AF388917

Modelling, Simulation and Control of Thermal Energy Systems

Modelling, Simulation and Control of Thermal Energy Systems

Editors

Kwang Y. Lee
Damian Flynn
Hui Xie
Li Sun

MDPI • Basel • Beijing • Wuhan • Barcelona • Belgrade • Manchester • Tokyo • Cluj • Tianjin

Editors
Kwang Y. Lee
Baylor University
USA

Damian Flynn
University College Dublin
Ireland

Hui Xie
Tianjin University
China

Li Sun
Southeast University
China

Editorial Office
MDPI
St. Alban-Anlage 66
4052 Basel, Switzerland

This is a reprint of articles from the Special Issue published online in the open access journal *Energies* (ISSN 1996-1073) (available at: https://www.mdpi.com/journal/energies/special_issues/Thermal_Energy_System).

For citation purposes, cite each article independently as indicated on the article page online and as indicated below:

LastName, A.A.; LastName, B.B.; LastName, C.C. Article Title. *Journal Name* **Year**, *Article Number*, Page Range.

ISBN 978-3-03943-360-5 (Hbk)
ISBN 978-3-03943-361-2 (PDF)

Contents

About the Editors

Kwang Y. Lee is a Full Professor and Chair of the Department of Electrical and Computer Engineering at Baylor University in Waco, Texas, U.S.A. He received his B.S. degree in Electrical Engineering from Seoul National University, his M.S. degree in Electrical Engineering from North Dakota State University, and the Ph.D. degree in Systems Science from Michigan State University. Before joining Baylor, Dr. Lee was a faculty member of electrical engineering at Michigan State University, Oregon State University, University of Houston, and the Pennsylvania State University. For 21 years at Penn State, he served as Director of Power Systems Control Laboratory and the Intelligent Distributed Controls Research Laboratory. Dr. Lee is the author of over 700 technical publications and book chapters. His current research interests include power system control and optimization, economic operation, generation expansion planning, reactive power planning, load forecasting, power plant control, fuel cell power generation, and intelligent system applications to power systems. Dr. Lee is a Life Fellow of IEEE, active in the Intelligent Systems Subcommittee and Working Group on Modern Heuristic Optimization Techniques at the IEEE Power and Energy Society. He served as the Chair of the Technical Committee on Power and Energy Systems (TC 6.3) in the International Federation of Automatic Control (IFAC) from 2014 to 2020. He has been a frequent panelist on intelligent control, distributed simulation, and combined research and curriculum development, and a tutorial speaker/organizer on artificial neural networks, fuzzy systems and evolutionary computation. Dr. Lee served as an Editor of IEEE Transactions on Energy Conversion and Associate Editor of IEEE Transactions on Neural Networks, Associate Editor of IFAC Journal on Control Engineering Practice. He edited two IEEE-Press/Wiley Books: *Modern Heuristic Optimization Techniques* (2008) and *Applications of Modern Heuristic Optimization Methods in Power and Energy Systems* (2020); and co-authored a Springer Book: *Intelligent Network Integration of Distributed Renewable Generation* (2017) and an *IEEE-Press/Wiley Book: Probabilistic Power System Expansion Planning with Renewable Energy Resources and Energy Storage Systems* (2020).

Damian Flynn is an Associate Professor in the School of Electrical and Electronic Engineering at University College Dublin, Ireland. He obtained M.Eng. and Ph.D. degrees in Electrical and Electronic Engineering from The Queen's University of Belfast, N. Ireland, respectively, in 1987 and 1991, before being appointed as a lecturer at the same university. He later joined University College Dublin in 2009. Dr Flynn's research interests involve grid integration of renewables, power system operation and stability analysis, energy systems integration and power generation control. He has published 200+ research papers, 3 books and various technical studies, supported through various joint research and consulting projects with industry, and national and European research awards. He is a member of International Energy Agency (IEA) Wind Task 25: Design and Operation of Power Systems with Large Amounts of Wind Power, International Federation of Automatic Control (IFAC) Technical Committee on Power and Energy Systems (TC 6.3) and European Energy Research Alliance (EERA) Joint Programme on Energy Systems Integration. He is an Editor of IEEE Transactions on Energy Conversion, a Subject Editor of IET Renewable Power Generation and Associate Editor for Wiley Interdisciplinary Reviews: Energy and Environment. He has been an Editor for two books: *Thermal Power Plant Simulation and Control* (IEE, 2003) and *Advances in Energy Systems: The Large-Scale Renewable Energy Integration Challenge* (Wiley, 2019), and he has co-authored a book: *Wind Power Integration: Connection and System Operational Issues* (IET, 2007, 2014).

Hui Xie received his PhD in propulsion machine and engineering at Tianjin University, China in 1998, and now he holds a position as Professor and Vice Director in State Key Laboratory of Engines at Tianjin University and also as Director of Autonomous Driving Cross-research Platform. His research interests include intelligent control of engine, powertrain and vehicle, autonomous driving vehicle and big data analysis. His research achievements include advanced intelligent control algorithms for engines, multi-core hardware control architecture and self-optimization energy management methods. He received the 2014 National Educating Achievement Award, the 2018 China Machinery Industry Science and Technology Award, and the 2019 Tianjin Government Science and Technology Award.

Li Sun is currently an Associate Professor at School of Energy and Environment, Southeast University, Nanjing, China. He received his B.S. degree from Southeast University in 2012 and Ph.D. degree from Tsinghua University, Beijing, China in 2017. He was a Visiting Scholar at Baylor University, Waco, Texas, U.S.A. from 2014 to 2015, and a Visiting Assistant Professor at Robert Frederick Smith School of Chemical and Biomolecular Engineering, Cornell University, Ithaca, New York, U.S.A. from 2019 to 2020. His research interests include: 1) fuel cell control, 2) battery management systems, and 3) microgrid operation. His professional service includes: 1) Guest Editor: *IFAC Journal on Control Engineering Practice*, 2) Guest Editor: *Energies*, 3) Guest Editor: Sustainable Energy Technologies and Assessments (Elsevier), and 4) Associate Editor, American Control Conference (ACC), 2016–present, 4) Associate Editor, IFAC World Congress 2020.

Preface to "Modelling, Simulation and Control of Thermal Energy Systems"

Against a background of ever-increasing resource scarcity and environmental regulations, recent decades have experienced rapid growth in the development of various renewable energy sources, such as wind, tidal and solar power. The variable and uncertain nature of these energy sources is well known, and greater utilization of power electronic converters presents new challenges for power system stability. Consequently, across both industry and the research community, a range of control and operational strategies have been proposed and implemented, recognizing the increasing flexibility, ramping and load regulation requirements being placed on conventional thermal power generation. Similarly, thermal engines, particularly those based on diesel and gasoline, are also facing greater environmental concerns, such that efficient control, which fulfills requirements of high efficiency, low pollution and high durability, is an emerging requirement.

It follows that modelling, simulation and control are key aspects for thermal energy systems to provide innovative and effective solutions. A detailed understanding of the thermal conversion mechanism(s) can be achieved through detailed dynamic modelling, which can then inform the design of advanced control strategies to improve system performance, both in economic and environmental terms. Simulation studies, and experimental test beds, can play a very important role here before proceeding to field tests. Consequently, this Special Issue is dedicated to exploring the state of the art in modelling and control of thermal energy systems, and to provide a practical and comprehensive forum for exchanging novel research ideas and empirical practices. After a detailed review process, 12 papers were accepted for publication, covering general themes of enhanced process control, optimization of performance against cost and emissions-based objectives, and modelling of thermally sensitive equipment performance, with applications in thermal power generation, district heating systems and electric vehicle batteries, to name but a few. A summary of the accepted papers is provided as follows.

The paper "On the flexible operation of supercritical circulating fluidized bed: burning carbon based decentralized active disturbance rejection control" by Fan Zhang et al. [1] investigates the ability of a supercritical circulating fluidized bed (CFB), as a prominent clean coal technology, to improve its operational flexibility using decentralized active disturbance rejection control. A CFB plant presents the advantages of high efficiency, fuel flexibility, and low-cost emission control, but the large inertia, strong nonlinearity, and multivariable coupling make it a challenging task to harmonize the boiler's slow dynamics with the turbine's fast dynamics. Since burning carbon in the furnace responds faster than throttle steam pressure when the fuel flow rate changes, this is utilized to provide dynamic compensation. Simulations are presented for a 600-MW supercritical CFB unit, to verify the load following and disturbance rejection merits of the proposed method: https://www.mdpi.com/1996-1073/12/6/1132.

The paper "Optimization of dispatching electricity market with consideration of a solar-assisted coal-fired power generation system" by Rongrong Zhai et al. [2] investigates multi-objective optimization for load dispatch of a solar-assisted coal-fired power generation system, considering coal consumption, NOx emissions, and power purchase cost. Based upon a test system consisting of one coal-fired unit which is retrofitted to be solar-assisted, it is concluded that the loading of the solar-assisted unit is increased, particularly with increased solar irradiation. In addition, coal consumption, NOx emissions and power costs are reduced, and based upon on-grid power tariff variations, a reduction in electricity purchase costs can be achieved: https://www.mdpi.com/1996-1073/12/7/1284.

The paper "Transient analysis and execution-level power tracking control of the concentrating solar thermal power plant" by Xiufan Liang and Yiguo Li [3] studies an execution-level power tracking control strategy for a concentrating solar plant (CSP), primarily aimed at coordinating control of the sluggishly responding steam generator and the fast responding steam turbine. Concentrating solar power is a promising technology for exploiting solar energy, given its ability to integrate with thermal energy storage, and so offer similar operability to that of fossil-fired power plants. The power output of such plants is generally scheduled to maximize operating revenue, but this can result in frequent changes in power reference signal and introduces challenges for real-time power tracking. After analyzing the transient behavior of a CSP plant to gain insight into the system dynamic characteristics and control difficulties, two control strategies are studied through simulation experiments, based on heuristic PI control, and advanced model predictive control: https://www.mdpi.com/1996-1073/12/8/1564.

The paper "Thermal fatigue modelling and simulation of flip chip component solder joints under cyclic thermal loading" by Liangyu Wu et al. [4] develops a model, based on the Darveaux energy method, to investigate influential factors which affect thermal fatigue of flip chip component solder joints in thermal energy systems. Under cyclic thermal loading, a theoretical heat transfer and thermal stress model is developed, which is then used to show the effects of environmental and power parameters on thermal fatigue life. It is indicated that solder joints located at the outer corner point of an array tend to have the shortest life, and increments in either power density or ambient temperature, or reductions in power conversion time or ambient pressure, will result in a reduced thermal fatigue life for the key solder joints: https://www.mdpi.com/1996-1073/12/12/2391.

The paper "Development of engine efficiency characteristic in dynamic working states" by Piotr Bera [5] applies an artificial neural network (ANN) method to represent the efficiency characteristic of a combustion engine in dynamic working states. The model comprises engine speed, angular acceleration, engine torque, torque change intensity, and fuel mass flow parameters, measured on a test bed of a spark ignition engine in static and dynamic working states. Detailed analysis is presented of the ANN design, data preparation, and training method, with a simplified ANN subsequently created to represent a two-dimensional efficiency characteristic: https://www.mdpi.com/1996-1073/12/15/2906.

The paper "Temperature dependent parameter estimation of electrical vehicle batteries" by Anna Pózna et al. [6] presents a parametric temperature dependent battery model, based upon static relationships. The proposed method is intended as a computationally effective way of determining key battery parameters at a given temperature from actual estimated values, and previously determined static temperature dependence. A two-step method is employed which includes a parameter estimation step of the key parameters at different temperatures, followed by a static optimization step that determines the temperature coefficients of the corresponding parameters. The effectiveness of the method was verified by simulation experiments on a more complex battery model, which also describes the detailed dynamic thermal behavior of the battery: https://www.mdpi.com/1996-1073/12/19/3755.

The paper "Stacked auto-encoder modeling of an ultra-supercritical boiler-turbine system" by Hao Zhang et al. [7] presents a deep neural network framework using stacked auto encoders (SAEs) to model an ultra-supercritical (USC) coal-fired boiler turbine unit. Such units are widely used in modern power plants due to their high efficiency and low emissions, but it is also a multivariable system with large inertia, severe nonlinearity, and strong coupling, such that building an accurate system model using traditional identification methods is almost impossible. Maximum correntropy is chosen as the loss function for training of the SAE, with real-time measurement data from an USC unit used for training and validation. When compared against a traditional multilayer perceptron network model, the SAE model is seen to be superior in both forecasting plant dynamic behavior as well as eliminating the influence of outliers: https://www.mdpi.com/1996-1073/12/21/4035.

The paper "Modelling and analysis of plate heat exchangers for flexible district heating systems" by Serafym Chyhryn [8] develops a temperature dependent overall heat transfer coefficient (OHTC) model of a plate heat exchanger, based on a linear approximation of thermophysical components of the forced convection coefficient (FCC). Heat transfer in a plate heat exchanger is regulated by mass flows, but flexible operation and demand variations cause shifts in temperature levels, which need to be accounted for as part of efficient system operation. The presented modelling approach accounts for temperature variations, avoids the iterative lookup of thermophysical properties and requires fewer inputs. Experimental model verification is performed on a laboratory plate heat exchanger, and a practical estimation procedure is proposed based on component data. Finally, operational optimization test cases for a basic district heating system are used to demonstrate the superior performance of the proposed approach against existing models: https://www.mdpi.com/1996-1073/12/21/4141.

The paper "Supplementary control of air–fuel ratio using dynamic matrix control for thermal power plant emission" by Taehyun Lee et al. [9] applies dynamic matrix control (DMC) to the supplementary control of existing combustion control loops within thermal power plants in order to reduce environmental emissions. The conventional double cross limiting algorithm for combustion safety is formulated as constraints within the proposed DMC. Simulation results are presented for a 600-MW drum-type power plant and a 1000-MW ultra-supercritical once-through boiler power plant, and in both cases tight control of the air–fuel ratio is seen to be effective in reducing emissions: https://www.mdpi.com/1996-1073/13/1/226.

The paper "Uncertainties in the testing of the coefficient of thermal expansion of overhead conductors" by Miren Bedialauneta et al. [10] analyzes the effect of some of the uncertainty sources in the testing of the thermal expansion coefficient of overhead line conductors. Utilities and conductor manufacturers usually carry out verification of the thermal expansion coefficient of overhead conductors on an actual size span, based on observation of changes in conductor length as a result of conductor temperature variations. However, additional factors can affect the coefficient value. Subsequently, the thermal expansion process for line conductors is described and sources of uncertainty are identified, before their effect on coefficient testing for high temperature low sag (HTLS) conductors is quantified: https://www.mdpi.com/1996-1073/13/2/411.

The paper "Multiresolution GPC-structured control of a single-loop cold-flow chemical looping testbed" by Shu Zhang et al. [11] develops a self-tuning controller design methodology, encompassing a spatiotemporal, multi-resolution, deadbeat control loop for a chemical looping process. Chemical looping is a near-zero emission process for coal-fired generation, based on multi-phase gas–solid flow, but with extremely challenging nonlinear, multi-scale dynamics with jumps. Consequently, traditional robust control techniques are largely inapplicable, so temporal and spatiotemporal multi resolution modelling is applied to address the process complexity, along with corresponding model-based control laws. A nonlinear autoregressive with exogenous input model structure, nonlinear in the wavelet basis, but linear in parameters, is used to identify the dominant temporal process dynamics. The effectiveness of the proposed methodology in producing fast recursive real-time algorithms for controlling highly uncertain nonlinear multi-scale processes is then demonstrated for a difficult chemical looping cold flow tracking control problem: https://www.mdpi.com/1996-1073/13/7/1759.

The paper "Multi-objective optimal operation for steam power scheduling based on economic and exergetic analysis" by Yu Huang et al. [12] evaluates the thermodynamic efficiency of steam supply, based on exergetic analysis, in order to formulate a mixed-integer linear programming (MILP) optimal scheduling model. Steam supply scheduling plays an important role in providing reliable energy supply to meet heat and electricity demand in both the industrial and residential sectors, but system complexity makes it challenging to operate efficiently. Contradictory operational objectives, in terms of economic cost and thermodynamic efficiency, make online scheduling even more intractable. An epsilon constraint-based method is used to obtain the Pareto front of the multi-objective optimization model, while a fuzzy approach is introduced to determine the actual operational strategy. Results from single-period and multi-period multi-objective optimal scheduling are used to verify the effectiveness of the model, and the proposed solution, relative to single objective approaches: https://www.mdpi.com/1996-1073/13/8/1886.

These papers collectively offer a wide range of approaches for modelling, simulation and control development, supported by detailed analysis and discussion, as applied to a range of applications within the domain of thermal energy systems. The Editors of this Special Issue would like to express thanks to all the authors who submitted their work, for the careful attention of all the reviewers in assessing the papers, and to the MDPI staff for their diligent support in preparing this Special Issue.

Kwang Y. Lee, Damian Flynn, Hui Xie, Li Sun
Editors

Article

On the Flexible Operation of Supercritical Circulating Fluidized Bed: Burning Carbon Based Decentralized Active Disturbance Rejection Control

Fan Zhang, Yali Xue *, Donghai Li, Zhenlong Wu and Ting He

State Key Lab of Power System, Department of Energy and Power Engineering, Tsinghua University, Beijing 100084, China; zhfan@tsinghua.edu.cn (F.Z.); lidongh@tsinghua.edu.cn (D.L.); wu-zl15@mails.tsinghua.edu.cn (Z.W.); he-t14@mails.tsinghua.edu.cn (T.H.)
* Correspondence: xueyali@tsinghua.edu.cn

Received: 17 February 2019; Accepted: 18 March 2019; Published: 22 March 2019

Abstract: Supercritical circulating fluidized bed (CFB) is one of the prominent clean coal technologies owing to the advantages of high efficiency, fuel flexibility, and low cost of emission control. The fast and flexible load-tracking performance of the supercritical CFB boiler-turbine unit presents a promising prospect in facilitating the sustainability of the power systems. However, features such as large inertia, strong nonlinearity, and multivariable coupling make it a challenging task to harmonize the boiler's slow dynamics with the turbine's fast dynamics. To improve the operational flexibility of the supercritical CFB unit, a burning carbon based decentralized active disturbance rejection control is proposed. Since burning carbon in the furnace responds faster than throttle steam pressure when the fuel flow rate changes, it is utilized to compensate the dynamics of the corresponding loop. The parameters of the controllers are tuned by optimizing the weighted integrated absolute error index of each loop via genetic algorithm. Simulations of the proposed method on a 600 MW supercritical CFB unit verify the merits of load following and disturbance rejection in terms of less settling time and overshoot.

Keywords: supercritical circulating fluidized bed; boiler-turbine unit; active disturbance rejection control; burning carbon; genetic algorithm

---

1. Introduction

Circulating fluidized bed (CFB) technology has demonstrated its ability to efficiently utilize a wide variety of fuels, including high sulfur coal to coal gangue and coal slurries [1]. Taking coal-water slurries containing petrochemicals fuels for example, through experiments and calculations the advantages are much lower anthropogenic emissions and ash residue, low cost of the components, positive economic performance indicators of storage, transportation, and combustion, as well as higher fire and explosion safety [2,3]. It is believed that a combination with supercritical steam cycle to increase the efficiency of energy conversion is one of the futures of CFB combustion technology [4]. The first 600 MW supercritical CFB boiler demonstration project was put into commercial operation in 2013 [5]. By the end of 2017, more than eighty-two supercritical CFB boilers were in operation or under construction in China [6].

As Lyu pointed out [4], in the demonstration the control of the 600 MW supercritical CFB boiler was the heart of the matter. Some factors attribute to the difficulties of the coordinated control of the unit:

1. Higher requirement for operational flexibility. With increasing intermittent renewable energy integrated in the grid, thermal power plants are required to operate in a wider range [7]. Supercritical CFB boilers can regulate their load from 30% to 100%, which extends 20% more in the low load region compared with pulverized coal-fired boilers [8]. However, the considerable quantities of bed materials in furnace result in large inertia of the CFB boiler. In addition, dynamics of the boiler vary at different operation conditions, leading to strong nonlinearity. Both of these factors make it hard to design controllers of coordinated control system (CCS) to harmonize the boiler's slow dynamics with the turbine's fast dynamics so as to follow the command from grid promptly.

2. Capability to reject disturbance in fuel. Since the CFB boiler works with a variety of fuels, the variability of fuel brings in disturbance for the unit operation. In addition, the amounts of fuel that enter the boiler sometimes fluctuate due to mechanical reasons. Consequently, it is necessary to design advanced controllers so as to suppress the influence of disturbance from fuel.

3. Complex dynamics of the supercritical CFB unit. Besides the thermal inertia, strong nonlinearity, and time delay of supercritical CFB unit, multivariable coupling has a significant impact on the controller design [9]. The adjustments of manipulated variables would cause changes in all controlled variables. Furthermore, the unit would become more complicated when the bed temperature of the CFB boiler is taken into consideration [10].

As can be anticipated, a well-designed control system of the supercritical CFB unit can yield potential environmental and economic benefits.

Much of the literature has paid particular attention to this problem. Among them, investigation of the dynamic characteristic and mechanism-based modeling for supercritical CFB boilers lays the groundwork. Prior knowledge about subcritical CFB boilers is essential to the modeling research. Majanne and Köykkä presented a dynamic model which consisted of the air-flue gas and the water-steam systems [11]. The model was based on the first principles mass, energy, and momentum balances and experimental correlations about reaction kinetics and heat transfer, and was finally tested against measured process data. Furthermore, a mechanism-based control model in the form of transfer functions for the CCS of the subcritical coal-fired 300 MW CFB unit was established based on the dynamic characteristics in [12]. The research has been extended to the supercritical CFB unit. In [13] a hybrid dynamic model was developed to characterize the main physical and chemical processes in a supercritical CFB boiler. Steady-state verification was made to evaluate the accuracy of the model while step responses of different manipulate variables were tested. Through some reasonable simplification, a nonlinear control model of supercritical CFB unit was established in [14], and the parameters of the system model were identified by steady-state derivation, function fitting, and optimization algorithm. The correctness of the model structure and validity of the identification method were verified by operation data of a 600 MW supercritical CFB unit in service.

Based on the derived dynamic model of CFB unit, a variety of control methods are adopted for the coordinated control purpose. The proportional-integral-derivative (PID) control is the most widely used control strategy in real control engineering. Hultgren and Hao et al. analyzed the relative gain and designed a decentralized PID control structure for the CFB unit [15,16]. The controllers in [16] were devised based on desired dynamic equation (DDE) while an heuristic algorithm was used to optimize the PID controllers for the CFB unit in [17]. In [18], dynamic feedforward was employed to improve the performance of PID controllers. on basis of decentralized PI control, two disturbance observers (DOBs) were designed to estimate and compensate the effect of coupling in the CFB unit [19]. However, higher requirements are imposed for the supercritical CFB unit which the conventional decentralized PID control can hardly satisfy. Some advanced control methods have been discussed, such as fuzzy control [20–22], neural network based control [23,24], etc. Although good performances were observed in simulations, these methods could seldom be used in the real CFB power plant due to the restrictions in the distributed control system.

Due to the ability to deal with uncertainty, active disturbance rejection control (ADRC) has received increased attention across a number of disciplines in recent years, such as gasoline engines [25], proton exchange membrane fuel cell [26], electromechanical actuator [27], pendulum cart system [28], piezo-driven positioning stage [29], flight control [30], and so on. The basic principle of ADRC is the estimation and actively compensation via extended state observer (ESO). The stability analysis of ADRC and the convergence of ESO have been studied by many researchers [31,32]. ADRC has also been employed for the process control of CFB unit, such as the superheated steam temperature control [33], primary air control [34], secondary air control [35], combustion system control [36], and control of boiler-turbine unit [37]. However, none of them have taken control problems of supercritical CFB unit into consideration. The increase in inertia aggravates the difficulty of the supercritical CFB unit. Recent studies show that combination of burning carbon in the furnace of CFB with heat signal can improve the control performance [9]. To make use of the burning carbon, Gao et al. added the rate of heat released in combustion of burning carbon as the feedforward signal on basis of decentralized PI controllers. However, detailed analysis of the control structure should be undertaken.

The primary aim of this research is to provide reasonably consistent evidence of an association between burning carbon and operation performance and to explore the enhancement of the operation performance of the supercritical CFB unit, making the following contributions:

- Burning carbon is integrated into the control framework to accelerate the load following;
- The disturbance rejection performance is improved via the design of decentralized ADRC controllers;
- Genetic algorithm (GA) is employed to tune the parameters of the ADRC controllers.

The remaining part of the paper proceeds as follows: Section 2 introduces the 600 MW supercritical CFB boiler-turbine unit and analyzes its dynamics. In Section 3, the decentralized ADRC framework is proposed for the supercritical CFB unit, in which burning carbon information is utilized. In order to achieve satisfying performance, GA is used to tune the controllers for both traditional decentralized controllers and the proposed method in Section 4. In Section 5, simulation results are given to verify the merits of the proposed method. Finally, some conclusions are drawn in Section 6.

2. Performance Analysis of Supercritical Circulating Fluidized Bed Boiler-Turbine Unit Model

Gao et al. investigated the dynamics of supercritical CFB unit and established the nonlinear model for control purposes [14]. The model represented the behavior of the boiler-turbine unit in the 600 MW supercritical CFB power plant located at Baima, China, and was validated by its operation data.

Simplified working process of the supercritical CFB boiler-turbine unit is illustrated in Figure 1. The essential working principle of the boiler-turbine unit is energy conversion. The chemical energy stored in coal is transformed into steam thermal energy by the boiler, then it is transformed into rotational mechanical energy by the turbine, and finally it is transformed into electric energy by the turbogenerator.

In the derived model, the manipulate variables are the fuel flow rate command u_B (u_1, kg/s), feedwater flow rate D_{fw} (u_2, kg/s), and turbine throttle valve opening u_t (u_3, %); the controlled variables are the throttle steam pressure P_{st} (y_1, MPa), separator steam enthalpy h_m (y_2, kJ/kg), and active electric power generated by the turbogenerator N_e (y_3, MW), respectively. Some typical operating conditions are shown in Table 1.

Table 1. Typical steady-state operation conditions of the supercritical circulating fluidized bed (CFB) unit.

	y_1 (MPa)	y_2 (kJ/kg)	y_3 (MW)	u_1 (kg/s)	u_2 (kg/s)	u_3 (%)
High (100%)	23.93	2609.53	600	32.79	485.98	91.51
Medium (70%)	19.30	2669.29	420	24.54	335.60	79.40
Low (40%)	12.53	2804.35	240	15.28	184.10	69.91

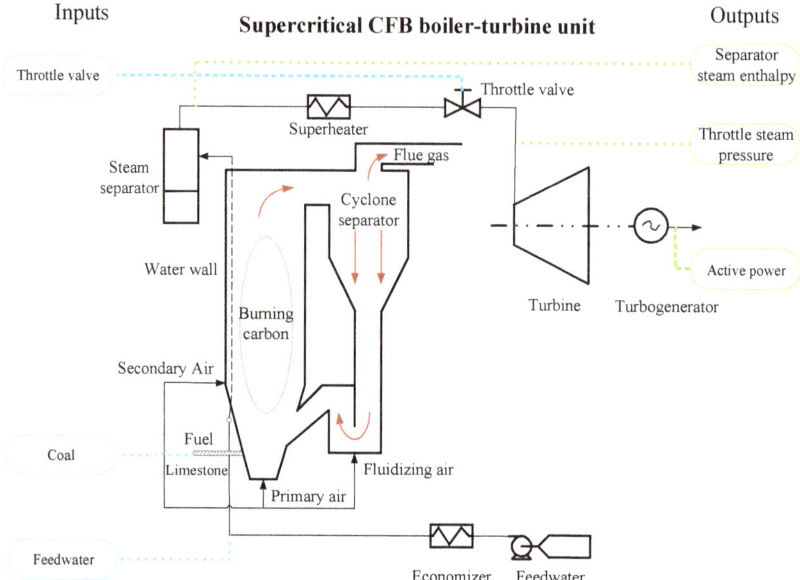

Figure 1. Simplified diagram of a supercritical circulating fluidized bed (CFB) boiler-turbine unit.

Since the purpose of modeling is to design and test advanced control algorithms suited for the supercritical CFB boiler-turbine unit, many simplifications are made during the modeling, however, for the most important variables of the unit, the model can capture both the steady-state and dynamic properties of the unit well thus is very suited for controller design and test.

Large inertial, nonlinear, and strong multivariable coupling behavior of the unit can be clearly indicated through step response tests. Taking u_B for example, Figure 2 shows the step response at typical operation conditions. The evidence from this test suggests that unit takes more than 4000 s to reach steady-state when an increase in fuel command u_B is occurred. Meanwhile, it is also of interest to note u_B has an effect on all the outputs. Multivariable coupling should be taken into consideration when the control system is designed. Furthermore, the unit exhibits signs of nonlinearity since the amplitudes and time constants at different operation conditions are quite different from each other.

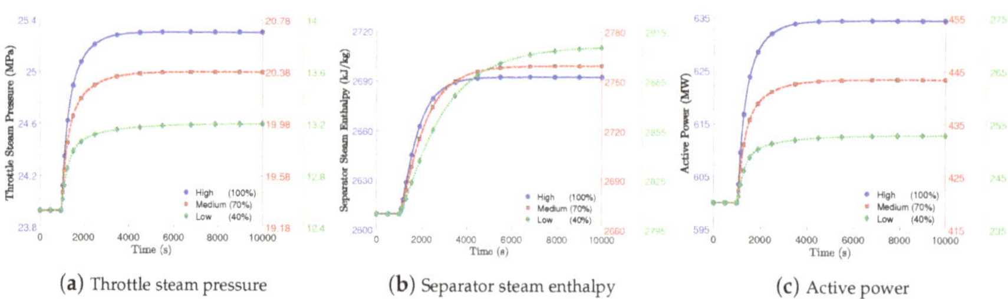

(**a**) Throttle steam pressure   (**b**) Separator steam enthalpy   (**c**) Active power

Figure 2. Outputs of a supercritical CFB unit when u_B increases by 5%.

We also investigate the character of burning carbon when the unit changes its load, as shown in Figure 3. Researchers have discussed the model of burning carbon in CFB unit [14,38]. Here we follow the model proposed in [14] because it can reflect the change tendency and is simple enough to satisfy the fast calculation requirement in real-time operation. Figure 3 indicates that burning carbon in the furnace is strongly related to the operation condition of the unit. Once the fuel flow command changes, the burning carbon varies accordingly. The second major finding is that burning carbon will reach the steady-state earlier than the throttle steam pressure when the fuel flow rate changes. This fact motivates us to make use of the burning carbon information to design the control system for the unit.

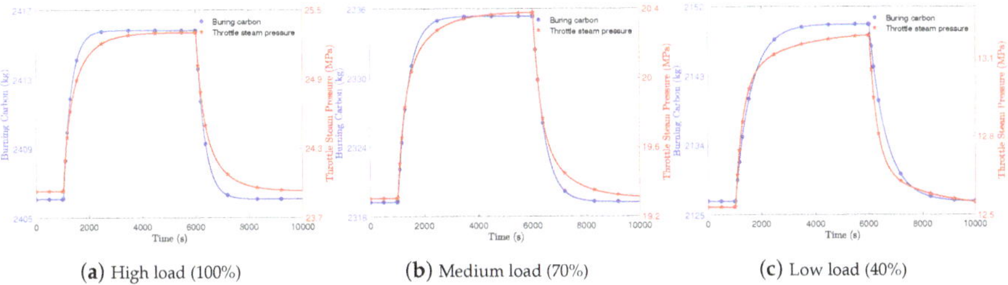

(a) High load (100%)   (b) Medium load (70%)   (c) Low load (40%)

Figure 3. Performance of burning carbon in a supercritical CFB unit when u_B increases by 5% at 1000 s and decreases by 5% at 6000 s.

3. Burning Carbon Based Decentralized Active Disturbance Rejection Control of a Supercritical CFB Unit

As mentioned previously, large inertial and unknown disturbances are the two main problems in the operation of the supercritical CFB boiler-turbine unit, therefore we propose a burning carbon based decentralized ADRC method to deal with both issues simultaneously.

3.1. Linear Active Disturbance Rejection Control

Without loss of generality, a class of nonlinear plant can be depicted by the following equation:

$$y^{(n)} = bu + f\left(y^{(n-1)}, y^{(n-2)}, \cdots, y\right) + d \tag{1}$$

where y is the measurable system output, u is the measurable control input, d is the unknown external disturbance, $f(\cdot)$ is the unknown internal (state-dependent and potentially nonlinear) dynamics of the process, and b is the unknown input scaling factor.

In order to design the input signal to make the output track the desired reference regardless of the unmodeled/unknown disturbance, the above system can be firstly augmented using a virtually extended state:

$$\begin{cases} y^{(n)} = b_0 u + \sigma \\ \dot{\sigma} = h \end{cases} \tag{2}$$

where b_0 is the approximation of b, σ represents the lumped disturbance including the unmodeled dynamics and external disturbance [32]:

$$\sigma = f(\cdot) + d + (b - b_0)\,u \tag{3}$$

It is assumed that σ is differentiable.

If the lumped disturbance is regarded as one dimensional state, we can define $x = [x_1, \cdots, x_n, x_{n+1}]^T = \left[y, y^{(1)}, \cdots, y^{(n)}, \sigma\right]^T$. So plant (2) can be described in the state space representation as

$$\begin{cases} \dot{x} = Ax + Bu + E\sigma \\ y = Cx \end{cases} \tag{4}$$

where

$$A = \begin{bmatrix} 0 & 1 & 0 & \cdots & 0 \\ 0 & 0 & 1 & \cdots & 0 \\ \vdots & \vdots & \vdots & \ddots & \vdots \\ 0 & 0 & 0 & \cdots & 1 \\ 0 & 0 & 0 & \cdots & 0 \end{bmatrix}_{(n+1)\times(n+1)}$$

$$B = \begin{bmatrix} 0 \\ \vdots \\ 0 \\ b_0 \\ 0 \end{bmatrix}_{(n+1)\times 1}, \quad E = \begin{bmatrix} 0 \\ \vdots \\ 0 \\ 0 \\ 1 \end{bmatrix}_{(n+1)\times 1}$$

$$C = \begin{bmatrix} 1 & 0 & \cdots & 0 \end{bmatrix}_{1\times(n+1)}$$

For the extended system model (4), $(n+1)$-th order extended state observer (ESO) is designed to estimate the unknown lumped disturbance σ [39]:

$$\begin{cases} \dot{z} = Az + Bu + L(y - \hat{y}) \\ \hat{y} = Cz \end{cases} \tag{5}$$

where $z = [z_1, \cdots, z_n, z_{n+1}]^T = [\hat{x}_1, \cdots, \hat{x}_n, \hat{\sigma}]^T$ is the estimate of state x, $L = [\beta_1, \cdots, \beta_{n+1}]^T$ is the observer gain. If the gains $\beta_1, \cdots, \beta_{n+1}$ are chosen properly, the lumped disturbance σ can be estimated as well as the states x.

Then, the control law is designed as

$$u = \frac{-z_{n+1} + u_0}{b_0} \tag{6}$$

where u_0 is to be determined to meet the specific type of application. Since the estimate of the extended state z_{n+1} approximates the lumped disturbance σ, i.e., $z_{n+1} \approx \sigma$, when combining control law (6) with the plant (4) we get

$$y^{(n)} \approx u_0 \tag{7}$$

which reduces the uncertain plant (4) to a cascade form of integrators. The canonical form of cascade of integrators makes the system trivial to govern due to inherent robustness against any perturbation in the system [40,41]. One should notice that the plant is still under the influence of uncertainties, however, the impact on output is removed [40].

The derived model (7) can be effectively controlled by state feedback law [42]:

$$u_0 = k_1(r - z_1) + k_2(\dot{r} - z_2) + \cdots + k_n(r^{(n-1)} - z_n) \tag{8}$$

where r is the desired reference.

For convenience in industrial applications, the 1st and 2nd order linear ADRC are commonly used [33]. Taking the 1st order ADRC for example, it can be reduced from the general form (5), (6), and (8) to the following:

$$\begin{cases} \dot{z}_1 = z_2 + b_0 u + \beta_1(y - z_1) \\ \dot{z}_2 = \beta_2(y - z_1) \end{cases} \tag{9}$$

$$u = \frac{u_0 - z_2}{b_0} \tag{10}$$

$$u_0 = k_p(r - y) \tag{11}$$

Thus, for the 1st order ADRC we have four parameters to be determined, namely k_p, b_0, β_1, and β_2. Once the ADRC is well tuned, the system output can track the reference while overcoming the uncertainties. The structure of the 1st order ADRC is illustrated by Figure 4, where G_p denotes the transfer function of the controlled plant.

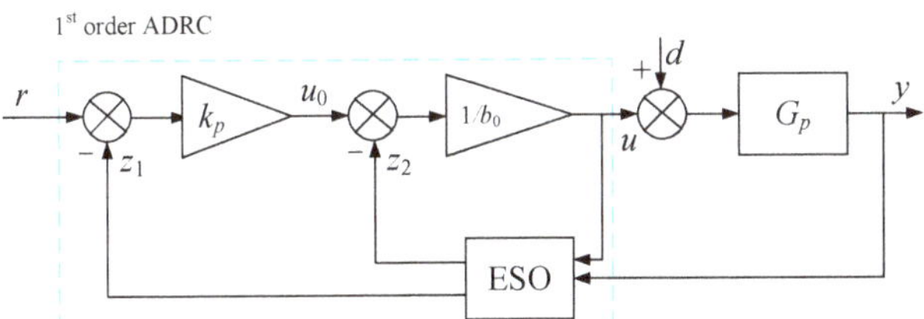

Figure 4. Structure of 1st order active disturbance rejection control.

3.2. Burning Carbon Based Decentralized ADRC for Supercritical CFB Boiler-Turbine Unit

Decentralized control is widely used in the industry process due to its simplicity. Instead of decentralized PID control, we propose the decentralized ADRC control for the supercritical CFB unit to improve the capability to follow load command in large-scale and enhance the capacity to unknown fuel variation. Figure 5 shows the designed control structure.

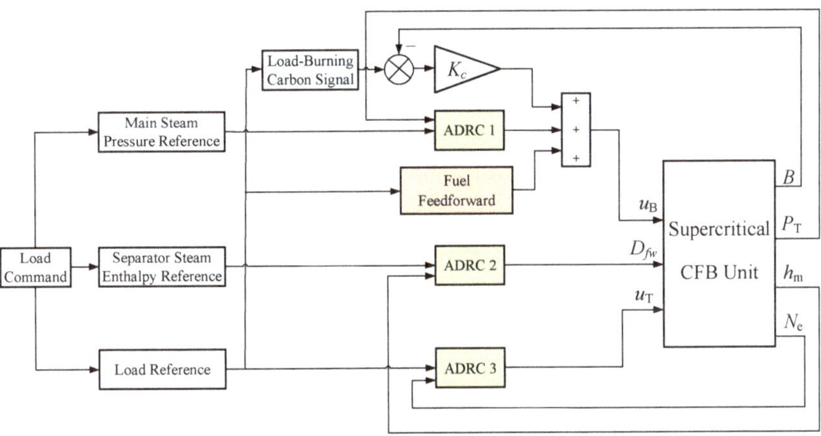

Figure 5. Structure of the burning carbon based decentralized active disturbance rejection control (ADRC) for a supercritical CFB unit.

In the proposed control structure, three ADRC controllers are devised for the individual loops. The ADRC controllers can not only alleviate the coupling of different loops, but also enhance the disturbance rejection performance. Fuel feedforward is included in this structure so that the output power can track the load command promptly. It is noted that we design the dynamic compensation for the fuel-main steam pressure loop based on the load-burning carbon signal. In the following, control of fuel-main steam pressure loop is analyzed. Its structure is shown in Figure 6.

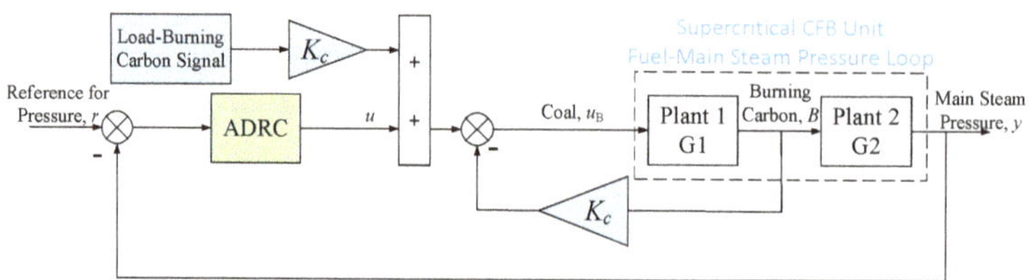

Figure 6. Control structure of fuel-main steam pressure loop based on load-burning carbon signal.

To illustrate the effectiveness of the dynamic compensation, the transfer function from coal to burning carbon is identified and used for analysis. The identified model $G_1 = \dfrac{K}{Ts+1}e^{-\tau s}$ at 100% and 40% operation conditions are:

$$G_{1(100)} = \frac{6.0567}{306.1587s+1}e^{-28s}$$
$$G_{1(40)} = \frac{29.6893}{591.5113s+1}e^{-28s} \tag{12}$$

The dynamics of G_1 varies considerably at different operation conditions, including the gain coefficient and time constant. The large time constant in G_1 indicates the huge inertial in CFB boiler, especially when it is at low load operation condition. In addition, since $\dfrac{\tau}{T+\tau} < 0.08$ the effect of time delay can be ignored when we analyze the control structure. The load-burning carbon signal is precalculated based on past operation data.

The equivalent transfer function from ADRC's output to burning carbon in (12) is

$$B'(s) \approx \frac{K\left(K_cK+1\right)}{Ts+K_cK+1}u \tag{13}$$

Compared with the original transfer function from burning carbon to ADRC's output,

$$B(s) = \frac{K}{Ts+1}u \tag{14}$$

The pole is shifted left from $-\dfrac{1}{T}$ to $-\dfrac{K_cK+1}{T}$ when $K_c > 0$. Also, both (13) and (14) have the same static gain K. Thus, the compensated burning carbon could be accelerated. Figure 7 shows the unit step response of the compensated burning carbon and throttle steam pressure model at different loads, in which $K_c = 0.03$. It can be found that dynamics at low load is more compensated.

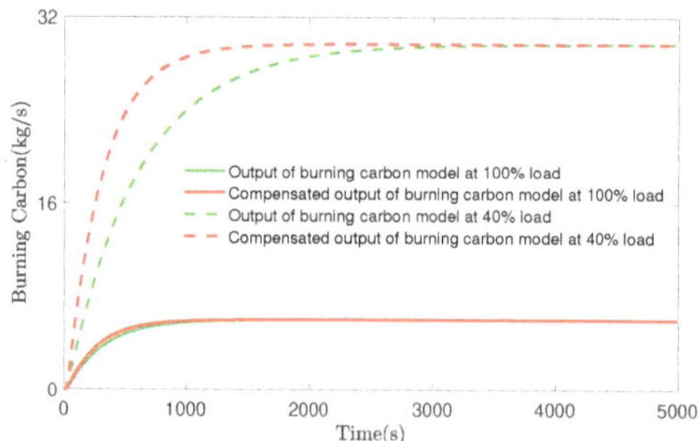

(**a**) Unit step response of the compensated burning carbon

Figure 7. *Cont.*

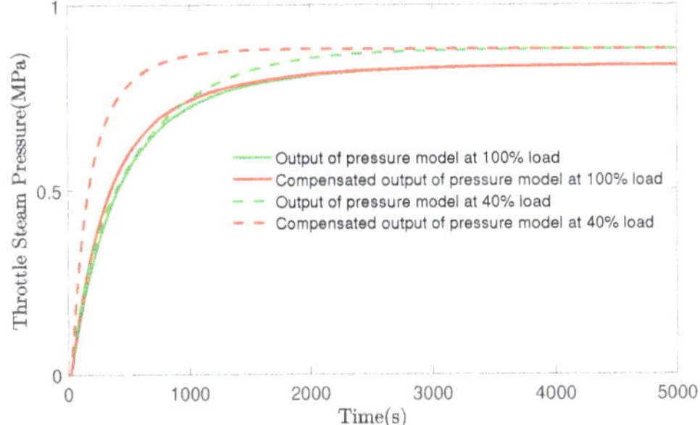

(**b**) Unit step response of the compensated throttle steam pressure

Figure 7. Unit step response of the compensated burning carbon and throttle steam pressure model at different loads.

The load-burning carbon signal is constructed according to the steady-state values of burning carbon at different operation conditions in advance. Based on the compensated fuel-main steam pressure loop, we design the decentralized ADRC controllers for the multi-input-multi-output (MIMO) supercritical CFB unit.

4. Tuning of ADRC Controllers

The performance of the ADRC controller is greatly affected by parameters b_0 in (10), k_p in (11), and β_1, β_2 in (9). Therefore, there are overall twelve parameters for the decentralized controllers to be optimized.

Genetic algorithm is a global search method that mimics the process of natural selection. It is one of the most well-known heuristic optimization methods, and has been used in various research areas [43]. In this research, the parameters in the decentralized ADRC controllers are optimized by GA.

To evaluate the control performance, the integrated absolute error (IAE) index is used,

$$\text{IAE} = \int_0^T |e(t)| dt \tag{15}$$

where $e(t)$ is the tracking error of the controlled variable. IAE tends to produce responses with less sustained oscillation. For the MIMO CFB unit, there are three controllers to be optimized. To obtain the trade-off between different loops, the fitness function for GA is defined as the weighted sum of each IAE,

$$J = \sum_{i=1}^{3} \omega_i \cdot \text{IAE}_i \tag{16}$$

where i denotes the i-th loop in CFB unit.

The optimization flowchart can be depicted as Figure 8.

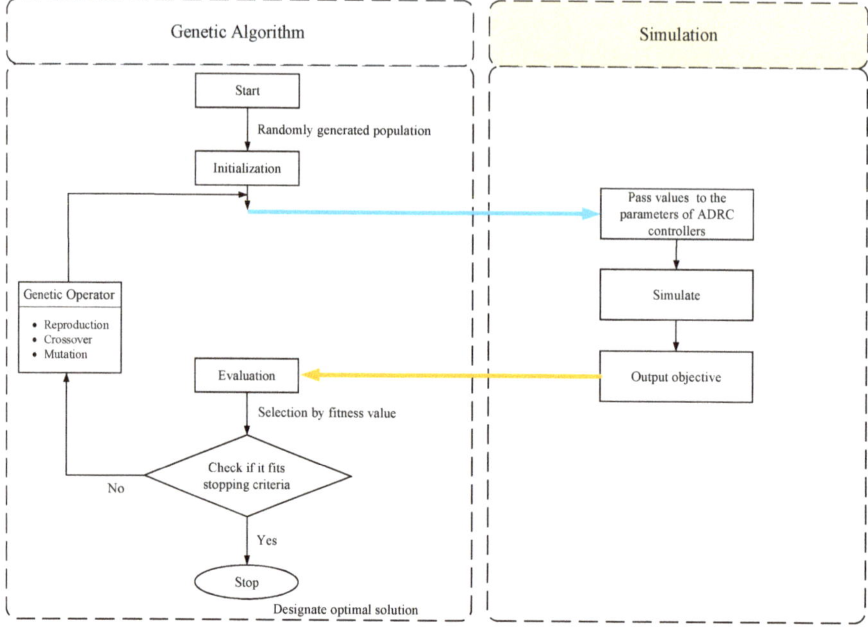

Figure 8. Flowchart of genetic algorithm based ADRC controller.

5. Simulations

In this section, the proposed burning carbon based decentralized ADRC is employed to control the supercritical CFB unit. Simulations under different scenarios, i.e., load tracking at different operation conditions and disturbance rejection, are performed to test the proposed method.

The simulation configuration of the burning carbon based decentralized ADRC of supercritical CFB unit is shown in Figure 5. The nonlinear dynamic model developed in [14] functions as the real plant since its dynamics has been tested by operation data. The fuel feedforward plays an important role in the boiler-turbine unit control. We design this feedforward signal based on the history steady-state data of the supercritical CFB unit so that it corresponds to the target load. The load-burning carbon signal is constructed based on the steady-state values of burning carbon at different operation conditions. The compensated gain K_c equals 0.03. As analyzed in Section 3.2, the introduced burning carbon can compensate the dynamics of the fuel-main steam pressure loop. The distributed ADRC controllers has been devised according to Equations (9) to (11). The parameters of the ADRC controllers are tuned by GA, the procedures of which are depicted in Section 4. The settings of GA are such that population size is chosen as 100, generation is 50, crossover fraction is 0.6, and individuals that are guaranteed to survive to the next generation are 10.

The proposed method is compared with two others, namely decentralized PI [16] and decentralized ADRC [37]. Both of them have the same fuel feedforward used in the proposed method. In addition, the controllers in these two methods are separately tuned by GA with the same settings. The differences in performance are expected to be found among the above three methods.

5.1. Load Tracking

The first case is designed to test the wide range load following performance of the controllers. The unit is assumed to decrease and increase at the rate of 2%/MCR/min at high (100%)/medium (70%)/low (40%) operation conditions, respectively. The results are shown in Figures 9 and 10.

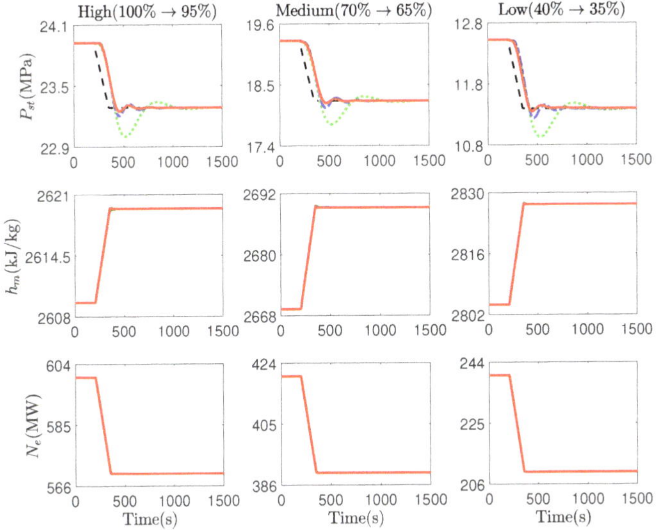

(**a**) Controlled variables (solid line: burning carbon based decentralized ADRC, dash-dotted line: decentralized ADRC, dotted line: decentralized PI, dashed line: reference).

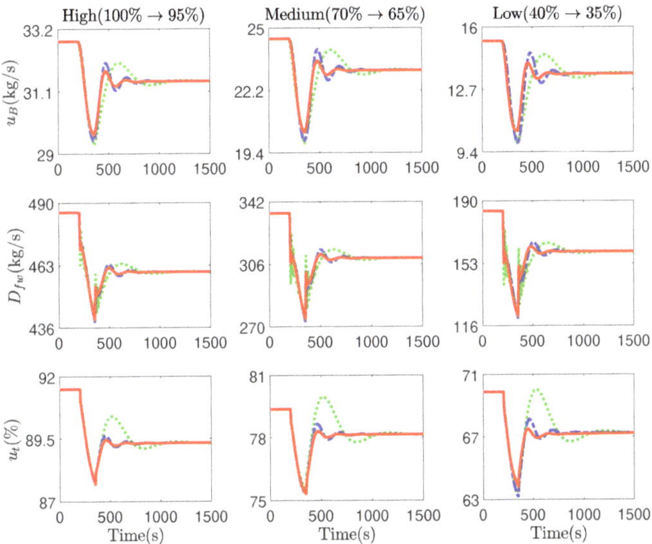

(**b**) Manipulated variables (solid line: burning carbon based decentralized ADRC, dash-dotted line: decentralized ADRC, dotted line: decentralized PI).

Figure 9. Load following performance of supercritical CFB unit at typical conditions (load decrease).

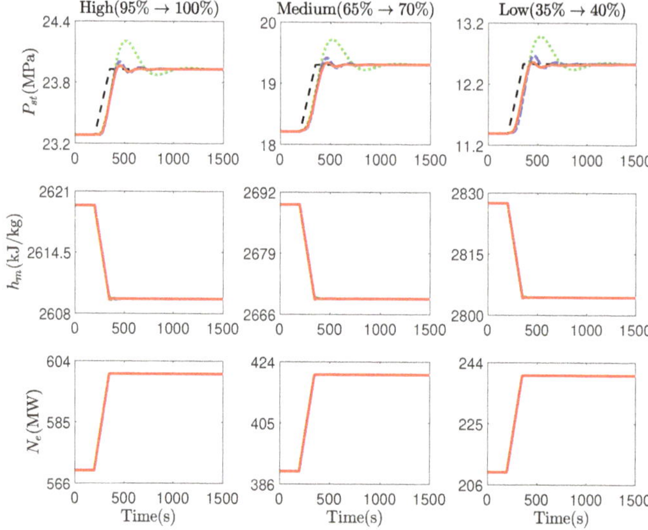

(**a**) Controlled variables (solid line: burning carbon based decentralized ADRC, dash-dotted line: decentralized ADRC, dotted line: decentralized PI, dashed line: reference).

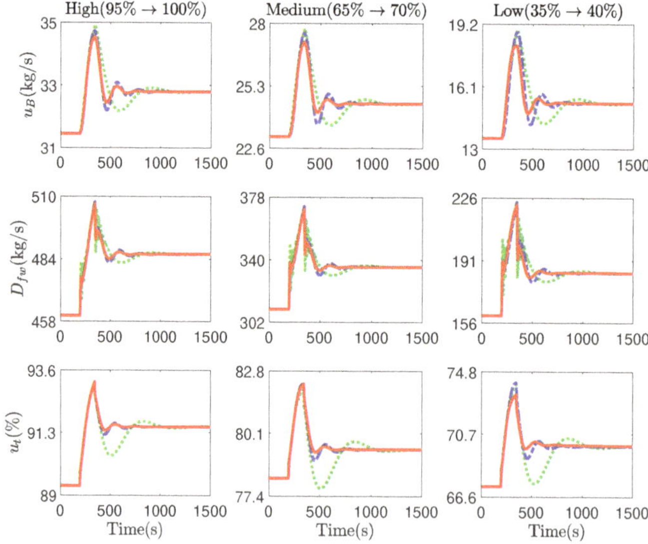

(**b**) Manipulated variables (solid line: burning carbon based decentralized ADRC, dash-dotted line: decentralized ADRC, dotted line: decentralized PI).

Figure 10. Load following performance of supercritical CFB unit at typical conditions (load increase).

The simulation results indicate that all three decentralized control methods can regulate the supercritical CFB unit and make the unit follow the command from grid. When the controllers are well tuned, both the output power N_e and separator steam enthalpy h_m can closely track the references. The major difference is the control performance in throttle steam pressure P_{st}, and the transient response criteria are listed in Tables 2 and 3.

Table 2. Transient response criteria of throttle steam pressure for the supercritical CFB unit at typical conditions (load decrease).

	Decentralized PI			Decentralized ADRC			Burning Carbon Based Decentralized ADRC		
	High	Medium	Low	High	Medium	Low	High	Medium	Low
Overshoot (%)	1.20	2.30	4.02	0.23	0.55	1.14	0.28	0.84	0.62
Settling time (s)	793.4	798.8	829.0	487.4	492.6	508.7	397.1	397.1	405.3

Table 3. Transient response criteria of throttle steam pressure for the supercritical CFB unit at typical conditions (load increase).

	Decentralized PI			Decentralized ADRC			Burning Carbon Based Decentralized ADRC		
	High	Medium	Low	High	Medium	Low	High	Medium	Low
Overshoot (%)	1.16	2.15	3.62	0.20	0.49	0.98	0.25	0.80	0.50
Settling time (s)	789.3	794.7	823.8	484.1	488.7	504.6	398.9	396.0	404.3

Decentralized PI control [16] with fuel feedforward is widely used in the real power plant. However, it takes the longest time for the throttle steam pressure to reach the steady-state. On the contrary, decentralized ADRC control [37] with fuel feedforward provides notable improvement. Since the coupling between the loops is regarded as disturbance and compensated by ADRC, its settling time is shortened while the overshoot is reduced.

The settling time of throttle steam pressure can be further improved under the proposed burning carbon based decentralized ADRC. As analyzed in Section 3.2, the burning carbon based compensated method ameliorates the dynamics of fuel-main steam pressure loop. The devised ADRC controller can result in a faster tracking performance. According to Tables 2 and 3, the settling time is about 50% less than that of decentralized PI method [16], and about 19% less than that of decentralized ADRC method [37]. In addition, it can be observed that the improvements of settling time at low load region is larger than that at high/medium load region. This is due to the fact that dynamics at low load are more compensated, as shown in Figure 7.

5.2. Disturbance Rejection

To further verity the disturbance rejection performance, a significant unknown step-type disturbance in fuel is considered in this case. This disturbance is common in coal-fired power plants because of the variability of coal. The unit is assumed to operate at 100% condition, and at $t = 500$ s the step-type disturbance $d = 3$ kg/s is acted on u_B, at $t = 2000$ s the step-type disturbance $d = -3$ kg/s is acted on u_B. The controllers are the same with those optimized in last case. The results are shown in Figure 11.

The results show that all these methods can remove the impact of the fuel disturbance, especially on the throttle steam pressure. However, decentralized PI [16] provides the slowest response because the integral action works on the bias between the output and the reference. Both decentralized ADRC [37] and the proposed burning carbon based decentralized ADRC method use the ESO to estimate the lumped disturbance and reject it according the principle of ADRC. Thus, the disturbance rejection is prompt and effective. The proposed burning carbon based decentralized ADRC is lightly better in terms of less settling time and overshoot.

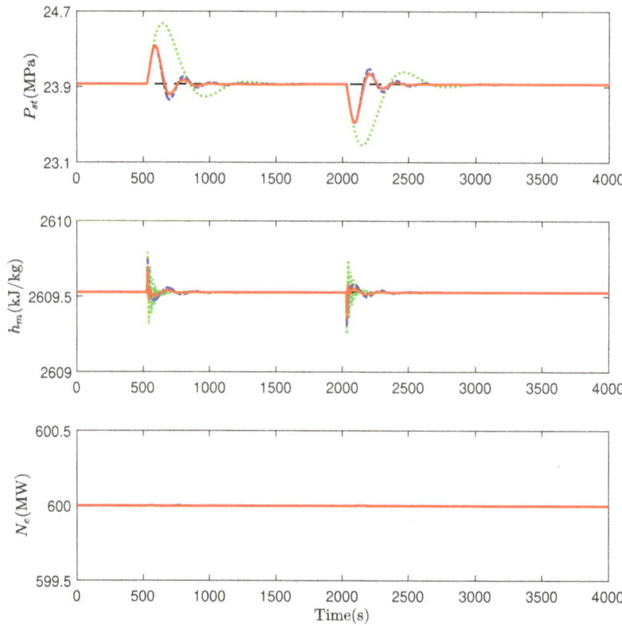

(**a**) Controlled variables (solid line: burning carbon based decentralized ADRC, dash-dotted line: decentralized ADRC, dotted line: decentralized PI, dashed line: reference).

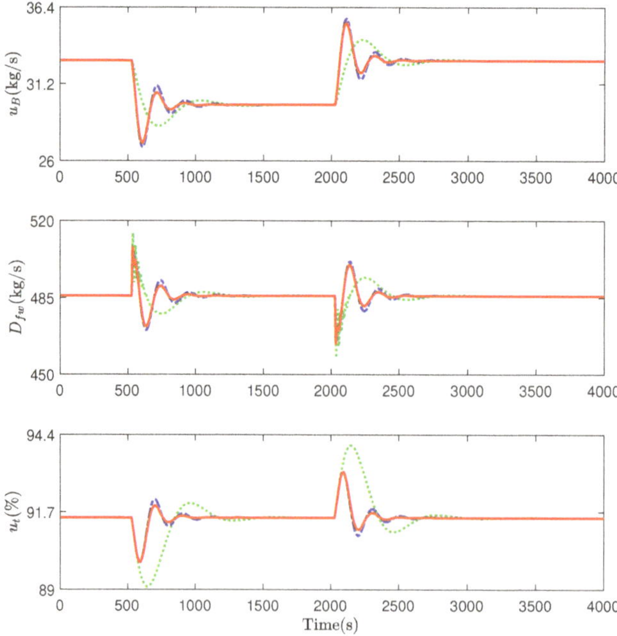

(**b**) Manipulated variables (solid line: burning carbon based decentralized ADRC, dash-dotted line: decentralized ADRC, dotted line: decentralized PI).

Figure 11. Disturbance rejection of the supercritical CFB unit under unknown fuel variation.

6. Conclusions

To achieve a sustainable future for renewable energy and integrate more renewable energy into the power grid, increasing the operational flexibility of supercritical power plants plays a crucial role in power systems.

The coordinated control system of the supercritical CFB unit is designed to harmonize the boiler's slow dynamics with the turbine's fast dynamics so as to meet the power grid's demand and maintain the parameters of the unit within the safe range. Since the burning carbon circulates in the boiler and releases the energy gradually, it can affect the change of the heat provided by the boiler and the impact of fuel variation on unit load. To this end, we make use of the burning carbon information to compensate the dynamics of the fuel-throttle pressure loop and reduce the regulating time.

In this research, we propose the burning carbon based decentralized ADRC for the operation of the supercritical CFB unit. The dynamics of the supercritical CFB unit, including the controlled variables and burning carbon in furnace, are analyzed on the basis of step response. One interesting finding is that burning carbon responds faster than the throttle steam pressure when the fuel flow rate changes. We utilize the burning carbon information to design the decentralized ADRC to reduce the influence of the large inertia of the supercritical CFB unit. A genetic algorithm is employed to optimize the controllers of the multivariable unit using a weighted integrated absolute error index. Through simulations of the supercritical CFB unit, it is demonstrated that the proposed method can notably reduce the settling time and maintain the disturbance rejection capability. These advantages benefit from the capacity of the well-tuned active disturbance rejection controllers and the utilization of burning carbon information. Consequently, a particular attention should be paid to the fast calculation of the burning carbon, as well as the steady-state values.

Applying the proposed method to the realistic supercritical CFB unit will be our future interest. Some practical issues should be taken into consideration, e.g., the implementation of the proposed method in the distributed control system of the power plant, and the tracing and undisturbed switching logic between controllers.

Author Contributions: Conceptualization, F.Z.; Funding acquisition, Y.X.; Methodology, F.Z., Y.X., D.L., Z.W. and T.H.; Project administration, Y.X. and D.L.; Software, F.Z., Z.W. and T.H.; Supervision, Y.X.; Validation, F.Z. and T.H.; Writing—original draft, F.Z.; Writing—review & editing, F.Z., Y.X. and D.L.

Funding: This research was funded by National Natural Science Foundation of China (NSFC) under Grant 51876096, and National Key Research and Development Program of China under Grant 2016YFB0600201.

Conflicts of Interest: The authors declare no conflict of interest.

Abbreviations

The following abbreviations are used in this manuscript:

ADRC	Active Disturbance Rejection Control
CFB	Circulating Fluidized Bed
CCS	Coordinated Control System
DDE	Desired Dynamic Equation
DOB	Disturbance Observer
ESO	Extended State Observer
GA	Genetic Algorithm
IAE	Integrated Absolute Error
MIMO	Multi-Input-Multi-Output
PID	Proportional-Integral-Derivative

References

1. Yue, G.; Cai, R.; Lyu, J.; Zhang, H. From a CFB reactor to a CFB boiler—The review of R&D progress of CFB coal combustion technology in China. *Powder Technol.* **2017**, *316*, 18–28. [CrossRef]
2. Nyashina, G.S.; Kurgankina, M.A.; Strizhak, P.A. Environmental, economic and energetic benefits of using coal and oil processing waste instead of coal to produce the same amount of energy. *Energy Convers. Manag.* **2018**, *174*, 175–187. [CrossRef]
3. Kurgankina, M.A.; Nyashina, G.S.; Strizhak, P.A. Advantages of switching coal-burning power plants to coal-water slurries containing petrochemicals. *Appl. Therm. Eng.* **2019**, *147*, 998–1008. [CrossRef]
4. Lyu, J.; Yang, H.; Ling, W.; Nie, L.; Yue, G.; Li, R.; Chen, Y.; Wang, S. Development of a supercritical and an ultra-supercritical circulating fluidized bed boiler. *Front. Energy* **2019**, *13*, 114–119. [CrossRef]
5. Chang, S.; Zhuo, J.; Meng, S.; Qin, S.; Yao, Q. Clean Coal Technologies in China: Current Status and Future Perspectives. *Engineering* **2016**, *2*, 447–459. [CrossRef]
6. Song, C.; Lyu, J.; Yang, H.; Wang, S.; Ling, W.; Yue, G. Research and Application of Supercritical and Ultra-supercritical Circulating Fluidized Bed Boiler Technology. *Proc. CSEE* **2018**, *38*, 338–347.
7. Brouwer, A.S.; van den Broek, M.; Seebregts, A.; Faaij, A. Operational flexibility and economics of power plants in future low-carbon power systems. *Appl. Energy* **2015**, *156*, 107–128, [CrossRef]
8. Liu, J.; Zeng, D.; Tian, L.; Gao, M.; Wang, W.; Niu, Y.; Fang, F. Control Strategy for Operating Flexibility of Coal-fired Power Plants in Alternate Electrical Power Systems. *Proc. CSEE* **2015**, *35*, 5385–5394.
9. Gao, M.; Hong, F.; Liu, J.; Chen, F. Investigation on the energy conversion and load control of supercritical circulating fluidized bed boiler units. *J. Process Control* **2018**, *68*, 14–22. [CrossRef]
10. Lv, Y.; Hong, F.; Yang, T.; Fang, F.; Liu, J. A dynamic model for the bed temperature prediction of circulating fluidized bed boilers based on least squares support vector machine with real operational data. *Energy* **2017**, *124*, 284–294. [CrossRef]
11. Majanne, Y.; Köykkä, P. Dynamic Model of a Circulating Fluidized Bed Boiler. *IFAC Proc. Vol.* **2009**, *42*, 255–260. [CrossRef]
12. Gao, M.; Hong, F.; Yan, G.; Liu, J.; Chen, F. Mechanism modelling on the coordinated control system of a coal-fired subcritical circulating fluidized bed unit. *Appl. Therm. Eng.* **2019**, *146*, 548–555. [CrossRef]
13. Wu, H.; Yang, C.; He, H.; Huang, S.; Chen, H. A hybrid simulation of a 600MW supercritical circulating fluidized bed boiler system. *Appl. Therm. Eng.* **2018**, *143*, 977–987. [CrossRef]
14. Gao, M.; Hong, F.; Zhang, B.; Liu, J.; Yue, G.; Yang, A.; Chen, F. Study on Nonlinear Control Model of Supercritical (Ultra Supercritical) Circulating Fluidized Bed Unit. *Proc. CSEE* **2018**, *38*, 363–372.
15. Hultgren, M.; Ikonen, E.; Kovács, J. Once-through Circulating Fluidized Bed Boiler Control Design with the Dynamic Relative Gain Array and Partial Relative Gain. *Ind. Eng. Chem. Res.* **2017**, *56*, 14290–14303. [CrossRef]
16. Hao, Y.; Li, D.; Tan, W. Design of the PID controller for circulating fluidized bed boiler combustion system. In Proceedings of the 31st Chinese Control Conference, Hefei, China, 25–27 July 2012; pp. 4580–4585.
17. Menhas, M.I.; Wang, L.; Hui, P.; Fei, M. CFBB PID Controller Tuning with Probability based Binary Particle Swarm Optimization Algorithm. In *Proceedings of the International Conference on Intelligent Computing for Sustainable Energy and Environment*; Springer: Berlin/Heidelberg, Germany, 2010; pp. 44–51.
18. Xue, Y.; Li, D.; Zhang, Y.; Gao, Q.; Wang, J.; Sun, Z. Decentralized nonlinear control of 300MWe circulating fluidized boiler power unit. In Proceedings of the 2012 UKACC International Conference on Control, Cardiff, UK, 3–5 September 2012; pp. 1019–1024. [CrossRef]
19. Sun, L.; Li, D.; Lee, K.Y. Enhanced decentralized PI control for fluidized bed combustor via advanced disturbance observer. *Control Eng. Pract.* **2015**, *42*, 128–139. [CrossRef]
20. Zhuo, X.; Lou, C.; Zhou, H.; Zhuo, J.; Fu, P. Hierarchical Takagi-Sugeno fuzzy hyperbolic tangent static model control for a circulating fluidized bed boiler thermal power unit. *Energy* **2018**, *162*, 910–917. [CrossRef]
21. Sun, H.; Sun, P. Study on fuzzy control of circulating fluidized bed temperature. In Proceedings of the 3rd International Conference on Control Science and Systems Engineering, Beijing, China, 17–19 August 2017; pp. 154–157. [CrossRef]
22. Wang, L.; Yang, R.; Pardalos, P.M.; Qian, L.; Fei, M. An adaptive fuzzy controller based on harmony search and its application to power plant control. *Int. J. Electr. Power Energy Syst.* **2013**, *53*, 272–278. [CrossRef]

23. Sethi, B.K.; Vinayak, J.R.; Sivakumaran, N. Internal model controller for CFBC boiler using neural networks. In Proceedings of the 2018 Indian Control Conference, Kanpur, India, 4–6 January 2018; pp. 276–281. [CrossRef]
24. Hu, M.; Ling, H.; Liu, D. The application of LM-BP Neural Network in the Circulating Fluidized Bed Unit. In *Proceedings of the 4th International Conference on Mechatronics, Materials, Chemistry and Computer Engineering 2015*; Liang, Z.; Li, X., Eds.; ACSR-Advances in Comptuer Science Research; Atlantis Press: Paris, France, 2015; Volume 39, pp. 2017–2020.
25. Xue, W.; Bai, W.; Yang, S.; Song, K.; Huang, Y.; Xie, H. ADRC With Adaptive Extended State Observer and its Application to Air–Fuel Ratio Control in Gasoline Engines. *IEEE Trans. Ind. Electron.* **2015**, *62*, 5847–5857. [CrossRef]
26. Sun, L.; Shen, J.; Hua, Q.; Lee, K.Y. Data-driven oxygen excess ratio control for proton exchange membrane fuel cell. *Applied Energy* **2018**, *231*, 866–875. [CrossRef]
27. Han, J.; Wang, H.; Jiao, G.; Cui, L.; Wang, Y. Research on Active Disturbance Rejection Control Technology of Electromechanical Actuators. *Electronics* **2018**, *7*, 174. [CrossRef]
28. Roman, R.C.; Radac, M.B.; Tureac, C.; Precup, R.E. Data-Driven Active Disturbance Rejection Control of Pendulum Cart Systems. In Proceedings of the 2018 IEEE Conference on Control Technology and Applications, Copenhagen, Denmark, 21–24 August 2018; pp. 933–938. [CrossRef]
29. Wang, G.; Xu, Q. Sliding Mode Control with Disturbance Rejection for Piezoelectric Nanopositioning Control. In Proceedings of the 2018 Annual American Control Conference, Milwaukee, WI, USA, 27–29 June 2018; pp. 6144–6149. [CrossRef]
30. Ma, S.; Sun, M.; Chen, Z. Interactive ADRC design for flight attitude control. In Proceedings of the 6th Data Driven Control and Learning Systems, Chongqing, China, 26–27 May 2017; pp. 611–616. [CrossRef]
31. Feng, H.; Guo, B.Z. Active disturbance rejection control: Old and new results. *Annu. Rev. Control* **2017**, *44*, 238–248. [CrossRef]
32. Huang, Y.; Xue, W. Active disturbance rejection control: Methodology and theoretical analysis. *ISA Trans.* **2014**, *53*, 963–976. [CrossRef] [PubMed]
33. Wu, Z.; He, T.; Li, D.; Xue, Y.; Sun, L.; Sun, L. Superheated steam temperature control based on modified active disturbance rejection control. *Control Eng. Pract.* **2019**, *83*, 83–97. [CrossRef]
34. Zhang, Y.; Xue, Y.; Li, D.; Wu, H.; Lee, K.Y. Active Disturbance Rejection Control for Bed-inventory Balancing of CFB Boilers with Pant-leg Structure. *IFAC-PapersOnLine* **2015**, *48*, 66–71. [CrossRef]
35. He, T.; Wu, Z.; Shi, R.; Li, D.; Sun, L.; Wang, L.; Zheng, S. Maximum Sensitivity-Constrained Data-Driven Active Disturbance Rejection Control with Application to Airflow Control in Power Plant. *Energies* **2019**, *12*, 231. [CrossRef]
36. Pan, F.; Liu, Q.; Sun, L.; Li, D.; Tan, W. A novel design of active disturbance rejection controller and its application in the Circulating Fluidized Bed Boiler combustion system. In Proceedings of the 2015 American Control Conference, Chicago, IL, USA, 1–3 July 2015; pp. 3950–3955. [CrossRef]
37. Zhang, Y.; Li, D.; Xue, Y. Active disturbance rejection control for Circulating Fluidized Bed Boiler. In Proceedings of the 12th International Conference on Control, Automation and Systems, JeJu Island, Korea, 17–21 October 2012; pp. 1413–1418.
38. Wu, H.; Xue, Y.; Guo, Z.; Wang, Z.; Li, Z. Dynamic Simulation Model of Combustion System of CFB Boilers. In *Proceedings of the 23rd International Conference on Fluidized Bed Conversion*; KIChE: Seoul, Korea, 2018; pp. 1–7.
39. Gao, Z. Scaling and bandwidth-parameterization based controller tuning. In Proceedings of the 2003 American Control Conference, Denver, CO, USA, 4–6 June 2003; Volume 6, pp. 4989–4996. [CrossRef]
40. Madoński, R.; Herman, P. Survey on methods of increasing the efficiency of extended state disturbance observers. *ISA Trans.* **2015**, *56*, 18–27. [CrossRef]
41. Xue, W.; Huang, Y.; Gao, Z. On ADRC for non-minimum phase systems: canonical form selection and stability conditions. *Control Theory Technol.* **2016**, *14*, 199–208. [CrossRef]

42. Tan, W.; Fu, C. Linear Active Disturbance-Rejection Control: Analysis and Tuning via IMC. *IEEE Trans. Ind. Electron.* **2016**, *63*, 2350–2359. [CrossRef]
43. Whitley, D.; Sutton, A.M. Genetic algorithms—A survey of models and methods. In *Handbook of Natural Computing*; Springer: Berlin/Heidelberg, Germany, 2012; Volume 2–4, pp. 637–671. [CrossRef]

Article

Optimization of Dispatching Electricity Market with Consideration of a Solar-Assisted Coal-Fired Power Generation System

Rongrong Zhai [1],*, Lingjie Feng [1], Hai Yu [2], Chao Li [1] and Yongping Yang [1]

[1] School of Energy Power and Mechanical Engineering, North China Electric Power University, Beijing 102206, China; fenglj1995@163.com (L.F.); Lichao201408@163.com (C.L.); yypncepu@163.com (Y.Y.)
[2] CSIRO Energy, 10 Murray Dwyer Circuit, Mayfield West, NSW 2304, Australia; hai.yu@csiro.au
* Correspondence: zhairongrong01@163.com

Received: 2 March 2019; Accepted: 28 March 2019; Published: 3 April 2019

Abstract: This study investigates the multi-objective optimization of load dispatch of a solar-assisted coal-fired power generation system. The improved environmental/economic load dispatch model considers coal consumption, NO_x emissions, and power purchase cost. The singular weighted method is utilized to solve this multi-objective and multi-constraint optimization problem. A power system that includes five power generators, one of which is retrofitted to a solar-assisted coal-fired unit, is also analyzed. It can be concluded that the loads of solar-assisted coal-fired units are higher than the original coal-fired unit, and with the increase of solar radiation, the gap between the loads of two units also increases. In addition, after retrofitting, the coal consumption, the NO_x emission, and power costs of units reduce by about 2.05%, 0.45%, and 0.14%, respectively. From the study on the on-grid power tariff, where the tariff drops from 16.29 cents/kWh to 3.26 cents/kWh, NO_x emissions drop from 12.31 t to 11.28 t per day, a reduction of about 8.38%. The cost of purchasing electricity decreases from $ 2,982,161.8 to $ 2,020,505.0 per day, a decrease of 32.25%. Therefore, when both coal-fired units and solar-assisted coal-fired units exist in a region, the use of solar-assisted coal-fired power generation units should be prioritized.

Keywords: Solar-assisted coal-fired power generation system; Singular weighted method; load dispatch

1. Introduction

The load dispatch of a power generation system is defined, on the basis of the efficiency, power, and other properties of each unit, as the arrangement of loads for the units [1]. The traditional load dispatch problem is to find the best way to arrange the power output from all types of generating plants to realize the maximum economic benefits. However, air pollution poses a serious threat to human health and the environment [2]. In China, the government has pressed various policies to address these problems. In terms of reducing pollutant emissions, the government has focused on industries. Pollutant emissions of the industries mostly come from coal-fired power generation. By the end of 2018, approximately 70% of the electricity in China was still derived from coal-fired power generation systems [3]. One way to deal with this situation is to exploit renewable energy to replace coal-fired power generation. Solar energy is a clean, renewable energy that can meet the energy demands of people. Solar-assisted coal-fired units that utilize various types of solar thermal systems for coupling traditional coal-fired power plants not only reduce pollutants and greenhouse gas emissions but also reduce the investment of solar energy facilities. Research by Wang et al. [4] showed the local influence on solar-assisted coal-fired power generation system (SCPGS) performance due to the substitution of chemical exergy by solar thermal exergy and the global influence on SCPGS performance jointly determines the final coal-savability for a SCPGS in a superposition manner. The superposition effect (the interaction of local influence and global influence) shows that the overall coal saving capability of SCPGS depends not only on the direct benefits brought by the local integration of solar energy, but also on the overall superposition effect of the integration on the system components. The comprehensive evaluation factor (f) is proposed on the premise of giving consideration to both the coal-savability and efficiency-promotability. By the end of 2018, the power generating capacity of grid-connected solar power was 174.63 million kW, an increase of 33.9% year-on-year [3].

Previous studies demonstrated that solar thermal technology is mature enough and that the combination of solar energy and coal-fired units is feasible. Desai et al. [5] proposed a method to determine a thermodynamic and cost-effective design for solar power plants without hybridization and storage. Some studies mainly focused on the use of thermal storage systems in solar thermal energy generation systems. Raul et al. [6] conducted experiments on a single spherical capsule and a lab-scale setup to investigate melting, and analyzed the effect of capsule diameter and porosity on the LHTES (based latent heat thermal energy storage). The results showed that the energy stored and extraction are faster for a smaller capsule diameter and higher porosity. Lakhani et al. [7] considered a multitube shell and tube latent heat thermal storage system, and found that the LHTES system with a lower overall diameter and longer length provides better overall performance for a solar thermal power plant. Corgnale et al. [8] studied solar power plants from the aspect of thermal energy storage systems and found that the selected storage systems with metal hydride high-temperature materials can achieve and exceed the requirements, such as volumetric energy density (25 kWhth/m^3) and operating temperature (600 °C). Yogev et al. [9] researched an operating strategy based on thermal storage with the latent heat of phase change materials (PCMs) and indicated that these PCMs can achieve stable power output without any heat transfer enhancement. Therefore, these materials can reduce the complexity and cost of the storage system. Ehrhart et al. [10] found a large gap between the design point with annual average performance (mainly due to the solar field and receiver subsystems) and energy losses because of the thermal energy storage being full to capacity. Eduard et al. [11] used life-cycle assessment to compare the environmental impacts of different thermal energy storage systems and found that the systems based on solid and liquid media exerted the lowest and the highest environmental impact per kWh stored, respectively. Other scholars studied the operating performance of solar-assisted coal-fired power generation systems. Eric et al. [12] proposed the concept of the solar-assisted coal-fired unit, which increases the efficiency and reduces the greenhouse gas emission of the traditional power plant. Yang et al. [13] took a 200 MW unit as an example and found that using medium- or low-temperature solar heat to replace the extraction can achieve a solar heat-to-power conversion efficiency of 36.5% for solar heat at 260 °C. Hong et al. [14] studied a 330 MW coal-fired power plant hybridized with solar heat in Sinkiang Province, China, and found that the levelized cost of electricity generation could be 20% to 30% lower than that of the solar-only thermal power plant.

Some studies proposed different methods to solve the problem of load distribution. Bhattacharjee et al. [15] proposed the real coded chemical reaction algorithm to solve the economic emission load distribution problem. In this method, they considered NO_x emissions, power demand equality constraint, and operating limit constraint. Jeddi et al. [16] proposed a new modified harmony search algorithm to solve the economic load dispatch problem and combined the economic emission load dispatch problem. Leena et al. [17] proposed use of Artificial Neural Networks, which are based on Levenberg Marquardt backpropagation Algorithm (LMBP), to make rapid decisions in the load scheduling center and solve the economic load dispatch problem. Coelho et al. [18] proposed particle swarm optimization approaches to solve the economic load dispatch problem. Another researcher studied the problem of electricity load distribution by considering wind power. Li et al. [19] proposed a stochastic multi-objective optimization method to solve the security-constrained optimal power flow problem with wind power and distributed load changing. This method considers the minimum expectation of fuel costs, the maximum wind power penetration with wind speed, and distributed load changes. Mondal et al. [20] proposed the gravitational search algorithm to solve the economic emission load dispatch problem by considering both thermal generators and wind turbines with the minimum amount of NO_x emissions and fuel costs. Aghaei et al. [21] presented a scenario-based stochastic programming framework to model the random nature of load demand and wind forecast errors. Taking empirical data as input, Modarresi et al. [22] formulated the look-ahead real-time economic dispatch problem using the scenario approach, which generated a quantifiable risk level in real-time economic scheduling and generated direct benefits to both real-time and intra-day decision making processes. Zhu et al. [23] proposed a decomposition-based multi-objective evolutionary algorithm to optimize the load distribution problem. Rizk et al. [24] proposed a new improved search algorithm based on parallel hurricane search optimization for non-smooth constrained economic emission dispatch problems to solve the nonlinear constrained economic emission dispatch problem. The results show that this method is superior to other optimization methods. Roy [25] gave the formulation of the maximum likelihood load dispatch, as well as Karush-Kuhn-Tucker conditions of its optimal solution, in order to reduce the impact of the instability of solar energy on grid connection; it is also a method for making certain operation reserves for solar energy. However, operating reserves may lead to a significant increase in solar grid connection costs. Modarresi et al. [26] introduced a new method of quantifying operation reserves in power systems with high intermittent resource utilization rate, and introduced an intuitive and rigorous risk diagram, which will help to obtain more cost-effective operation reserves while ensuring system reliability, load probability loss, and other performance indexes.

Only a few studies have analyzed solar-assisted coal-fired units from the aspect of load dispatch of the power generation system. In our previous work, we have done a lot of research on solar-assisted coal-fired power units on their unit features, performances analysis, integration optimization, and so on. Based on the previous work, how to make solar-assisted coal-fired power generation adapt to the complex and changeable environment of a power grid is further considered. In the present study, a coal-fired unit is retrofitted into a solar-assisted coal-fired unit, and the other remaining units are not changed. Under the given power demand, the multi-objective model considers coal consumption, NO_x emissions, and power purchase cost. The original and retrofitted units are compared and analyzed. This study may be used as a guide to propose appropriate policies that encourage and promote the development of solar-assisted coal-fired units.

2. Problem Description

The load dispatch of the power generation system is that the administrative office arranges loads for the power generating units on the basis of the efficiency, power, and other properties of each unit. Load distribution on the distribution of the energy and the operation of an electric power system is highly important. The load dispatch of the power generation changes with the operational characteristic of a unit. Therefore, the load dispatch of the power generation in this area changes when a solar-assisted coal-fired unit is introduced to the energy supply system. This study analyzes the changes in load dispatch when the coal-fired unit is retrofitted to a solar-assisted coal-fired unit. The diagram of the solar-assisted coal-fired system is shown in Figure 1. The system mainly contains "solar side" and "coal-fired side" parts. The "solar side" is composed of a solar collector system and heat exchange system. Solar collector plant is though collector plant form, in which the working medium is thermal oil. The heat exchanger is an oil–water heat exchanger. When the oil passes through the heat exchanger, its temperature decreases. When the water passes through the exchanger, its temperature increases. The solar collectors are produced by LUZ International, Ltd. A solar collector assembly is 47.1 m long with an aperture width of 5 m. The heat is absorbed in a steel absorber tube which is 66/70 mm in diameter, surrounded by an evacuated glass envelop which is 115/120 mm in diameter. The concentration ratio is about 71 and the collector focal length is 1.49 m. The parabolic trough axes are oriented due north-south to track the sun from east to west.

In the coal-fired side, the unsaturated feedwater from the condenser enters into the boiler after going through the condensate pump, four low-pressure reheaters (H5, H6, H7, and H8), a deaerator, a feedwater pump, and three high pressure reheaters (H1, H2, and H3). Then the feedwater absorbs heat in the boiler and becomes superheated steam. The superheated steam from the boiler is transported to the high-pressure cylinder in the turbine to produce power. The steam from the high-pressure cylinder goes into the boiler and is reheated in order to improve work capacity. Then, the reheated steam is transported to intermediate pressure and lower pressure cylinders to produce power. In the end, the final exhaust is condensed in the condenser.

As shown in the diagram, feed water from the deaerator enters the heat exchanger and absorbs the solar heat. Four modes of operation are available according to the heat exchanger outlet temperature of the feedwater.

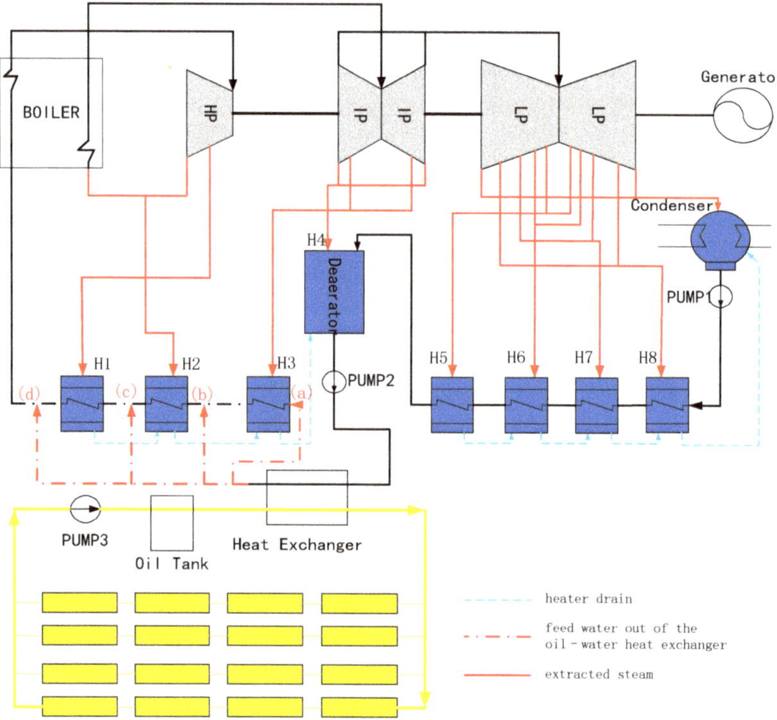

Figure 1. Diagram of solar-assisted coal-fired system.

(1) When the temperature of feed water out of the oil–water heat exchanger is lower than that of the original third heater. The feed water enters the third heater. This mode is shown in (a) position.

(2) When the temperature of the feed water out of the oil–water heat exchanger is lower than that of the original second heater but higher than that of the original third heater. The feed water enters the second heater. This mode is shown in (b) position.

(3) When the temperature of feed water out of the oil–water heat exchanger is lower than that of the original first heater but higher than that of the original second heater. The feed water enters the first heater. This mode is shown in (c) position.

(4) When the temperature of feed water out of the oil–water heat exchanger is higher than that of the original first heater. The feed water enters the boiler. This mode is shown in (d) position.

3. Methodology

Under certain constraints, multi-objective optimization aims to have more than one goal to achieve an optimal solution. One of the easiest ways to obtain the optimal solution is called standard quantization or the singular weighted method. A weight is assigned to each target. All the targets with their own weights are positioned together, and then the problem is solved under the same constraints of the original problem. The obtained solution is a solution to the original problem (called the Pareto solution). The weight arrangement is constantly changed; hence, different optimal solutions are obtained. The present study takes the units in a particular area as an example, simultaneously considering coal consumption, nitrogen oxide emissions, and electricity purchase. Then, the optimal solution and the load dispatch of each unit are obtained under the power demand in this region.

3.1. Objective Function

3.1.1. Minimum Coal Consumption

Coal consumption is a commonly used index in the evaluation of coal-fired power plants. Therefore, minimizing the coal consumption of coal-fired power plants is crucial in modern energy systems.

$$min f_1 = \sum_{j=1}^{N} \left[a_j P_j^2(t) + b_j P_j(t) + c_j \right] \tag{1}$$

where N is the number of units; a_j, b_j, and c_j are the coefficients of coal consumption; and $P_j(t)$ is the current output of the power plant.

3.1.2. Minimum Pollutant Emissions

Pollutant emission performance during power production is considered, and the function of minimum pollutant emissions is established in various energy systems by fitting the discharge curves of various energy systems. In this study, all the units are equipped with desulphurization equipment with an efficiency of above 98%. Therefore, the main consideration emissions of units are NO_x emissions.

$$min f_2 = \sum_{j=1}^{N} \left[u_j P_j^2(t) + v_j P_j(t) + w_j \right] \tag{2}$$

where u_j, v_j, and w_j are the coefficients of pollutant emission.

3.1.3. Minimum Power Costs

$$min f_3 = \sum_{j=1}^{N} \rho_j(t) P_j(t) \tag{3}$$

where $\rho_j(t)$ is the current power purchase cost.

3.2. Constraints

3.2.1. Load Balance between Supply and Demand

Load balance between supply and demand means that the power output and the electricity should be equal at any time.

$$\sum_{j=1}^{N} P_j(t) = P_D(t) + P_L(t) \tag{4}$$

where $P_L(t)$ stands for the grid loss and $P_D(t)$ stands for the grid power requirement.

3.2.2. Upper and Lower Limits for the Units

The power generation of the units is deeply influenced by the characteristics of the units. For example, a coal-fired power plant should work least at the load ratio of approximately 30%

to avoid extra oil consumption. The characteristics of different types of units should be considered to determine the upper and lower limits for the units.

$$P_{jmin} \leq P_j(t) \leq P_{jmax} \tag{5}$$

where P_{jmin} and P_{jmax} represent the lower and upper limits for the units, respectively.

3.3. Multi-Objective Optimization

The method for solving multi-objective function has rapidly developed. Some scholars proposed multi-objective differential evolution algorithms to solve problems [27–31]. Ghasemi [32] proposed improved multi-objective interactive honey bee mating optimization to solve the environmental/economic power dispatch problem. In this method, the fuel costs, pollutant emissions, and system loss should be minimized while satisfying certain system constraints. Boyhina et al. [33] proposed a trust region algorithm to solve the multi-objective economic emission load dispatch problem. The approach they proposed is efficient for solving ill-defined systems and non-convex multi-objective optimization problems.

The aim of conventional generation planning is to minimize the cost of power generation under conditions that meet the system constraints and electricity demand. Considering the lowest operating costs, fuel cost, or electricity purchase is a single objective function.

3.3.1. Singular Weighted Method

In accordance with the weight of the original objective function, a new objective function g(x) is constructed. The solution of the single-objective optimization function can be the efficient solution of the problem.

$$g(x) = w_1 \frac{f_1}{f_1^{min}} + w_2 \frac{f_2}{f_2^{min}} + \ldots + w_m \frac{f_m}{f_m^{min}} \tag{6}$$

where $f_i^{min} = \min\limits_{x \in s} f_i(x)$, $(i = 1,2, \ldots ,m)$ is the lowest solution of the single objective function; $f^{min} = \left(f_1^{min}, f_2^{min}, \ldots, f_m^{min}\right)$ is the optimal solution vector with all the elements denoting the optimal solution for each objective function; $w = (w_1, w_2, \ldots, w_m)$ is the weighted vector, and $\sum\limits_{j=1}^{N} w_i = 1$, $w_i \geq 0$, $i = 1,2, \ldots ,m$. In this model, $\frac{f_i}{f_i^{min}}$ is used to convert the different objective functions to the same magnitude.

The effective solution of the multi-objective function can be solved by solving the single objective function $\min\limits_{x \in s} g(x)$.

3.3.2. Realization of the Algorithm Using MATLAB

According to the singular weighted method, objective functions (1) to (3) can be converted to

$$g(x) = w_1 \frac{f_1}{f_1^{min}} + w_2 \frac{f_2}{f_2^{min}} + w_3 \frac{f_3}{f_3^{min}} \tag{7}$$

where f_1^{min}, f_2^{min}, and f_3^{min} stand for the optimal values of objective functions (1) to (3), respectively, and can be derived by solving the corresponding single-objective functions; and w_1, w_2 and w_3 stand for specified weight coefficients and known quantities. Equation (7) is simplified by removing the constant term to obtain

$$
\begin{aligned}
g(x) \quad &= w_1 \frac{\sum_{j=1}^{N} \left[a_j P_j^2(t) + b_j P_j(t)\right]}{f_1^{min}} + w_2 \frac{\sum_{j=1}^{N} \left[u_j P_j^2(t) + v_j P_j(t)\right]}{f_2^{min}} + w_3 \frac{\sum_{j=1}^{N} \rho_j(t) P_j(t)}{f_3^{min}} \\
&= \sum_{j=1}^{N} \left[\left(\frac{w_1 a_j}{f_1^{min}} + \frac{w_2 u_j}{f_2^{min}}\right) P_j^2(t)\right] + \sum_{j=1}^{N} \left[\left(\frac{w_1 b_j}{f_1^{min}} + \frac{w_2 v_j}{f_2^{min}} + \frac{w_3 \rho_j}{f_3^{min}}\right) P_j(t)\right]
\end{aligned} \tag{8}
$$

Therefore, the original problem can be solved by solving the objective function (8) under constraints (4) and (5). Quadprog function, a toolbox in MATLAB, can be used to solve quadratic programming problems.

General quadratic programming problems can be described as [34]

$$\text{Objective Function}: min\left(\frac{1}{2}x^T H x + f^T x\right) \tag{9}$$

$$\text{Constraints}: \begin{cases} Ax \leq B \\ A_{eq} = B_{eq} \\ x_m \leq x \leq x_M \end{cases} \tag{10}$$

Equation (9) can be solved by MATLAB.
Syntax:

$$[x, f_{opt}, \text{flag}, c] = \text{quadprog}\left(H, f, A, B, A_{eq}, B_{eq}, x_m, x_M, x_0\right) \tag{11}$$

where H is the Hessian matrix of the quadratic programming objective function, f is the vector of the coefficients of x, A is the matrix for linear inequality constraints, B is the vector for linear inequality constraints, A_{eq} is the matrix for the linear equality constraints, B_{eq} is the vector for the linear equality constraints, x_m is the vector of lower bounds, x_M is the vector of upper bounds, and x_0 is the initial point for x. If the constraint of matrix does not exist, an empty matrix is used to replace it. After the optimized operation finishes, the results will be returned in variable x, and the result of the optimization objective function will be returned in f_{opt}. If the flag is a positive number, the equation is solved successfully. The number of iterations and the number of function calls, algorithms, and other information will be returned in variable c.

4. Case Study

4.1. Basic Data

The area has five coal-fired units. No. 4 is retrofitted to solar-assisted coal-fired units. The retrofitted system is fictitious based on the real parameters. The parameters of five units are shown in Table 1.

Table 1. Parameters for the case study.

Unit No.	Coal Consumption Coefficients			No$_x$ Emission Coefficients			Lower and Upper Limits		On-grid Power Tariff
	(th^{-1})			(10e^{-7}th^{-1})			(MW)		(Cent/kWh)
	a_j	b_j	c_j	u_j	v_j	w_j	P_{jmin}	P_{jmax}	-
1	0.000175	0.11	3	6.49	5.554	4.091	310	570	4.40
2	0.00023	0.15	5	5.638	6.047	2.543	250	425	3.26
3	0.000116	0.07	7	4.586	5.094	4.257	350	700	4.07
4a	−0.000039	0.29	5.3	3.718	3.905	5.859	260	680	3.26
4b	a	b	6.43	3.624	−3.55	5.739			
5	0.00012	0.12	5	4.586	5.094	4.258	325	660	4.40

Note: a = −(0.000005662+0.00000003497DNI), b = 0.2707 − 0.00004014DNI + 0.00000003023DNI^2, DNI is the Direct Normal Irradiance, 4a stands for the original unit, 4b stands for the retrofitted unit.

On the basis of the parameters of No. 4 unit and collector and the values of DNI (direct normal insolation), the parameters of No. 4 unit are fitted to the function of load and DNI by using MATLAB. The results are presented in Table 1. The system of the retrofitted unit is the same as the solar-assisted coal-fired system in Figure 1. The parameters of No. 4 unit and collector are shown in Tables 2 and 3, respectively.

Table 2. Parameters of No. 4 unit.

Parameters	Units	Values				
Load	%	100	85	75	50	45
Capacity	MW	600.00	561.09	495.00	330.00	264.00
Parameters of main steam	Mpa/°C	24.2/566	24.2/566	23.5/566	15.9/566	13.3/566
Reheat temperature	°C	566	566	566	566	566
Feedwater mass flowrate	t/h	1836.5	1552.2	1360.1	899.0	733.3
Condenser pressure	kpa	4.9	4.9	4.9	4.9	4.9
Exhaust specific enthalpy	kJ/kg	2314.7	2347	2369.4	2440.1	2469.1
Designed coal consumption rate	g/kWh	270.9	275.7	278.8	293.3	298.3

Table 3. Parameters of collector.

Parameters	Units	Values
Width/Length	m	5/47.1
The number of Loop	-	70
The number of SCA (solar collector assembly) in Loop	-	8
Collector array pitch	m	15
Collector area	m^2	263,200
The total area	m^2	1,052,800

This study was based on the DNI value in Lhasa at a latitude of 29.60 N, longitude of 91.10 E, and elevation of 3658 m in Tibet Province, one of the highest solar irradiation areas in China. In this paper, the values of DNI in June 25 are used to calculate the coal consumption of the units. The DNI values are shown in Figure 2.

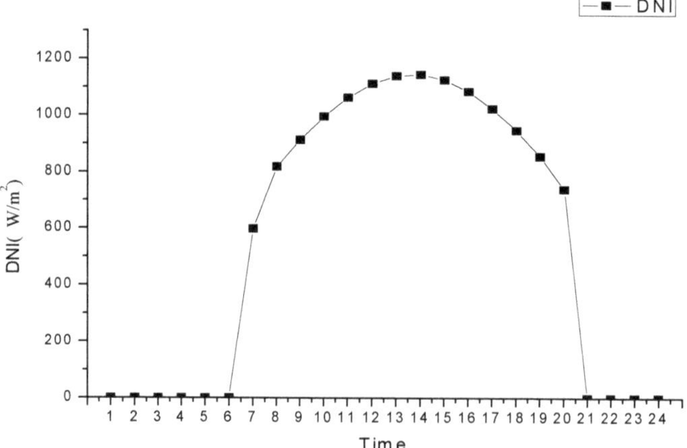

Figure 2. Values of DNI within 24 h.

The power load demands in this area are shown in Figure 3; w is chosen as w = (1/3, 1/3, 1/3). To simplify the calculation, the loss of grid is ignored.

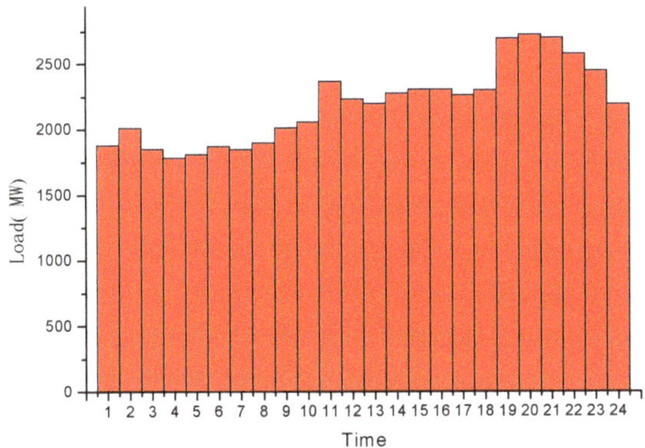

Figure 3. Power load demands within 24 h.

4.2. Results

When the results of w = (1/3, 1/3, 1/3), the parameters of this No. 4 original unit (OU) and retrofitted unit (RU) are shown in Table 4. More detailed values of each unit load allocation are shown in Table A1 in the Appendix A.

Table 4. Parameters of this No. 4 original unit (OU) and retrofitted unit (RU).

No. 4 Unit	Coal Consumption	NO_x Emission	Power Costs
	td^{-1}	td^{-1}	$\$d^{-1}$
OU	10,924.4	11.33	2,023,434.1
RU	10,699.9	11.29	2,020,505.0

Table 4 shows that the coal consumption of units is reduced by approximately 2.05% from 10924.4 t to 10699.9 t per day. The NO_x emission decreases by approximately 0.45% from 11.33 t to 11.29 t per day. The power costs are reduced by approximately 0.14% from $ 2,023,434.1 to $ 2,020,505.0 per day. Therefore, coal consumption has a greater degree of decline, while NO_x emissions and power cost decline are not obvious. This result demonstrates that the development of solar-assisted coal-fired units can effectively deal with resource exhaustion, environmental pollution, and other issues from energy, environmental, and economic aspects.

The loads of the original and retrofitted units are shown in Figure 4.

Figure 4 shows that the loads of units changing with time are both consistent with the power demand in this region. In periods 1 to 6, the loads of the original and retrofitted units are basically the same because the value of DNI is 0 in both original and retrofitted units. In periods 7 to 18, the loads of the retrofitted unit are higher than those of the original unit. With the increase or decrease in DNI, the difference between the loads of two units also shows a corresponding change. In period 11, the difference between the loads of the two units decreases because the load of the retrofitted unit has reached the upper limit of the unit. In period 14, DNI reaches a maximum of 1143.96 MW/m². The gap between two units also reaches a maximum of 40.64 MW. In periods 19 and 20, the loads of the original and retrofitted units have reached the upper limit of the unit. In periods 21 to 24, DNI is 0, and the loads of the original and retrofitted units are similar.

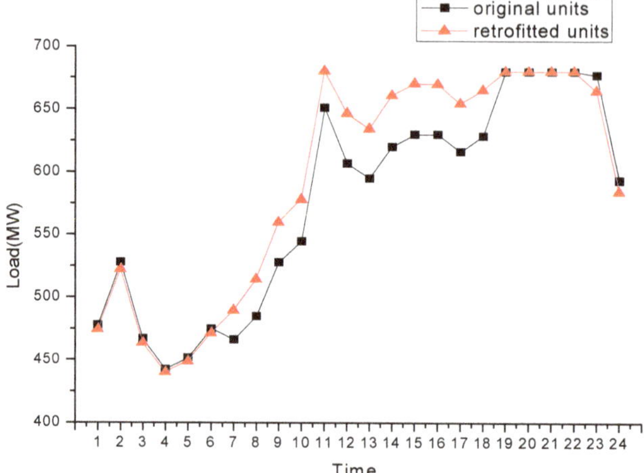

Figure 4. Load curves of No. 4 unit before and after reconstruction.

In periods 7 to 20, the loads for the units changing between the original and retrofitted units is shown in Figure 5.

Figure 5 shows that the load of No. 4 unit is higher after the coal-fired unit is retrofitted into the solar-assisted coal-fired unit in the presence of solar radiation. Nevertheless, the loads are lower for the remaining units. Thus, under the multi-objective optimal condition, the advantages of the solar-assisted coal-fired unit are higher than those of the coal-fired unit. Therefore, the use of solar-assisted coal-fired power generation units should be prioritized when both coal-fired units and solar-assisted coal-fired units are present in a region.

For periods 7, 11, and 14, load ratios of five units are shown in Figure 6.

As shown in Figure 6, in periods 7, 11, and 14, the loads of No. 4 unit have increased by 1.27%, 1.57%, and 1.79%, respectively. With increasing DNI values, the gap of loads between the coal-fired unit and solar-assisted coal-fired unit also increases.

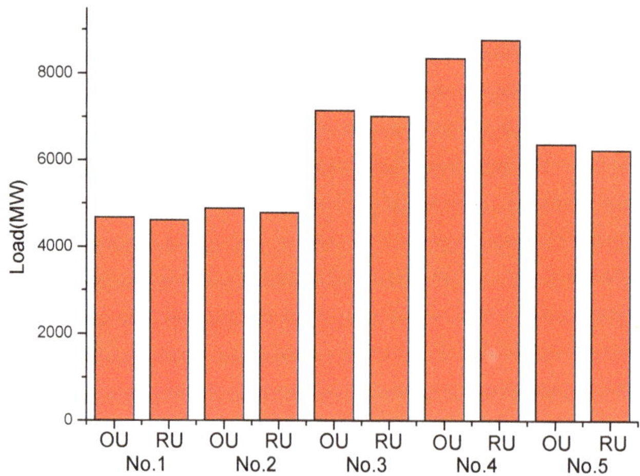

Figure 5. Load of units changing before and after reconstruction.

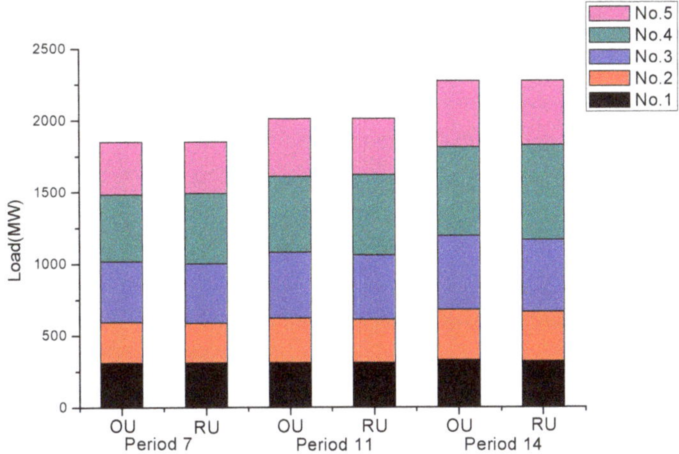

Figure 6. Load ratio of five units.

5. Uncertainty Analysis

Uncertainty analysis analyzes every factor that affects the project. Most of the basic data used in the evaluation process, such as investment, production, capital, revenue, and so on, are based on the calculation of forecasts and estimates, and it is assumed that these parameters remain unchanged in the calculation period. Decisions made on the basis of these conditions are called decisive decisions. However, due people's awareness of the limitations of predictions and the limited availability of information, these data and the actual situation may be different, thus leading to uncertainty. To reduce the risk of uncertainty and mistakes in the project, the impact of various uncertainties on the project must be analyzed. In the model building and solving process, some parameters are assumed. To match the model with the actual situation, the model requires uncertainty analysis.

5.1. Effects of Weighted Coefficients

The weighted coefficients are decided by the importance of different objectives. In this study, three weighted vectors are considered: (1/3, 1/3, 1/3), (0.5, 0.25, 0.25), and (0.8, 0.1, 0.1). The results are shown in Table 5.

Table 5. Effects of weighted vector on the results.

| Weighted Vector | Coal Consumption | | NO_x Emission | | Power Costs | |
| | td^{-1} | | td^{-1} | | $\$d^{-1}$ | |
	OU	RU	OU	RU	OU	RU
(1/3,1/3,1/3)	10,924.4	10,699.9	11.33	11.29	2,023,434.1	2,020,505.0
(0.5,0.25,0.25)	10,824.4	10,597.7	11.42	11.38	2,033,382.7	2,028,780.6
(0.8,0.1,0.1)	10,626.9	10,433.4	12.00	11.85	2,065,198.3	2,051,382.3

As shown in Table 5, the coal consumption, NO_x emissions, and power costs decline after the No. 4 unit is retrofitted under different weighted vectors. When the weighted vector transforms from (1/3, 1/3, 1/3) to (0.8, 0.1, 0.1), the coal consumption of the original unit declines by 2.72%, and the NO_x emissions and power costs increase by 5.91% and 2.06%, respectively. The coal consumption of the retrofitted unit declines by 2.49%, and the NO_x emissions and power costs increase by 5.00% and 1.53%, respectively. A comparison of the two data sets show that the change in the parameters of the retrofitted unit is less than that in the parameters of the original unit. This shows that small changes in the parameters of the solar-assisted coal-fired units are conducive to the operation stability of the unit.

The impact of weighting coefficients for the optimal solution is shown in Figure 7.

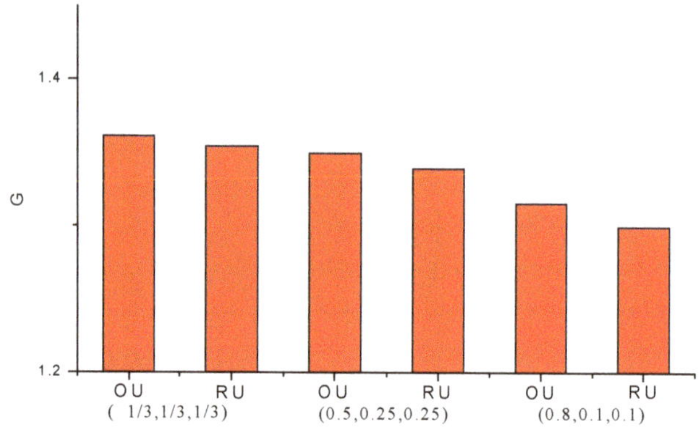

Figure 7. Impact of weighting coefficients for the optimal solution.

Figure 7 shows that the optimal solution of the solar-assisted coal-fired unit is less than that of the coal-fired unit with the same weighted vectors. The smaller the optimal solution, the better the performance of the unit. Thus, the performance of the solar-assisted coal-fired unit is better than that of the coal-fired unit. As the weight coefficients of coal consumption increase, the optimal solution decreases. This shows that among these three weighting coefficients, the weighting coefficients of coal consumption have the most important influence on the optimal solution.

5.2. Effects of on-Grid Power Tariff

In China, solar thermal power generation has not entered the commercial stage. The current on-grid power tariff of solar-assisted coal-fired units is not clearly defined. However, a large number of photovoltaic power generators have been put into use, so uncertainty analysis calculation refers to the PV (solar photovoltaic) on-grid power tariff. This paper selected 3.26, 6.52, 9.77, 13.03, and 16.29 cent/kWh as the on-grid power tariff for solar-assisted coal-fired units. The results are shown in Table 6.

Table 6. Effects of the on-grid power tariff on the results.

On-grid Tariff	Coal Consumption	NO_x Emission	Power Costs
Cent/kWh	td^{-1}	td^{-1}	$\$d^{-1}$
3.26	10,699.9	11.28	2,020,505.0
6.52	10,681.9	11.39	2,416,605.0
9.77	10,677.2	11.74	2,655,308.3
13.03	10,689.8	12.07	2,825,889.1
16.29	10,699.5	12.31	2,982,161.8

Table 6 shows a nonlinear relationship between coal consumption and on-grid tariff, but power purchase costs and NO_x emissions increase with increasing tariff amount. When the tariff decreases from 16.29 cent/kWh to 3.26 cent/kWh, NO_x emissions drop by approximately 8.38% from 12.31 t to 11.28 t per day. In addition, the cost of purchasing electricity decreases by 32.25% from \$ 2,982,161.8 to \$ 2,020,505.0 per day.

The impact of the on-grid power tariff on the optimal solution is shown in Figure 8.

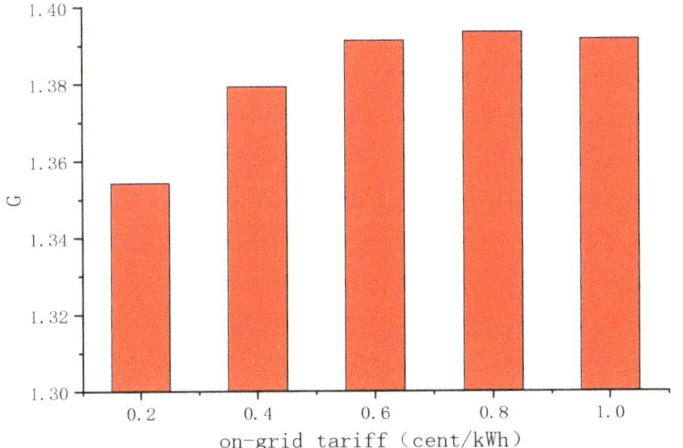

Figure 8. Impact of on-grid power tariff on the optimal solution.

As the tariff increases, the optimal solution also increases. The optimal solution finally stabilizes near 1.39. As the technology of the solar-assisted coal-fired units mature, their tariff gradually reduces. Therefore, the advantage of solar-assisted coal-fired units will become more apparent.

6. Conclusions

In this study, a coal-fired unit in a particular area is retrofitted to solar-assisted coal-fired units. The other remaining units are not changed. Under the given power demand, the multi-objective model considers coal consumption, NO_x emissions, and power purchase cost. The singular weighted method combined with the MATLAB toolbox is used to find the optimal solution in the multi-objective function. Thus, the changes in each unit's load before and after the No. 4 unit is retrofitted are compared and analyzed under optimal solution conditions.

The original and retrofitted unit loads are compared over 24 h. Results show that the loads of the solar-assisted coal-fired unit are higher than those of the coal-fired unit. With the increase in solar radiation, the gap between the loads of the two units also increases. The loads of five units before and after the No. 4 unit is retrofitted are compared under the solar radiation. The loads of the No. 4 unit increase, and the loads of the other units decrease.

In addition, uncertainty analysis in this section discusses the impact of weighting coefficients and on-grid power tariff changes on the results. With the increase in the weighted coefficient of coal consumption, the optimal solution decreases. Investigation of the on-grid power tariff shows that the optimal solution gradually declines as the tariff decreases.

Through this study, compared to coal-fired units, solar-assisted coal-fired units have an advantage in terms of energy saving and emission reduction. Therefore, the government should encourage and support the development of solar-assisted coal-fired units. The methods described in this paper may also be used to solve other problems. In real life, this method can be used to solve the load dispatch problem of this area according to the specific circumstances of units, DNI values, and power load. Then, recommendations on the regional distribution of load between units can be proposed.

Author Contributions: All of the authors have contributed to this research. The original ideas were provided by R.Z. and Y.H. The calculations and energy performance analysis were performed by L.F. and C.L. The revision of the draft was guided by R.Z. and Y.Y.

Funding: This research was funded by National Major Fundamental Research Program of China (No. 2015CB251505), China National Natural Science Foundation (No. 51776063), the Fundamental Research Funds for the Central Universities (2016YQ04), and China Scholarship Council.

Conflicts of Interest: The authors declare no conflict of interest.

Nomenclature

a_j, b_j, c_j	coal consumption coefficients	H	symmetric matrix
u_j, v_j, w_j	pollutant emission coefficients	A	matrix for linear inequality constraints
ρ_j	power purchase costs	B	vector for linear inequality constraints
P_j	power output of the existing power generating units	A_{eq}	matrix for linear equality constraints
		B_{eq}	vector for linear equality constraints
P_D	electricity demand for the grid	x_m	vector of lower bounds
P_L	grid loss	X_m	vector of upper bounds
P_{jmin}	lower limit for units	x_0	initial point for x
P_{jmax}	upper limit for units	OU	original unit
f	vector	RU	retrofitted unit

Appendix A Appendix

Table A1. Each unit load allocation of No. 4 original unit (OU) and retrofitted unit (RU).

Unit No.	P1(MW)		P2(MW)		P3(MW)		P4(MW)		P5(MW)	
Time	OU	RU	OU	RU	OU	RU	OU	RU	OU	RU
1	310	310	291	292	429	430	478	474	373	375
2	310	310	313	314	459	461	528	522	403	405
3	310	310	286	287	422	423	467	464	367	368
4	310	310	275	276	407	408	442	440	352	353
5	310	310	279	280	413	414	451	449	358	359
6	310	310	289	290	427	428	475	471	372	373
7	310	310	286	279	422	413	466	490	366	358
8	310	310	294	286	433	422	485	514	378	367
9	310	310	313	304	459	447	528	559	403	392
10	310	310	320	311	469	456	544	578	413	401
11	340	334	366	360	532	524	651	680	476	468
12	322	314	347	339	506	494	607	647	450	439
13	317	310	342	333	499	487	595	634	443	431
14	327	319	353	344	514	502	620	661	458	446
15	332	323	357	349	520	508	630	671	464	452
16	332	324	357	349	520	508	630	670	464	453
17	326	318	351	343	512	500	616	654	456	445
18	331	324	357	349	519	508	629	665	463	453
19	402	402	425	425	621	621	680	680	565	565
20	410	410	425	425	632	632	680	680	576	576
21	404	404	425	425	624	623	680	680	567	567
22	377	377	405	405	585	585	680	680	529	529
23	351	354	378	381	548	552	677	665	492	496
24	316	318	341	343	498	500	593	584	442	445

References

1. Yongping, Y.; Rongrong, Z.; Liqiang, D.; Masek, O.; Oakey, J. Study on multi-objective optimization of load dispatch including renewable energy and CCS technologies. *Int. J. Energy Res.* **2010**, *34*, 702–715. [CrossRef]
2. Bakos, G.C.; Tsechelidou, C. Solar aided power generation of a 300 MW lignite fired power plant combined with line-focus parabolic trough collectors field. *Renew. Energy* **2013**, *60*, 540–547. [CrossRef]
3. China Electric Council. 2018. Available online: http://www.cec.org.cn/d/file/guihuayutongji/tongjxinxi/niandushuju/2019-01-22/4fedb4c956f6059c5998913b10a6233a.pdf (accessed on 16 February 2019).
4. Ruilin, W.; Jie, S.; Hui, H.; Hongguang, J. Comprehensive evaluation for different models of Solar-aided coal-fired power generation system under common framework regarding both Coal-savability and efficiency-promotability. *Energy* **2018**, *143*, 151–167.
5. Desai, N.B.; Kedare, S.B.; Bandyopadhyay, S. Optimization of design radiation for concentrating solar thermal power plants without storage. *Sol. Energy* **2014**, *107*, 98–112. [CrossRef]

6. Raul, A.; Jain, M.; Gaikwad, S.; Saha, K. Modelling and experimental study of latent heat thermal energy storage with encapsulated PCMs for solar thermal applications. *Appl. Therm. Eng.* **2018**, *143*, 415–428. [CrossRef]

7. Lakhani, S.; Raul, A.; Saha, S. Dynamic modelling of ORC-based solar thermal power plant integrated with multitube shell and tube latent heat thermal storage system. *Appl. Therm. Eng.* **2017**, *123*, 458–470. [CrossRef]

8. Corgnale, C.; Hardy, B.; Motyka, T.; Zidan, R.; Teprovich, J.; Peters, B. Screening analysis of metal hydride based thermal energy storage systems for concentrating solar power plants. *Renew. Sustain. Energy Rev.* **2014**, *38*, 821–833. [CrossRef]

9. Yogev, R.; Kribus, A. Operation strategies and performance of solar thermal power plants operating from PCM storage. *Sol. Energy* **2013**, *95*, 170–180. [CrossRef]

10. Ehrhart, B.; Gill, D. Evaluation of Annual Efficiencies of High Temperature Central Receiver Concentrated Solar Power Plants with Thermal Energy Storage. *Energy Procedia* **2014**, *49*, 752–761. [CrossRef]

11. Oró, E.; Gil, A.; de Gracia, A.; Boer, D.; Cabeza, L.F. Comparative life cycle assessment of thermal energy storage systems for solar power plants. *Renew. Energy* **2012**, *44*, 166–173. [CrossRef]

12. Hu, E.; Yang, Y.; Nishimura, A.; Yilmaz, F.; Kouzani, A. Solar thermal aided power generation. *Appl. Energy* **2010**, *87*, 2881–2885. [CrossRef]

13. Yang, Y.; Yan, Q.; Zhai, R.; Kouzani, A.; Hu, E. An efficient way to use medium-or-low temperature solar heat for power generation–integration into conventional power plant. *Appl. Therm. Eng.* **2011**, *31*, 157–162. [CrossRef]

14. Hong, H.; Peng, S.; Zhao, Y.; Liu, Q.; Jin, H. A Typical Solar-coal Hybrid Power Plant in China. *Energy Procedia* **2014**, *49*, 1777–1783. [CrossRef]

15. Bhattacharjee, K.; Bhattacharya, A.; Halder nee Dey, S. Solution of Economic Emission Load Dispatch problems of power systems by Real Coded Chemical Reaction algorithm. *Int. J. Electr. Power Energy Syst.* **2014**, *59*, 176–187. [CrossRef]

16. Jeddi, B.; Vahidinasab, V. A modified harmony search method for environmental/economic load dispatch of real-world power systems. *Energy Convers. Manag.* **2014**, *78*, 661–675. [CrossRef]

17. Leena, D.; Krishna, T.C.; Mohan, L.K. Dynamic Economic Load Dispatch using Levenberg Marquardt Algorithm. *Energy Procedia* **2018**, *144*, 95–103.

18. Coelho, L.D.S.; Lee, C.S. Solving economic load dispatch problems in power systems using chaotic and Gaussian particle swarm optimization approaches. *Int. J. Electr. Power Energy Syst.* **2008**, *30*, 297–307. [CrossRef]

19. Li, M.S.; Wu, Q.H.; Ji, T.Y.; Rao, H. Stochastic multi-objective optimization for economic-emission dispatch with uncertain wind power and distributed loads. *Electr. Power Syst. Res.* **2014**, *116*, 367–373. [CrossRef]

20. Mondal, S.; Bhattacharya, A. Multi-objective economic emission load dispatch solution using gravitational search algorithm and considering wind power penetration. *Int. J. Electr. Power Energy Syst.* **2013**, *44*, 282–292. [CrossRef]

21. Aghaei, J.; Niknam, T.; Azizipanah-Abarghooee, R.; Arroyo, J.M. Scenario-based dynamic economic emission dispatch considering load and wind power uncertainties. *Int. J. Electr. Power Energy Syst.* **2013**, *47*, 351–367. [CrossRef]

22. Modarresi, M.S.; Xie, L.; Campi, M.G.; Simone, C.; Algo, T.; Anupam, K. PR Scenario-based Economic Dispatch with Tunable Risk Levels in High-renewable Power Systems. *IEEE Trans. Power Syst.* **2018**, 1. [CrossRef]

23. Zhu, Y.; Wang, J.; Qu, B. Multi-objective economic emission dispatch considering wind power using evolutionary algorithm based on decomposition. *Int. J. Electr. Power Energy Syst.* **2014**, *63*, 434–445. [CrossRef]

24. Rizk-Allah, R.M.; El-Sehiemy, R.A.; Wang, G.G. A novel parallel hurricane optimization algorithm for secure emission/economic load dispatch solution. *Appl. Soft Comput.* **2018**, *63*, 206–222. [CrossRef]

25. Roy, S. The maximum likelihood optima for an economic load dispatch in presence of demand and generation variability. *Energy* **2018**, *147*, 915–923. [CrossRef]

26. Modarresi, M.S.; Xie, L. An operating reserve risk map for quantifiable reliability performances in renewable power systems. In Proceedings of the 2014 IEEE PES General Meeting | Conference & Exposition, National Harbor, MD, USA, 27–31 July 2014.

27. Li, J.F.; Zhang, B.H.; Liu, Y.F.; Wang, K.; Wu, X.S. Spatial evolution character of multi-objective evolutionary algorithm based on self-organized criticality theory. *Phys. A Stat. Mech. Its Appl.* **2012**, *391*, 5490–5499. [CrossRef]
28. Lu, Y.; Zhou, J.; Qin, H.; Wang, Y.; Zhang, Y. Environmental/economic dispatch problem of power system by using an enhanced multi-objective differential evolution algorithm. *Energy Convers. Manag.* **2011**, *52*, 1175–1183. [CrossRef]
29. Wu, L.H.; Wang, Y.N.; Yuan, X.F.; Zhou, S.W. Environmental/economic power dispatch problem using multi-objective differential evolution algorithm. *Electr. Power Syst. Res.* **2010**, *80*, 1171–1181. [CrossRef]
30. Siddiqi, U.F.; Shiraishi, Y.; Dahb, M.; Sait, S.M. A memory efficient stochastic evolution based algorithm for the multi-objective shortest path problem. *Appl. Soft Comput.* **2014**, *14*, 653–662. [CrossRef]
31. Peng, C.; Sun, H.; Guo, J. Multi-objective optimal PMU placement using a non-dominated sorting differential evolution algorithm. *Int. J. Electr. Power Energy Syst.* **2010**, *32*, 886–892. [CrossRef]
32. Ghasemi, A. A fuzzified multi objective interactive honey bee mating optimization for environmental/economic power dispatch with valve point effect. *Int. J. Electr. Power Energy Syst.* **2013**, *49*, 308–321. [CrossRef]
33. El-sobky, B.; Abo-elnaga, Y. Multi-objective economic emission load dispatch problem with trust-region strategy. *Electr. Power Syst. Res.* **2014**, *108*, 254–259. [CrossRef]
34. Hunt, B.R.; Lipsman, R.L.; Rosenberg, J.M.; Coombes, K.R.; Osborn, J.E.; Stuck, G.J. *A Guide to MATLAB: For Beginners and Experienced Users*; Cambridge University Press: Cambridge, UK, 2006.

Article

Transient Analysis and Execution-Level Power Tracking Control of the Concentrating Solar Thermal Power Plant

Xiufan Liang and Yiguo Li *

Key laboratory of Energy Thermal Conversion and Control of Ministry of Education, Southeast University, Nanjing 210096, China; 676615246@qq.com
* Correspondence: lyg@seu.edu.cn; Tel.: +86-13913970596

Received: 25 March 2019; Accepted: 12 April 2019; Published: 25 April 2019

Abstract: Concentrating solar power (CSP) is a promising technology for exploiting solar energy. A major advantage of CSP plants lies in their capability of integrating with thermal energy storage; hence, they can have a similar operability to that of fossil-fired power plants, i.e., their power output can be adjusted as required. For this reason, the power output of such CSP plants is generally scheduled to maximize the operating revenue by participating in electric markets, which can result in frequent changes in the power reference signal and introduces challenges to real-time power tracking. To address this issue, this paper systematically studies the execution-level power tracking control strategy of an CSP plant, primarily aiming at coordinating the control of the sluggish steam generator (including the economizer, the boiler, and the superheater) and the fast steam turbine. The governing equations of the key energy conversion processes in the CSP plant are first presented and used as the simulation platform. Then, the transient behavior of the CSP plant is analyzed to gain an insight into the system dynamic characteristics and control difficulties. Then, based on the step-response data, the transfer functions of the CSP plant are identified, which form the prediction model of the model predictive controller. Finally, two control strategies are studied through simulation experiments: (1) the heuristic PI control with two operation modes, which can be conveniently implemented but cannot coordinate the control of the power tracking speed and the main steam parameters, and (2) advanced model predictive control (MPC), which overcomes the shortcoming of PI (Proportional-Integral) control and can significantly improve the control performance.

Keywords: CSP plant model; transient analysis; power tracking control; two-tank direct energy storage

---

1. Introduction

In recent years, solar energy has become the second-largest energy source after wind energy among the renewable energy sources that are used for electricity production [1]. The concentrating solar power (CSP), which uses either organic oil or molten salt as its heat transfer fluid (HTF) to absorb and transfer solar energy, is currently the most commercially attractive solar thermal-based power generation technology [2].

To fill the generation gaps in intermittent solar energy, the CSP plant is generally integrated with thermal energy storage (TES), which enables the CSP plant to control its power output flexibly in the presence of solar uncertainty [3]. Actually, the TES enables the CSP plant to be partly independent from constantly changing solar radiation [4], reducing the short-term load variation and extending or shifting the power supply period [5]. Therefore, the CSP plant is potentially capable of supplying the power on demand, participating in electricity markets by scheduling the power production throughout each day [6], and providing ancillary services such as regulating the grid frequency [7]. By participating in electricity markets, the revenue of the CSP plant can be significantly improved [1]. For this reason, power scheduling is required to maximize the revenue, which is known as the decision-level power generation control [8]. Additionally, when the CSP plant serves as the load-following power plant, it must be able to regulate the power output rapidly in response to the changing demand for power supply [9].

The demand for the flexible operation of the CSP plant brings a high requirement to its execution-level dynamic control. The CSP plant must be able to rapidly adjust its power output in a wide operation range and simultaneously maintain the stability of the main steam pressure and temperature for safety and economic reasons. To achieve this goal, it is necessary to perform a thorough investigation of the system dynamic behavior, and on this basis find an appropriate execution-level control strategy for controlling power tracking.

Existing research works related to the execution-level control of CSP plants integrated with TES mainly focused on the control of the HTF temperature in collector field. Cirre et al. used a feedback linear control scheme to control the HTF temperature, which can reduce the influence of process nonlinearity [10]. Gallego and Camacho proposed a state-space model predictive control to reject external disturbance on the HTF temperature [11]. Alsharkawi and Rossiter developed an improved gain scheduling predictive control, incorporating a feed-forward strategy to improve the temperature control performance [12]. Nevado Reviriego, Hernández-del-Olmo, and Álvarez-Barcia studied a nonlinear adaptive control scheme for the HTF temperature, which can cope with the time-varying nature of the process [13].

However, few of these research works have focused on the execution-level power tracking control of CSP plants, which is even more challenging than the HTF temperature control. In fact, power tracking is not a stand-alone control problem. The control action that changes the power output can bring significant disturbances to the main steam (steam flowing into the turbine) pressure and temperature. Accelerating the power tracking rate can easily result in significant fluctuation in the main steam parameters, which imposes a negative effect on the safe and economic operation of the CSP plant. Furthermore, in CSP plants, the steam generator dynamics are much slower than the steam turbine dynamics, and it is challenging to coordinate the control of two systems with completely different response speeds.

Considering these issues, this paper proposes an execution-level power tracking control strategy for CSP plants that aims at achieving the dual tasks of fast power tracking and small fluctuation in the main steam parameters by coordinating the operation of the steam generator and the turbine. The heuristic PI control strategy with two operation modes, i.e., the fast power tracking mode (FT mode) and the smooth operation mode (SO mode), is first studied based on our knowledge about the process. However, each operation mode of heuristic PI can only satisfy one of the two control tasks. Therefore, the advanced model predictive control (MPC) strategy is further proposed to enhance the coordinating strategy and achieve both of the control tasks.

The remainder of this paper is organized as follows. The description and modeling of the CSP plant are presented in Section 2. Section 3 includes the transient analysis and process model identification. In Section 4, the heuristic PI control strategy and the MPC strategy are formulated, and their control performance is evaluated. Section 5 contains the conclusions.

2. Simulation Model of the CSP Plant

This section presents the simplified simulation model of a CSP plant with two-tank direct TES. Figure 1 shows the simplified scheme of the plant considered in this study, and its working principle is described as follows. The HTF (orange line in Figure 1) from the cold molten salt tank absorbs the solar radiation in the solar field, which comprises a set of single-axis tracking parabolic trough concentrators, and is then stored in the hot molten salt tank. In the meantime, the hot tank releases the stored HTF, which sequentially flows through the superheater, the boiler, and the economizer. In the economizer, the working fluid in the Rankine cycle is in liquid phase (blue line) and preheated close to its saturation temperature. In the boiler, the working fluid undergoes a phase change and evaporates from liquid to saturated steam. In the superheater, the saturated steam is further heated into superheated steam. Finally, the superheated steam (main steam) passes through the main steam valve and drives the steam turbine to produce electric power. To control the main steam temperature, an attemperator, which can spray feed water, is installed before the last-level superheater.

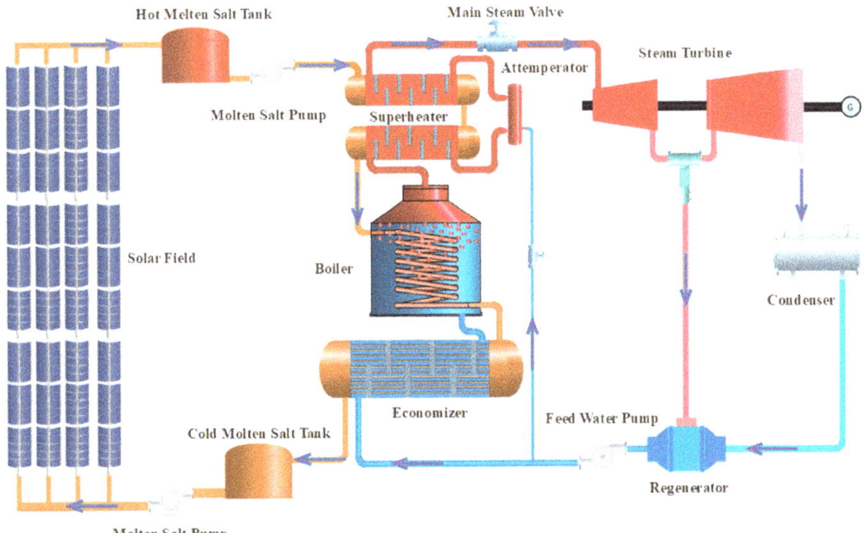

Figure 1. System configuration of the concentrating solar power (CSP) plant with two-tank direct thermal energy storage (TES) (the arrows represent the fluid flow direction).

2.1. Solar Collector Model

The solar collector consists of a parabolic mirror, a glass envelope, and an absorber tube. The sunlight entering the mirror aperture is focused on the focal line, where the absorber tube is positioned. The absorber tube is enclosed by an evacuated glass envelope to prevent the heat loss to the environment. The dynamics of the HTF, the absorber tube, and the glass envelope can be described using a piecewise lumped parameter model discretized along the focal line (see Figure 2).

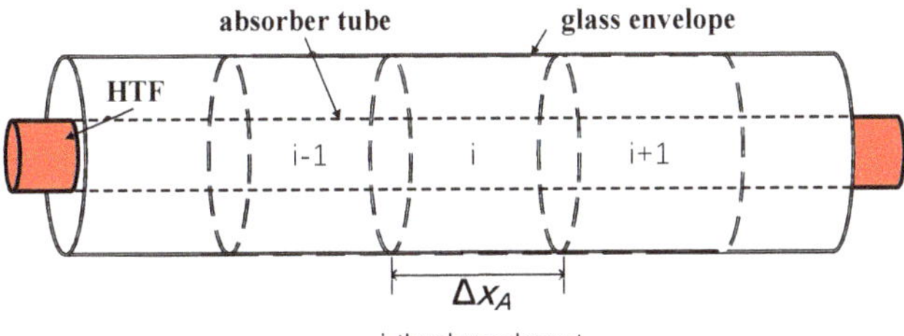

Figure 2. A side view of the solar collector.

A.   Energy balance of the *i*-th HTF volume element:

$$\rho_H C_H \hat{A}_H \Delta x_A \frac{dT_{HA}^{(i)}}{dt} = q_{H,cold} C_H (T_{HA}^{(i-1)} - T_{HA}^{(i)}) + h_{H,T} P_{TA,i}(T_{TA}^{(i)} - T_{HA}^{(i)}), \tag{1}$$

where Δx_A is length of the volume element, $q_{H,cold}$ is the mass flowrate of the HTF outflowing from the cold tank, and $h_{H,T}$ is the heat transfer coefficient between the HTF and the tube, which can be determined using empirical correlations developed in [14].

B.   Energy balance of the i-th absorber tube volume element:

$$\rho_T C_T \hat{A}_T \frac{dT_{TA}^{(i)}}{dt} = h_{H,T} P_{TA,i}\left(T_{HA}^{(i)} - T_{TA}^{(i)}\right) - \frac{\sigma}{\frac{1}{\varepsilon_{TA}} + \frac{1-\varepsilon_G}{\varepsilon_G}\left(\frac{r_{TA,o}}{r_{G,i}}\right)} P_{TA,o}\left(\left(T_{TA}^{(i)}\right)^4 - \left(T_G^{(i)}\right)^4\right) + I_C \eta_{optical} w \tag{2}$$

where $\eta_{optical}$ is the optical efficiency of the solar collector.

C.   Energy balance of the i-th glass envelope volume element:

$$\rho_G C_G \hat{A}_G \frac{dT_G^{(i)}}{dt} = \frac{\sigma}{\frac{1}{\varepsilon_{TA}} + \frac{1-\varepsilon_G}{\varepsilon_G}\left(\frac{r_{TA,o}}{r_{G,i}}\right)} P_{G,i}\left(\left(T_{TA}^{(i)}\right)^4 - \left(T_G^{(i)}\right)^4\right) - \sigma \varepsilon_G P_{G,o}\left(\left(T_G^{(i)}\right)^4 - \left(T_{sky}\right)^4\right) - h_{G,air} P_{G,o}\left(T_G^{(i)} - T_{air}\right) \tag{3}$$

where T_{sky} is the temperature of the sky, and T_{air} is the ambient air temperature.

2.2. Storage Tank Model

The dynamics of the hot tank and the cold tank are modeled using the same equations. The mass and energy balance of the storage tanks are established as:

$$\rho_H \frac{dV_{St}}{dt} = q_{H,in} - q_{H,out} \tag{4}$$

$$\rho_H C_H \frac{d(V_{St} T_{St})}{dt} = C_H T_{H,in} q_{H,in} - C_H T_{St} q_{H,out} \tag{5}$$

where the derivative of $V_{St} T_{St}$ is used instead of T_{St}, because the volume of the molten salt in the storage tanks can be varying.

2.3. Economizer Model

The function of the economizer is to preheat the feed water. The economizer in the CSP plant is generally a cross-flow shell-and-tube heat exchanger of one-shell pass and one-tube pass [15], and its overall heat transfer rate can be calculated as [16]:

$$Q_E = FA_E h_E \cdot LTMD \tag{6}$$

where $LTMD$ is the logarithmic mean temperature difference, F is the correction factor, which can be calculated analytically based on the geometry and fluid temperature of the heat exchanger [16], and h_E can be calculated using the empirical correlations presented in [16]. The dynamics of the outlet feed water and HTF temperature are determined from the overall heat transfer rate Q_E:

A.   Feed water temperature:

$$\rho_L C_L V_L \frac{dT_{LE}}{dt} = q_{LE}(H_{Fw} - H_{LE}) + Q_E, \tag{7}$$

B.   HTF temperature:

$$\rho_H C_H V_H \frac{dT_{HE}}{dt} = C_H q_{H,hot}(T_{HB} - T_{HE}) - Q_E \tag{8}$$

2.4. Boiler Model

The boiler is the place where the feed water is heated into saturated steam. It has been demonstrated that the energy stored in the metal and liquid water dominates the dynamics of the boiler pressure. The boiler model can be developed as [17]:

$$K_B \frac{dP_B}{dt} = Q_B - q_{SB}(H_{SB} - H_{LE})$$ (9)

where:

$$K_B = \rho_{Sw} V_{Sw} \frac{\partial H_{Sw}}{\partial P_B} + C_{MB} M_{MB} \frac{\partial T_B}{\partial P_B}$$ (10)

V_{Sw} can be assumed to be constant because the liquid level in the boiler is generally well controlled [17]. The dynamics of the HTF in the boiler tube are described using the same piecewise lumped parameter model developed for the HTF in the absorber tube. The boiler tube temperature T_{TB} can be assumed to be equal to the boiler temperature T_B [18]; then, the heat exchange between the HTF in the boiler tube and the boiler tube is calculated as:

$$Q_B = \sum_{i=1}^{NB} h_{H,T} P_{HB,i}(T_{HB}^{(i)} - T_B)\Delta x_B$$ (11)

where NB is the number of the boiler tube's volume elements. The saturated steam outflowing from the boiler mainly depends on the boiler pressure and the downstream superheater pressure. It can be calculated using the empirical formula developed in [19]:

$$q_{SB} = \kappa_B P_B^{0.9} \sqrt{P_B - P_{Sh}},$$ (12)

where κ_B is the fitting constant.

2.5. Superheater Model

Generally, the superheaters in CSP plants are also tube-and-shell heat exchangers, and their heat transferring analysis is similar to that of the economizer. In superheaters, the overall heat transfer rate Q_{Sh} and the dynamics of the HTF at the tube side can be determined using the same models developed for the economizer. However, the dynamics of the superheated steam (the main steam) at the shell side must be reconsidered, because it is a compressible non-ideal gas featuring a completely different characteristic from the incompressible liquid in the economizer. The general mass and energy balance equation of the superheated steam can be formulated as [17]:

$$V_{Sh} \frac{\partial \rho_S}{\partial p_S} \frac{dp_{Sh}}{dt} + V_{Sh} \frac{\partial \rho_S}{\partial T_{Sh}} \frac{dT_{Sh}}{dt} = q_{SB} + q_{LA} - q_{SSh},$$ (13)

$$V_{Sh} \frac{\partial(\rho_S H_{SSh})}{\partial p_{Sh}} \frac{dp_{Sh}}{dt} + \left(V_{Sh} \frac{\partial(\rho_S H_{SSh})}{\partial T_{Sh}} + M_{MSh} C_{MSh}\right)\frac{dT_{Sh}}{dt} = q_{SB} H_{SB} + q_{LA} H_{Fw} + Q_{Sh} - q_{SSh} H_{SSh}$$ (14)

where the partial derivatives of the steam properties are solved using the XSteam Packages developed for Matlab.

2.6. Turbine Model

The mass flowrate of the main steam is mainly affected by its thermal–physical properties and the main steam valve opening at the governing stage. Their relationship can be described by the following equation [20]:

$$q_{SSh} = \kappa_{Sh} \delta(p_{Sh})^{1-\gamma}(\rho_{Sh})^{\gamma} \quad 0 \le \gamma \le 0.5$$ (15)

where κ_{Sh} and γ are fitting constants. The range of γ goes from 0, meaning that the steam is just saturated, to 0.5, meaning that the steam is highly superheated and behaves similar to ideal gas. Generally, $\gamma = 0.3$ can have a good fit to the real process [20]. The power output of the steam turbine can be calculated based on the practical enthalpy drop of the main steam. The ideal enthalpy drop and the practical enthalpy drop are bridged by the turbine's relative internal efficiency η_{Tu}:

$$H_{SSh} - H_{Ex}^* = \eta_{Tu}(H_{SSh} - H_{Ex}), \tag{16}$$

$$H_{Ex}^* - H_{Tu,out}^* = \eta_{Tu}(H_{Ex} - H_{Tu,out}), \tag{17}$$

where H_{Ex} is the enthalpy of the extracted steam that drops along the isentropic enthalpy curve, $H_{Tu,out}$ is the enthalpy of the turbine exhaust steam that drops along the isentropic enthalpy curve, and the superscript "*" represents the enthalpy of the actual process. Considering the Rankine cycle with a single-stage regenerator, the final power output of the turbine can be calculated using the product of the ideal enthalpy drop, turbine relative internal efficiency, and the main steam mass flowrate:

$$N_e = \eta_{Tu}q_{SSh}[\alpha(H_{SSh} - H_{Ex}) + (1 - \alpha)(H_{Ex} - H_{T,out})], \tag{18}$$

where α is the ratio of the extracted steam mass flowrate to the main steam mass flowrate. Assuming that the feed water temperature and condenser pressure are constant, α can be calculated as:

$$\alpha = \frac{H_{Fw} - H_{Cond,out}}{H_{Ex}^* - H_{Cond,out}} \tag{19}$$

2.7. Parameter Settings of the Simulation Model

The parameters of the simulation model are set according to the parameters of a real 5-MW CSP plant reported in [15]. The geometric design data as well as the steady-state operating points of the CSP plant, which are related to the construction and simulation of the model, are presented in Table 1. The nominal operating points of the model show good agreement with the plant data, indicating that the parameter setting is reasonable and the model can be used to simulate the real process.

Table 1. Partial data of the model and the CSP plant reported in [15]. HTF: heat transfer fluid.

Parameters	Model	Plant Data	Parameters	Model	Plant Data
Absorber tube diameter	0.07 m	0.07 m	Total length installed (collector)	5400 m	5400 m
Parallel collector assemblies	9	9	Collector number in each assembly	6	6
Heat transfer area of the economizer	145 m^2	150 m^2	Heat transfer coefficient of the economizer	0.96 kW/m^2/K	1.04 kW/m^2/K
Heat transfer area of the boiler	322 m^2	330 m^2	Heat transfer coefficient of the boiler	1.21 kW/m^2/K	1.04 kW/m^2/K
Heat transfer area of the superheater	72 m^2	31 m^2	Heat transfer coefficient of the superheater	0.79 kW/m^2/K	0.88 kW/m^2/K
Direct solar radiation	1.9 kW/m^2	1.9 kW/m^2	Power output	4.80 MW	4.77 MW
HTF temperature in cold tank	304.6 °C	290.0 °C	HTF temperature in hot tank	567.8 °C	555.2 °C
Main steam temperature	403.6 °C	404.6 °C	Main steam pressure	8.84 MPa	Not given
HTF mass flowrate outflowing the hot tank	135 kg/s	135.8 kg/s	HTF mass flowrate outflowing the cold tank	135 kg/s	135.8 kg/s
Main steam mass flowrate	19.7 kg/s	18.5 kg/s	Feed water mass flowrate	18.8 kg/s	18.6 kg/s
Boiler pressure	9.37 MPa	9.03 MPa	Boiler temperature	306.37 °C	311.4 °C
HTF volume in hot tank	2000 m^3	1990 m^3	HTF volume in cold tank	6000 m^3	6158 m^3
Feed water temperature	240 °C	243.2 °C	Economizer temperature (outlet liquid water)	300.1 °C	298.5 °C

3. Transient Analysis and Process Model Identification of the CSP Plant

To identify the control difficulties and find appropriate strategies for the power tracking control problem, the transient behavior of the CSP plant should be analyzed in order to understand how the manipulating variables and external disturbances can influence the controlled variables.

Since the step response can present the dynamic information of the CSP plant in a clear manner, including the settling time of the process, the coupling effect between the operating variables, and the influence of external disturbances, etc., the step response experiment of the key input variables that have a significant influence on the operation of the CSP plant is performed for transient analysis. In addition, using the step-response data, we can identify the transfer function of the CSP plant as the prediction model of the MPC controller.

3.1. Open-Loop Step Response Analysis

Step response simulation is carried out on the exogenous input $d(I_C)$ and manipulating variables u_i through five cases (u_1: the mass flowrate of the HTF outflowing from the hot tank $q_{H,hot}$; u_2: opening of the main steam valve δ; u_3: the mass flowrate of the spraying water q_{LA}; u_4: the mass flowrate of the HTF outflowing from the cold tank $q_{H,cold}$).

The step increase value of the step variables are shown as follows: in case I, u_1 steps increase from 135 kg/s to 150 kg/s; in case II, u_2 steps increase from 80% to 90%; in case III, u_3 steps increase from 0.8 kg/s to 1.0 kg/s; in case IV, u_4 steps increase from 135 kg/s to 150 kg/s; and in case V, d steps increase from 1.9 kW/m^2 to 2.1 kW/m^2. In each case, there is only one signal step, and the step signal starts at 200 s. The simulation results are plotted in Figure 3.

Case I represents the situation where the hot tank releases the stored energy to generate more electricity. The increment in the mass flowrate of the HTF outflowing from the hot tank (u_1) significantly enhances the heat transfer from the HTF to the working fluid in the Rankine cycle and produces more superheated steam. Meanwhile, the feed water mass flowrate will increase with the steam mass flowrate to maintain a constant boiler liquid level. Although the main steam temperature initially increases because of the enhanced heat transfer, it is later cooled down by the increasing feed water and finally falls below the initial value. The main steam pressure also increases drastically as more steam is generated. In this case, the settling time of the main steam pressure and power output is approximately 150 s, which is shorter than that of the main steam temperature (about 500 s).

In case II, increasing the opening of the main steam valve (u_2) causes an instant boost in the power output, which is much faster than manipulating the mass flowrate of the HTF outflowing from the hot tank (u_1). This indicates that the turbine dynamics is faster than the steam generator and can be operated to enforce an immediate change in the power output. However, manipulating u_2 does not change the thermal energy flowing into the power generation system; therefore, the final power output almost drops to its initial value. As u_2 increases, the flow resistance of the main steam reduces rapidly, causing a significant drop in the main steam pressure with a similar setting time as that in case I (about 150 s).

In case III, the increment in the mass flowrate of spraying water (u_3) greatly lowers the main steam temperature, while it only has a minor influence on the main steam pressure, because the increased amount of the spraying water is very small compared to the main steam flow. Since the spraying water increases the irreversible loss and reduces the thermal efficiency, the turbine power output is slightly reduced, as shown in Figure 3. In this case, the setting time of the main steam temperature is approximately 500 s.

Case IV shows that increasing the HTF flowing through the collector (u_4) significantly reduces the HTF temperature at the collector outlet: from 567.5 °C to 514.5 °C within 400 s. However, it has little influence on the temperature of the HTF stored in the hot tank, owing to the large heat capacity of the stored HTF. Therefore, both the temperature and the mass flowrate of the HTF flowing into the steam generator remain unchanged, demonstrating that the manipulation of u_4 almost has no influence on the steam generation side.

In case V, with the increase of direct solar irradiance incident (d), the HTF temperature in the solar collector rapidly rises from 567.5 °C to 595 °C in 600 s; however, the temperature of the hot storage tank only increases by 0.4 °C in the same timescale, which indicates that the influence from the solar irradiance on the power generation side is also significantly attenuated by the storage tanks.

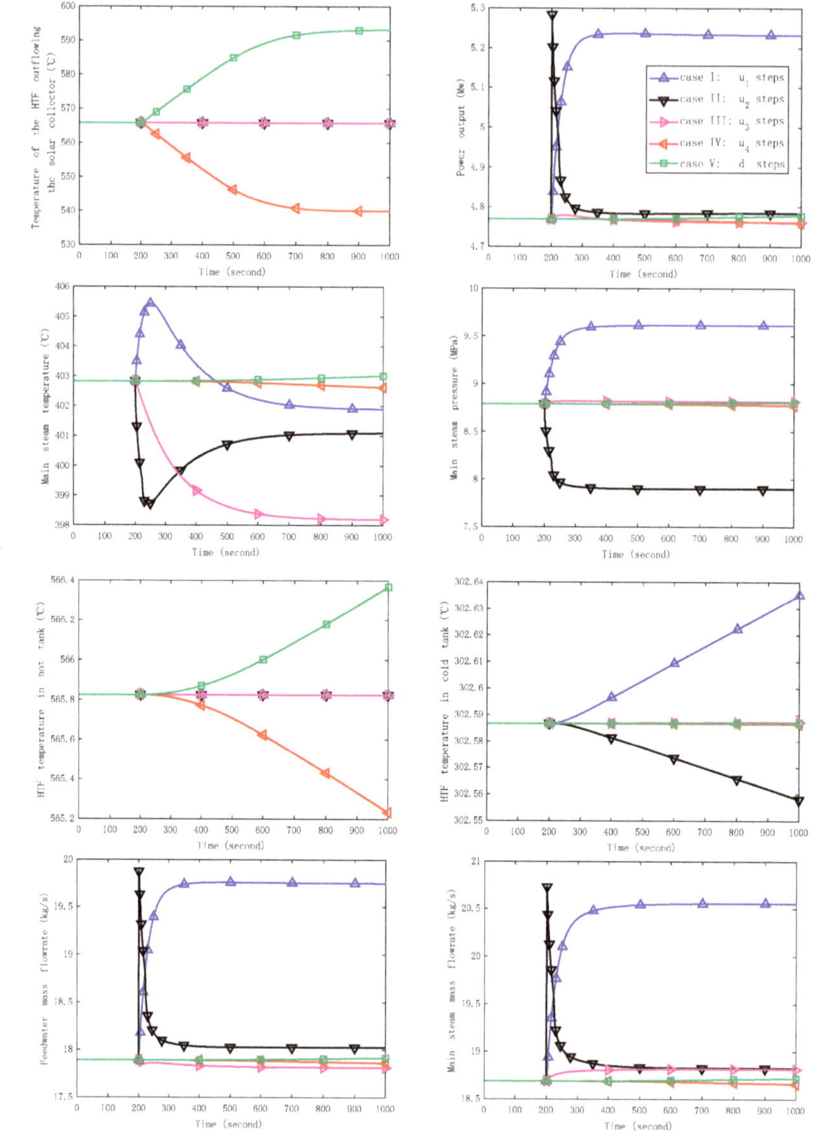

Figure 3. Simulation results of the step response (the legend on the top right side is applied to all the pictures in Figure 3).

3.2. Process Model Identification

From the step-response analysis, we find a strong coupling effect between the manipulating variables and the controlled variables. To quantify this effect, we introduce the maximal relative deviation of the controlled variable:

$$\mu_{ij} = \frac{\left|\Delta y_j^{max}\right|/\left|y_j^*\right|}{\left|\Delta u_i^{step}\right|/\left|u_i^*\right|},\tag{20}$$

where u_i^* and y_j^* are the initial value of the step variable u_i and the controlled variable y_j, respectively. Δu_i^{step} is the step increase value of u_i. Δy_j^{max} is the maximal deviation of y_j from y_j^* in the step response. All the μ_{ij} values are listed in Table 2 (y_1: power output; y_2: main steam pressure; y_3: main steam temperature; y_4: HTF temperature at the collector outlet). In the same row in Table 2, a higher value of μ_{ij} means that u_i has a relatively stronger influence on y_j than the other input variables.

Table 2. The value of μ_{ij} ("-" means μ_{ij} is too small and can be ignored).

Controlled Variables	u_1	u_2	u_3	u_4	d
y_1	0.8832	0.8639	-	-	-
y_2	0.8437	0.8118	-	-	-
y_3	0.0581	0.0814	0.0457	-	-
y_4	-	-	-	0.4300	0.4838

According to the results in Table 2, we can ignore the dynamics between the weakly interacted input variables and the controlled variables. Then, the following process model structure can be used to describe the dynamics of the CSP plant:

$$\begin{bmatrix} y_1 \\ y_2 \\ y_3 \\ y_4 \end{bmatrix} = \begin{bmatrix} g_{11}(s) & g_{12}(s) & 0 & 0 \\ g_{21}(s) & g_{22}(s) & 0 & 0 \\ g_{31}(s) & g_{32}(s) & g_{33}(s) & 0 \\ 0 & 0 & 0 & g_{44}(s) \end{bmatrix} \begin{bmatrix} u_1 \\ u_2 \\ u_3 \\ u_4 \end{bmatrix} + \begin{bmatrix} 0 \\ 0 \\ 0 \\ g_d(s) \end{bmatrix} d \tag{21}$$

where $g_{ij}(s)$ is the transfer function from the j-th manipulating variable to the i-th controlled variable. This model structure indicates that the original five-input–four-output control system can be decomposed into two independent control systems: a three-input–three-output power control system (u_1, u_2, u_3 and y_1, y_2, y_3) and a two-input–one-output HTF temperature control system (u_4, d, and y_4). Therefore, the control of the solar collector side and the power generation side are unrelated to each other, and we can design the power tracking controller regardless of the HTF temperature control in the solar collector. Using the step-response data, the transfer function models are identified in the System Identification toolbox in MATLAB (R2017a, The MathWorks, Inc., Natick, MA, USA) and the result is presented in Equation (22):

$$\begin{cases} \begin{bmatrix} y_1 - y_1^* \\ y_2 - y_2^* \\ y_3 - y_3^* \end{bmatrix} = \begin{bmatrix} \dfrac{0.001}{s+0.035} & \dfrac{5.2(s+0.0015)}{s+0.053} & 0 \\ \dfrac{0.0018}{s+0.032} & \dfrac{-0.526}{s+0.06} & 0 \\ \dfrac{0.01(s-0.0017)}{s^2+0.043s+0.0003} & \dfrac{-3.199(s+0.0025)}{s^2+0.0705s+0.0005} & \dfrac{-0.18}{s+0.0077} \end{bmatrix} \begin{bmatrix} u_1 - u_1^* \\ u_2 - u_2^* \\ u_3 - u_3^* \end{bmatrix} \\ y_4 - y_4^* = \dfrac{-0.00014}{s^2+0.017s+7.8\times10^{-5}} \cdot (u_4 - u_4^*) + \dfrac{0.00027(s+0.013)}{s^2+0.0068s+2.7\times10^{-5}} \cdot (d - d^*) \end{cases} \tag{22}$$

where the terms with superscript "*" represent the initial values of the step variables. The model fitness to the step-response data is shown in Table 3. The high fitness value indicates that the identified model captures the key dynamics of the CSP plant.

Table 3. The model fitness to the step-response data.

	$g_{11}(s)$	$g_{12}(s)$	$g_{21}(s)$	$g_{22}(s)$	$g_{31}(s)$	$g_{32}(s)$	$g_{33}(s)$	$g_{44}(s)$	$g_d(s)$
Fit to data	94.18%	87.19%	93.54%	91.44%	93.71%	90.31%	94.2%	91.19%	94.06%

4. Power Tracking Control System Design

This section investigates the power tracking strategy of the CSP plant. The heuristic PI control with two operation modes is first developed; then, the advanced MPC strategy is studied to further improve the control performance.

4.1. Heuristic PI Control

The PI control strategy is widely used in industrial processes owing to its convenience in parameter tuning, low implementation cost, and good robustness. In practice, PI controllers are implemented in discrete form:

$$u(t_k) = u(t_{k-1}) + K_p\left[\left(1 + \frac{\Delta t}{T_i}\right)(y_r(t_k) - y(t_k)) - (y_r(t_{k-1}) - y(t_{k-1}))\right] \tag{23}$$

where Δt is the sample time, t_k is the present sample time instance, t_{k-1} is the last sample time instance, y_r is the reference signal, y is the measurement of the controlled variable, K_p is the controller gain, and T_i is the integral time constant.

Based on heuristic knowledge, the PI control strategy with two operation modes is proposed for the CSP plant, taking account of the power tracking speed and the main steam parameter fluctuations.

A. Fast power tracking mode

In FT mode, the turbine maintains the power demands and the steam generator maintains the main steam pressure. This mode gives a fast power tracking rate, because the turbine has very fast dynamics and can immediately respond to the change of the power reference signal. However, this leads to a violent manipulation on the main steam valve, which brings substantial disturbance to the main steam pressure. This disturbance cannot be well compensated, because the steam generator has a large process inertia, and cannot generate enough steam in a timely manner to maintain the main steam pressure. In this mode, the power output, the main steam pressure, and the temperature are regulated by the opening of the main steam valve, the mass flowrate of the HTF outflowing from the hot tank, and the mass flowrate of the spraying water, respectively, via three independent PI controllers.

B. Smooth operation mode

In SO mode, the turbine maintains the main steam pressure and the steam generator maintains the power demands. This mode gives a slow power tracking rate, but a smooth operation of the main steam parameters. On one hand, it is not feasible to force aggressive control action on the sluggish steam generator to accelerate the power tracking speed, because this will cause the control system to be oscillatory and even unstable. Hence, the power tracking rate will inevitably reduce. On the other hand, when the control action on the steam generator is not strong, the main steam pressure and temperature are less disturbed and can be more easily controlled by manipulating the turbine and the attemperator. In this mode, the power output, the main steam pressure, and the temperature are regulated by the mass flowrate of the HTF outflowing from the hot tank, the opening of the main steam valve, and the mass flowrate of the spraying water, respectively, via three independent PI controllers.

4.2. Model Predictive Control Strategy

MPC is an advanced control technique that employs a model to predict the future response of the plant to the manipulating variables and minimizes the error between the plant response and the reference signal. At each sample time, the MPC solves a finite horizon optimization problem yielding a finite sequence of control actions, and only the first control action in the sequence is applied to the plant. Since MPC can automatically coordinate several control loops with strong interactions, it is recommended for the control of multivariable systems.

In this section, the state-space model-based MPC is employed to enhance the power tracking performance, and the controller design consists of three steps.

A. Construction of prediction model

A state-space model that can facilitate the design of a multivariable controller is used as the prediction model:

$$\begin{cases} x_{k+1} = Ax_k + Bu_k \\ y_k = Cx_k + Du_k \end{cases} \tag{24}$$

where x_k, $u_k = \begin{bmatrix} u_1(k) & u_2(k) & u_3(k) \end{bmatrix}^T$ and $y_k = \begin{bmatrix} y_1(k) & y_2(k) & y_3(k) \end{bmatrix}^T$ are the state vector, the input vector, and the output vector at time k, respectively. A, B, C, and D are the system matrixes. The state-space matrixes can be obtained by converting the identified transfer function model using the MATLAB command "tf2ss".

Then, the prediction model (24) is transformed into the augmented style to impose integral action on the MPC, so that an offset-free tracking performance can be obtained in the presence of model–plant mismatch [21]:

$$\begin{cases} \widehat{x}_{k+1} = \widehat{A} \cdot \widehat{x}_k + \widehat{B} \cdot \Delta u_k \\ y_k = \widehat{C} \cdot \widehat{x}_k + D \cdot \Delta u_k \end{cases} \tag{25}$$

where $\widehat{A} = \begin{bmatrix} A & 0 \\ C & I_3 \end{bmatrix}$, $\widehat{B} = \begin{bmatrix} B \\ D \end{bmatrix}$, $\widehat{C} = \begin{bmatrix} C & I_3 \end{bmatrix}$, $\widehat{x}_k = \begin{bmatrix} x_k - x_{k-1} \\ y_{k-1} \end{bmatrix}$, $\Delta u_k = u_k - u_{k-1}$, I_3 is the three-order unit matrix.

By stacking up Equation (25) for N_y steps, the prediction of future output sequences can be obtained: $y_p = \begin{bmatrix} y_{k+1}^T & y_{k+2}^T & \cdots & y_{k+N_y}^T \end{bmatrix}^T$, which can be expressed using $N_u(N_u < N_y)$ future control sequences $\Delta u_p = \begin{bmatrix} \Delta u_k^T & \Delta u_{k+1}^T & \cdots & \Delta u_{k+N_u-1}^T \end{bmatrix}^T$:

$$y_p = \Phi_x \widehat{x}_k + \Phi_u \begin{bmatrix} \Delta u_p & 0_{1\times3} & \cdots & \underbrace{0_{1\times3}}_{(N_y-N_u)\,items} \end{bmatrix}^T \tag{26}$$

where:

$$\Phi_x = \begin{bmatrix} \widehat{C} \cdot \widehat{A} \\ \widehat{C} \cdot \widehat{A}^2 \\ \vdots \\ \widehat{C} \cdot \widehat{A}^{N_y} \end{bmatrix} \tag{27}$$

$$\Phi_u = \begin{bmatrix} \widehat{C} \cdot \widehat{B} & D & 0 & 0 & \cdots & 0 \\ \widehat{C} \cdot \widehat{A} \cdot \widehat{B} & \widehat{C} \cdot \widehat{B} & D & 0 & \cdots & 0 \\ \vdots & \vdots & \vdots & \vdots & \ddots & 0 \\ \widehat{C} \cdot \widehat{A}^{N_u-1} \cdot \widehat{B} & \widehat{C} \cdot \widehat{A}^{N_u-2} \cdot \widehat{B} & \cdots & \widehat{C} \cdot \widehat{A} \cdot \widehat{B} & \widehat{C} \cdot \widehat{B} & D \\ \widehat{C} \cdot \widehat{A}^{N_u} \cdot \widehat{B} & \widehat{C} \widehat{A}^{N_u-1}\widehat{B} & \cdots & \widehat{C} \cdot \widehat{A}^2 \cdot \widehat{B} & \widehat{C} \cdot \widehat{A} \cdot \widehat{B} & \widehat{C} \cdot \widehat{B} \\ \vdots & \vdots & \vdots & \vdots & \vdots & \vdots \\ \widehat{C} \widehat{A}^{N_y-1}\widehat{B} & \widehat{C} \widehat{A}^{N_y-2}\widehat{B} & \cdots & \widehat{C} \cdot \widehat{A}^{N_y-N_u+1} \cdot \widehat{B} & \widehat{C} \cdot \widehat{A}^{N_y-N_u} \cdot \widehat{B} & \widehat{C} \cdot \widehat{A}^{N_y-N_u-1} \cdot \widehat{B} \end{bmatrix} \tag{28}$$

Note that the future control sequence beyond N_u (the control horizon) is assumed to be constant, i.e., $\Delta u_{k+Nu}^T \cdots \Delta u_{k+Ny-1}^T = 0$.

B. Estimation of immeasurable states

Since the state-space model is developed via data identification, its state vector does not have physical meanings, and cannot be measured. Therefore, it is necessary to estimate its value via a state observer on the basis of measured inputs and outputs:

$$\hat{x}_{k+1} = \widehat{A}\cdot\hat{x}_k + \widehat{B}\cdot u_k + K\left(y_k - \widehat{C}\cdot\hat{x}_k - Du_k\right) \tag{29}$$

where the superscript "^" means the estimated value. The observer gain K can be calculated if the matrix H and G and a symmetric positive definite matrix X exist, such that the following LMI (linear matrix inequality) problem is feasible [22]:

$$\begin{bmatrix} H^T + H - X & \left(\widehat{HA} + \widehat{GC}\right)^T \\ \widehat{HA} + \widehat{GC} & X \end{bmatrix} > 0 \tag{30}$$

And the observer gain is $K = H^{-1}G$. At each sample time, we replace the states of x_k^{aug} in Equation (26) with the estimated states to update the prediction model.

C. Calculation of optimal control moves

The objective function of MPC is designed to achieve an optimal trade-off between the rapidity of set-point tracking and the intensity of control actions; hence, the MPC controllers can achieve good performance more easily than conventional PI controllers that cannot ensure the optimality of control actions. By minimizing the quadratic objective function with the consideration of actuator constraints, the control moves of the MPC can be calculated:

$$\begin{aligned} J(\Delta u_p) &= \left(y_p - r\right)^T Q\left(y_p - r\right) + \Delta u_p^T R \Delta u_p \\ s.t. \quad &\Delta u_{min} \leq \Delta u_p \leq \Delta u_{max} \\ &u_{min} \leq u_p \leq u_{max} \end{aligned} \tag{31}$$

where $r = \begin{bmatrix} r_{k+1}^T & r_{k+2}^T & \cdots & r_{k+N_y}^T \end{bmatrix}^T$ is the reference signal of the controlled variables, Δu_{min} and Δu_{max} are the rate constraints of the actuators, and u_{min} and u_{min} are the amplitude constraints of the actuators. Q and R are the adjustable weighting matrixes for the tracking error and the control actions, respectively.

4.3. Case Study

This section presents the simulation study of power tracking control. The heuristic PI tuned using the conventional Ziegler–Nichols method is compared with MPC. The sample time for the PI controller and MPC is set at 1 s. The prediction horizon and control horizon of the MPC are $N_y = 500$ and $N_u = 10$, respectively. The weighting matrixes in MPC are given as follows: $Q = diag\{\ 1 \quad 30 \quad 35\ \}$ and $R = diag\{\ 0.08 \quad 15 \quad 15\ \}$. The physical constraints of the actuator are: $\Delta u_{max} = -\Delta u_{min} = \begin{bmatrix} 1.5 & 0.01 & 0.02 \end{bmatrix}$, $u_{min} = \begin{bmatrix} 0 & 0.4 & 0 \end{bmatrix}$, and $u_{max} = \begin{bmatrix} 200 & 1 & 2 \end{bmatrix}$. The tuning parameters of the heuristic PI are shown in Table 4.

Table 4. Tuning parameters of the heuristic PI.

Modes of the Heuristic PI	Power Control Loop		Main Steam Pressure Control Loop		Main Steam TEMPERATURE Control Loop	
	K_p	T_i	K_p	T_i	K_p	T_i
Smooth operation	0.3	2.3	0.05	27	0.04	18
Fast power tracking	1.2	2.1	0.26	15	0.04	18

The simulation case is designed as follows: initially, the power reference signal (regarded as the power generation schedule) is set at 4.78 MW. At 200 s and 1000 s, it changes from 4.78 MW to 5.5 MW and to 3.0 MW, respectively. The direct solar irradiance incident on the solar collector is initially set at 1.9 kW/m^2; at 1500 s, it reduces to 1.0 kW/m^2. The reference signal of the main steam temperature and pressure are set as 402.8 °C and 8.79 MPa, respectively. The simulation results are shown in Figure 4. Note that since the control of the HTF temperature at the collector outlet is not studied in this paper, we use a well-tuned PI controller to control it in the simulation cases. The status of the TES system is also presented to analyze the energy flow of the CSP plant.

As shown in Figure 4a–c, the MPC has the best performance: a fast power tracking rate and the least fluctuation in main steam parameters. When the power reference signal increases/decreases, the MPC increases/decreases the mass flowrate of the HTF outflowing from the hot tank and the main steam valve opens on time, controlling the power output of the CSP plant rapidly to follow the power generation schedule, as shown in Figure 4a. Since the MPC can anticipate the interactions between the different control loops, the change rate of the main steam valve opening and spraying the water mass flowrate is well coordinated with the change rate of the mass flowrate of the HTF from the hot tank, so that the pressure and temperature of the main steam are closely maintained to their set points.

However, the PI controllers cannot achieve a fast power tracking and a smooth operation simultaneously. In FT mode, the heuristic PI attains a similar power tracking rate to the MPC, while there are significant fluctuations in the main steam parameters. As shown in Figure 4b, the main steam valve quickly opens to instantly generate the power required by the generation schedule. This brings significant disturbance to the steam pressure. However, unlike the MPC, the heuristic PI (FT mode) cannot predict the influence of such incoming disturbance, and it only uses the sluggish steam generator to regulate the rapidly changing pressure, which inevitably results in an untimely adjustment. Additionally, the main steam temperature is also disturbed by the frequent regulation in the main steam pressure, which results in an oscillatory transient and is difficult to be compensated by the single-looped PI.

In SO mode, the heuristic PI has a smooth transient of main steam parameters but a much lower power tracking rate (approximately 200 s slower than the PI of the FT mode and the MPC). The control moves for power tracking are quite conservative so that it does not cause substantial variations in the main steam parameters. This low-level disturbance can be timely eliminated by the PI controller. However, even with such a mild control action, the fluctuation of the main steam parameters is still more intensive than that of the MPC.

In summary, although the trend of the control actions is reasonable, the single loop-based heuristic PI does not consider the strong interactions between the multiple loops, and hence is unable to give the best control action. Additionally, owing to the past error-based mechanism, it is difficult for the PI controller to satisfy the prompt control requirement of the CSP plant with slow dynamics. Therefore, the MPC can achieve an improved performance to the heuristic PI.

Figure 4d,e, presents the status of the TES system. The temperature of the HTF in the storage tanks almost has no variation; hence, the volume of the HTF stored in the hot tank alone can be regarded as the indicator of the amount of the stored thermal energy. In the starting 200 s, the received solar energy (1.9 kW/m^2) is in perfect balance with the generated power (4.78 MW), and hence, the HTF volume in the hot tank is kept constant. Then, from 200 s to 1000 s, the power reference signal increases to 5.5 MW while the received solar energy is still equivalent to 4.78 MW. The hot tank must release the stored energy to compensate for the insufficient solar energy input, which results in a gradual reduction in the HTF volume. From 1000 s to 1500 s, the power output changes to 3 MW, which is smaller than the received solar energy (4.78 MW), and the HTF volume of the hot tank increases in order to store the excessive solar energy. After 1500 s, the direct solar irradiance incident reduces to 1.0 kW/m^2, which is insufficient to provide 3 MW power output, leading to a reduction in the HTF volume in hot tank. The results show that via manipulating the storage system, both the heuristic PI and MPC can automatically adjust the energy balance between the received solar energy, the stored thermal energy, and the power output.

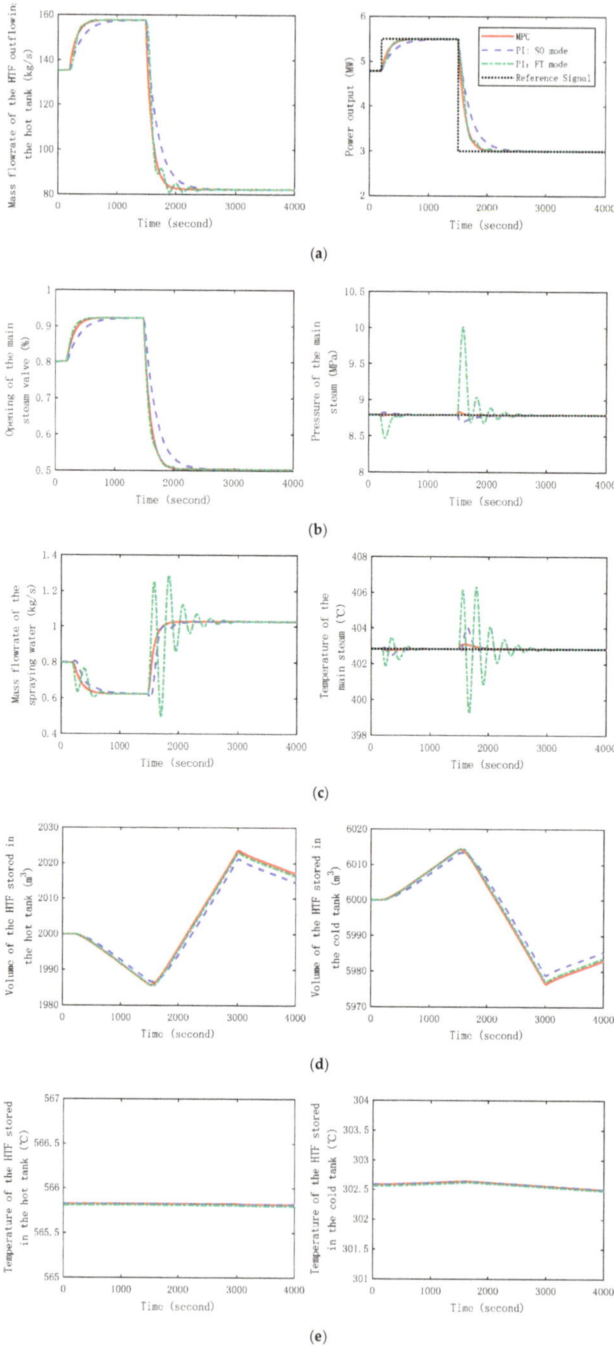

Figure 4. Simulation results of the control system (the legend on the top right side is applied to all the pictures in Figure 4). (**a**) Manipulating variable u_1 and controlled variable y_1; (**b**) manipulating variable u_2 and controlled variable y_2; (**c**) manipulating variable u_3 and controlled variable y_3; (**d**) HTF stored in the hot tank and HTF stored in the cold tank; (**e**) HTF temperature in the hot tank and HTF temperature in the cold tank.

5. Conclusions

The execution-level power tracking control is significant to the economic, safe, and flexible operation of the CSP plant, which is challenging and had not yet been studied in previous research.

To address this issue, this paper proposes two control strategies: heuristic PI control with two operation modes, and advanced model predictive control. Step response analysis of the CSP plant is first performed, and we find that: (1) there is a strong coupling effect between the system inputs and outputs, (2) the external disturbance (the change of solar radiation) has little influence on the power generation side, and (3) in the sense of execution-level control, the design of the power tracking controller is completely independent from the HTF temperature control at the solar collector side. Then, the performance of the heuristic PI control and MPC control are compared through simulation study. The results show the following. (1) The heuristic PI with FT mode can achieve a fast power tracking rate; however, this causes a considerable fluctuation in the main steam parameters. (2) The heuristic PI with SO mode can achieve a smooth operation of the main steam parameters; however, it has a much slower power tracking rate. (3) The MPC can well handle the strong interactions between the control loops, and hence achieves the dual task of fast power tracking and smooth operation of the main steam parameters. (4) Both the heuristic PI and MPC can automatically balance the energy flow of the CSP plant.

Within the proposed MPC control scheme, the power generation schedule from the decision-level controller can be timely tracked, which guarantees the economic operation of the CSP plant. Additionally, the main steam parameters are maintained around the designed operating range, ensuring the safe operation of the plant.

Future works on the power tracking control of the CSP plant will be extended to the widely used two-tank indirect TES system, which is even more challenging because the change of solar radiation can directly influence the power output, and it is difficult to compensate for this disturbance.

Author Contributions: Conceptualization, X.L. and Y.L.; Methodology, Y.L.; Software, X.L.; Validation, X.L.; Formal Analysis, X.L.; Investigation, X.L.; Resources, Y.L.; Data Curation, Y.L.; Writing-Original Draft Preparation, X.L.; Writing-Review & Editing, Y.L.; Supervision, Y.L.; Project Administration, Y.L.; Funding Acquisition, Y.L.

Funding: This research was funded by National Key R&D Program of China grant number [2018YFB1502904], and National Natural Science Foundation of China (NSFC) grant number [51476027], [51576041], [51506029].

Conflicts of Interest: The authors declare no conflict of interest.

Abbreviations

A	overall area, m^2
$\hat{A}$	cross sectional area, m^2
C	specific heat, $kJ/(kg \cdot K)$
H	specific enthalpy, kJ/kg
h	heat transfer coefficient, $W/(m^2 \cdot K)$
I	direct solar irradiance incident, W/m^2
k	thermal conductivity, $W/(m \cdot K)$
M	mass, kg
P	perimeter, m
PI	proportional-integral
p	pressure, MPa
Q	transferred heat, kJ
q	mass flowrate, kg/s
T	temperature, $^\circ C$
V	volume, m^3
w	aperture width of the solar collector, m

Greek letters

δ	main steam valve opening
ε	emissivity
η	efficiency
μ	dynamic viscosity, Pa·s
ρ	density, kg/m^3
σ	the Stefan-Boltzmann constant, W/(m2·K^4)

Subscripts

B	boiler
C	collector
Cond	condenser
cold	cold tank
E	economizer
Ex	extracted steam
Fw	feed water
G	glass envelope
H	heat transfer fluid (HTF)
HA	HTF outflowing the absorber
HB	HTF outflowing the boiler
HE	HTF outflowing the economizer
HSh	HTF outflowing the superheater
hot	hot tank
i	inner
in	inlet
L	liquid water
LA	liquid water outflowing the attemperator
LE	liquid water outflowing the economizer
MB	metal in the boiler
MSh	metal in the superheater
o	outer
out	outlet
S	steam
SB	steam outflowing the boiler
St	storage tank
Sh	superheater
Sw	saturated water
SSh	steam outflowing the superheater
T	tube
TA	tube in the absorber
TB	tube in the boiler
TE	tube in the economizer
Tu	turbine

References

1. Petrollese, M.; Cocco, D.; Cau, G.; Cogliani, E. Comparison of three different approaches for the optimization of the CSP plant scheduling. *Sol. Energy* **2017**, *150*, 463–476. [CrossRef]
2. Vasallo, M.J.; Bravo, J.M. A MPC approach for optimal generation scheduling in CSP plants. *Appl. Energy* **2016**, *165*, 357–370. [CrossRef]
3. Vasallo, M.J.; Bravo, J.M.; Cojocaru, E.G.; Gegúndez, M.E. Calculating the profits of an economic MPC applied to CSP plants with thermal storage system. *Sol. Energy* **2017**, *155*, 1165–1177. [CrossRef]
4. Cirocco, L.R.; Belusko, M.; Bruno, F.; Boland, J. Controlling stored energy in a concentrating solar thermal power plant to maximise revenue. *Renew. Power Gener. IET* **2015**, *9*, 379–388. [CrossRef]

5. Liu, M.; Tay, N.H.S.; Bell, S.; Belusko, M.; Jacob, R.; Will, G.; Saman, W.; Bruno, F. Review on concentrating solar power plants and new developments in high temperature thermal energy storage technologies. *Renew. Sustain. Energy Rev.* **2016**, *53*, 1411–1432. [CrossRef]

6. Casati, E.; Casella, F.; Colonna, P. Design of CSP plants with optimally operated thermal storage. *Sol. Energy* **2015**, *116*, 371–387. [CrossRef]

7. Usaola, J. Participation of CSP plants in the reserve markets: A new challenge for regulators. *Energy Policy* **2012**, *49*, 562–571. [CrossRef]

8. Camacho, E.F.; Gallego, A.J. Optimal operation in solar trough plants: A case study. *Sol. Energy* **2013**, *95*, 106–117. [CrossRef]

9. Franchini, G.; Barigozzi, G.; Perdichizzi, A.; Ravelli, S. Simulation and performance assessment of load-following CSP plants. In Proceedings of the 3rd Southern African Solar Energy Conference, Kruger National Park, South Africa, 11–13 May 2015.

10. Cirre, C.M.; Berenguel, M.; Valenzuela, L.; Camacho, E.F. Feedback linearization control for a distributed solar collector field. *Control Eng. Pract.* **2007**, *15*, 1533–1544. [CrossRef]

11. Gallego, A.J.; Camacho, E.F. Adaptative state-space model predictive control of a parabolic-trough field. *Control Eng. Pract.* **2012**, *20*, 904–911. [CrossRef]

12. Alsharkawi, A.; Rossiter, J.A. Towards an improved gain scheduling predictive control strategy for a solar thermal power plant. *IET Control Theory Appl.* **2017**, *11*, 1938–1947. [CrossRef]

13. Nevado Reviriego, A.; Hernández-del-Olmo, F.; Álvarez-Barcia, L. Nonlinear Adaptive Control of Heat Transfer Fluid Temperature in a Parabolic Trough Solar Power Plant. *Energies* **2017**, *10*, 1155. [CrossRef]

14. Rolim, M.M.; Fraidenraich, N.; Tiba, C. Analytic modeling of a solar power plant with parabolic linear collectors. *Sol. Energy* **2009**, *83*, 126–133. [CrossRef]

15. Manenti, F.; Ravaghi-Ardebili, Z. Dynamic simulation of concentrating solar power plant and two-tanks direct thermal energy storage. *Energy* **2013**, *55*, 89–97. [CrossRef]

16. Caputo, A.C.; Pelagagge, P.M.; Salini, P. Heat exchanger design based on economic optimisation. *Appl. Therm. Eng.* **2008**, *28*, 1151–1159. [CrossRef]

17. Åström, K.J.; Bell, R.D. Drum-boiler dynamics. *Automatica* **2000**, *36*, 363–378. [CrossRef]

18. Powell, K.M.; Edgar, T.F. Modeling and control of a solar thermal power plant with thermal energy storage. *Chem. Eng. Sci.* **2012**, *71*, 138–145. [CrossRef]

19. De Mello, F.P. Boiler models for system dynamic performance studies. *IEEE Trans. Power Syst.* **1991**, *6*, 66–74. [CrossRef]

20. Leva, A.; Maffezzoni, C.; Benelli, G. Validation of drum boiler models through complete dynamic tests. *Control Eng. Pract.* **1999**, *7*, 11–26. [CrossRef]

21. Wu, X.; Shen, J.; Li, Y.; Wang, M.; Lawal, A. Flexible operation of post-combustion solvent-based carbon capture for coal-fired power plants using multi-model predictive control: A simulation study. *Fuel* **2018**, *220*, 931–941. [CrossRef]

22. Wu, X.; Wang, M.; Shen, J.; Li, Y.; Lawal, A.; Lee, K.Y. Reinforced coordinated control of coal-fired power plant retrofitted with solvent based CO2 capture using model predictive controls. *Appl. Energy* **2019**, *238*, 495–515. [CrossRef]

Article

Thermal Fatigue Modelling and Simulation of Flip Chip Component Solder Joints under Cyclic Thermal Loading

Liangyu Wu [1], Xiaotian Han [1], Chenxi Shao [2], Feng Yao [3] and Weibo Yang [1,*]

[1] School of Hydraulic, Energy and Power Engineering, Yangzhou University, Yangzhou 225127, Jiangsu, China; lywu@yzu.edu.cn (L.W.); xthan@microflows.net (X.H.)

[2] Key Laboratory of Energy Thermal Conversion and Control of Ministry of Education, School of Energy and Environment, Southeast University, Nanjing 210096, Jiangsu, China; 220140482@seu.edu.cn

[3] Jiangsu Key Laboratory of Micro and Nano Heat Fluid Flow Technology and Energy Application, School of Environmental Science and Engineering, Suzhou University of Science and Technology, Suzhou 215009, Jiangsu, China; yaofeng@usts.edu.cn)

* Correspondence: wbyang@yzu.edu.cn; Tel.: +86-514-8797-1315

Received: 29 May 2019; Accepted: 19 June 2019; Published: 21 June 2019

Abstract: Thermal Fatigue of flip chip component solder joints is widely existing in thermal energy systems, which imposes a great challenge to operational safety. In order to investigate the influential factors, this paper develops a model to analyze thermal fatigue, based on the Darveaux energy method. Under cyclic thermal loading, a theoretical heat transfer and thermal stress model is developed for the flip chip components and the thermal fatigue lives of flip chip component solder joints are analyzed. The model based simulation results show the effects of environmental and power parameters on thermal fatigue life. It is indicated that under cyclic thermal loading, the solder joint with the shortest life in a package of flip chip components is located at the outer corner point of the array. Increment in either power density or ambient temperature or the decrease in either power conversion time or ambient pressure will result in short thermal fatigue lives of the key solder joints in the flip chip components. In addition, thermal fatigue life is more sensitive to power density and ambient temperature than to power conversion time and ambient air pressure.

Keywords: electronic device; flip chip component; thermal stress; thermal fatigue; life prediction

1. Introduction

With the rapid development of microelectronics, microelectronic devices with variable thermal loads have been extensively used in the thermal energy system [1–3], chemical production [4–6], biomedical detection [7–9], deep space exploration fields [10–12], and in microelectromechanical systems (MEMS) [13–16]. The flip chip technology is widely applied in MEMS, in which solder joints are used as both mechanical supports and interconnections between electronic components [17,18]. However, under cyclic thermal loads, solder joints always tend to fatigue failure resulting from the mismatched thermal expansion of the materials in the solder joints [14,19]. Therefore, the thermal fatigue life of a solder joint under cyclic thermal loads is directly related to the safe operation and reliability of the entire electronic device [20]. In this context, the thermal fatigue mechanism and prediction of thermal fatigue life are essential to the designs and reliabilities of electronic products.

Recently, several theoretical efforts to investigate the thermal fatigue mechanism and predict the thermal fatigue life have been undertaken [21,22]. Cheng et al. [17] examined the reliability of the solder interconnect of an advanced ultrafine-pitch integrated circuit chip through three-dimensional finite element (FE) numerical simulations and experimental tests. Incorporating the fatigue criteria based on energy, Chen et al. [23] conducted the improvement of the geometric simplification methods. The accuracy of simulation results depends on the parameters of the local model, including the mesh density, step size, as well as cut boundaries. Jiang et al. [24] compared the prediction results of the fatigue life solder balls in ball grid array (BGA) packaging obtained by the Darveaux model and the Coffin–Manson model, which are energy-based and strain-based, respectively. In either model, the fatigue life increases with height and decreases with diameter. However, the thermal stress distributions for the solder joints in a BGA product is usually nonuniform, which gives rise to uneven thermal fatigue lives for them. It is of significance finding the key solder joint that determines the thermal fatigue life of a BGA product.

In addition, the thermal fatigue life of an electronic device depends on the ambient environment [25–27] and cooling condition [28,29]. Considering the environmental conditions of Mars, the reliability of the plastic BGA was experimentally studied under four different thermal cycles (−55 to 100 °C, −55 to 125 °C, −65 to 150 °C, and −120 to 85 °C) [30]. Ghaffarian has taken the optical photomicrographs of BGA to record the progression and characteristics of damage with numerous thermal cycle intervals. Under the thermal cycles with extreme-temperature conditions, the reliabilities of surface-mounted electronic package test boards were assessed for future long-term deep space missions in extreme-temperature environments [31]. The highly accelerated life testing (HALT) technique was applied by Ramesham [32] to assess electronic packaging during long-term deep-space explorations with the extreme temperature ranging from −150 °C to +125 °C in order to achieve optimized design. During this accelerated test within 12 h, an abnormal electrical continuity occurred in the plastic BGA. However, the underlying relationship between the extreme ambient environment and thermal fatigue life of an electronic device is still unclear. In particular, limited attempts have been made to reveal the coupled effects of extreme ambient environments and cyclic thermal loads on thermal fatigue lives.

Although there are several attempts to interpret the thermal fatigue life of an electronic device with the extreme ambient environment via experimental investigations, numerical attempts in exploring the thermal fatigue lives of electronic devices is less available, especially when the electronic devices are operating under low temperature and pressure. Therefore, based on the Darveaux energy method, a theoretical heat transfer and thermal stress model for flip chip components in the cavity of an initial static air flow field and the finite boundary temperature under cyclic thermal loading was developed in an attempt to predict the thermal fatigue lives of flip chip component solder joints. The location of the critical solder joint in a flip chip component with the shortest thermal fatigue life that determines the safety of the entire electronic device is examined. The effects of ambient environment conditions and thermal loads on the thermal fatigue life of the critical solder joint are discussed. This work can provide a further insight into the thermal reliability of an electronic device under extreme ambient environments.

2. Mathematical Model

In this paper, a model consist of a 4×4 solder joint array of a flip chip component is developed, as shown in Figure 1. The model mainly includes a silicon chip, a circuit board, a copper sheet, and solder joints, as shown in Figure 2. There are 16 solder joints in the 4×4 array. Each solder joint has a height of 0.3 mm, a surface area of 4.847×10^{-7} m^2, and a volume of 3.035×10^{-11} m^3. The shape of the silicon chip is a rectangular parallelepiped with a length of 2 mm, a width of 2 mm, and a height of 0.5 mm. The shape of the copper sheet is a rectangular parallelepiped with a length of 2.7 mm, a width of 2.7 mm, and a height of 0.07 mm. The shape of the circuit board is a rectangular parallelepiped with a length, width and height of 2.7 mm, 2.7 mm, and 1.23 mm, respectively. In this paper, the following assumptions are made: (1) the influences of the printed copper wire, filler and other components on the model are neglected; (2) any residual stress and strain that may occur during the manufacturing process are ignored; (3) the materials in each part are ideally connected; and (4) when the temperature varies, the heat conduction between each kind of material in the model is considered.

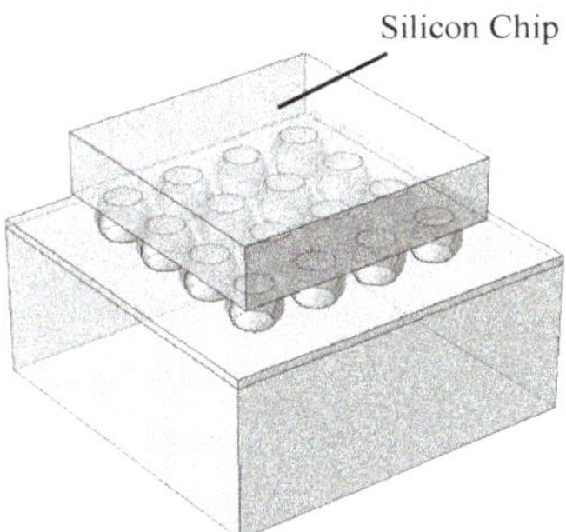

Figure 1. Geometry model of a flip chip component.

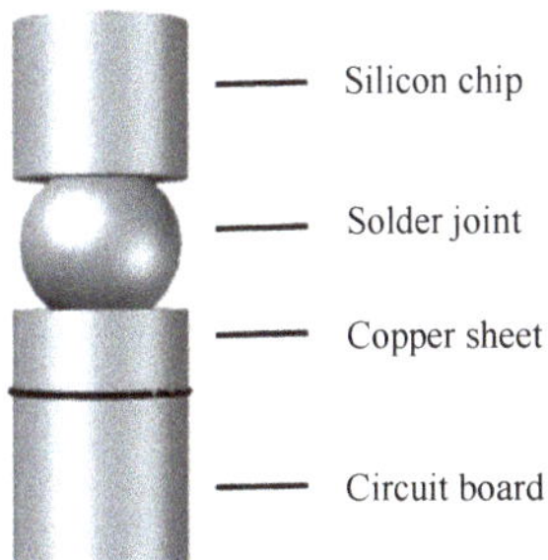

Figure 2. Schematic of the flip chip component assembly material.

2.1. Governing Equations

In this paper, the circuit board, copper sheet, and silicon chip are regarded as isotropic, linear elastic materials. The stress and strain have a simple one-to-one correspondence, and the constitutive equation can be expressed as follows [24]:

$$\{\sigma\} = [D]\{\varepsilon_{el}\}, \tag{1}$$

where $\{\sigma\}$ is the stress component, $[D]$ is the elastic stiffness matrix, and $\{\varepsilon_{el}\}$ is the elastic strain component.

When the ambient temperature of the material is greater than half of its own melting point, viscoplastic behaviors should be considered. Since the mechanical parameters and properties of tin-lead solder are affected by temperature and time, the unified viscoplastic Anand constitutive equation [24] gives the stress-strain response of tin-lead solder under thermal loading.

To further study the influence of the working environment of on the thermal fatigue life, the coupled effects of the air flow, thermal convection, heat conduction, and structural mechanics are considered to investigate the thermal characteristics and failure mechanisms of electronic devices in extreme environments. In this paper, cyclic thermal loading of electronic devices is used to simulate the normal working/idle process of an electronic package. In addition, the Darveaux energy method is applied in predicting the thermal fatigue lives. To study the influences of ambient temperature on the thermal fatigue lives of electronic devices, a silicon chip with a cyclic thermal load is considered as a heat source, which conducts heat between each part of the electronic devices. The time-dependent function of the cyclic thermal load is:

$$P = P_0(t), \tag{2}$$

where P_0 is the power per unit volume of the flip chip component chip heat source. The cyclic thermal load under typical conditions is shown in Figure 3, where the maximum power density of the silicon chip is $P_{max} = 5 \times 10^7$ W·m^{-3} and the minimum power density is $P_{min} = 1 \times 10^7$ W·m^{-3}. The conversion time of the chip power is $t_{trans} = 600$ s.

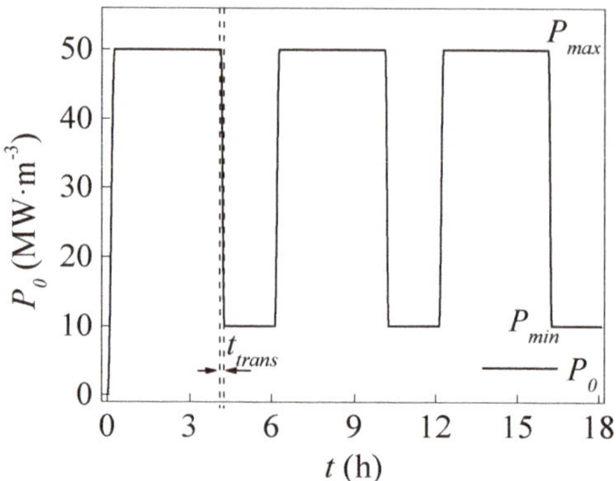

Figure 3. Power density of a cyclic thermal load.

2.2. Boundary Conditions

At the beginning of every process, the whole zone of the computational domain is at thermal equilibrium state with a constant temperature and pressure. As shown in Figure 4, the boundaries of the computational domain are assumed to be open faces of constant pressure and temperature through which air can flow, which means the boundaries has the identical temperature and pressure with the ambient. Thus, the boundary conditions of the computational domain can all be written as:

$$T_w = T_a;\; P_w = p_a, \tag{3}$$

where, the subscripts a and w represent the ambient and the boundaries of computational domain. In this simulation, the ambient temperature and pressure are fluctuant parameters, and the typical ambient temperature and pressure of the whole computational domain is $T_a = 20\ °C$ and $p_a = 1$ bar, respectively. Additional, the boundary of the solid mechanics field is set to limit the rigid displacement:

$$\omega(x, y, z) = 0\ (x = 2.8\ \text{mm},\; y = 0.1\ \text{mm},\; z = 1.41\ \text{mm}), \tag{4}$$

where ω is the displacement of the copper sheet.

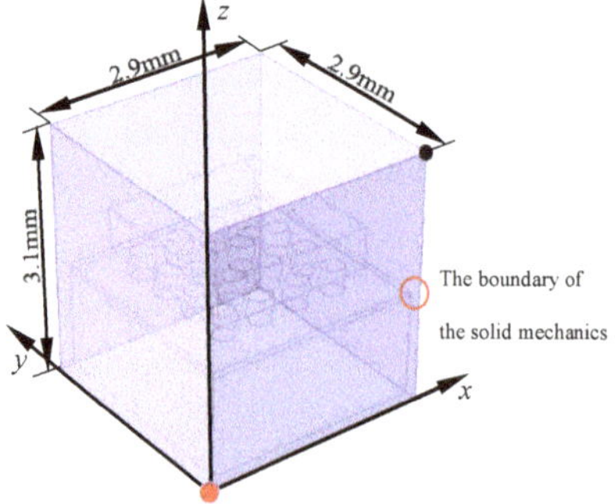

Figure 4. Power density of the cyclic thermal load.

The number of breakdown cycles of the electronic component is calculated to predict the thermal fatigue life of the flip chip component. Therefore, the thermal fatigue life of a solder joint in the flip chip component is predicted based on the Darveaux failure model [24].

$$N = K_1 \Delta W_{ave}^{K_2} + \frac{a}{K_3 \Delta W_{ave}^{K_4}}, \tag{5}$$

$$\Delta W_{ave} = \frac{\int \Delta W_d \delta V}{\int \delta V}, \tag{6}$$

$$\Delta W_d = W_d' - W_d = \frac{1}{2}(\varepsilon' \sigma' - \varepsilon \sigma), \tag{7}$$

where N denotes the cyclic number. In this study, the duration of one cycle is 6 h. ΔW_{ave} is the increment in the density of average viscoplastic strain energy. ΔW_d is the viscoplastic strain energy density increase. W_d' and W_d are the strain energy densities of adjacent cycles. ε' and ε are the equivalent strains of adjacent cycles. σ' and σ are the equivalent stresses of adjacent cycles. $a = 2.65 \times 10^{-4}$ m is the characteristic fracture length, which can be regarded as the diameter of the interface between the solder joint and other materials. K_1, K_2, K_3, and K_4 are related physical property parameters, and their specific values are shown in Table 1.

Table 1. Material failure parameters.

Parameter	Value	Description
K_1	3.25 cycles·Pa^{-1}	Initial crack energy coefficient
K_2	−1.52	Initial crack energy index
K_3	4.04×10^{-3} Pa·cycle^{-1}	Crack growth energy coefficient
K_4	0.98	Crack growth energy index

In this paper, a grid independence test is performed by calculating the thermal fatigue lives at different grid numbers. The specific grid data and calculation results are shown in Table 2.

Table 2. Computational grid data of the flip chip component model.

Number of Grids	Thermal Fatigue Life (h)
12981970	17484
2783059	17520
1072916	17598
543249	17706
219631	17880
103691	18744
56625	19248
34138	20124

As shown in Table 2, after the number of grids reaches 1×10^6, the predicted thermal fatigue life is basically stable. Considering the calculation accuracy and calculation time, it is suitable to select the calculation grid with a grid number of 1,072,916.

2.3. Model Validation

To validate the model, the finite element numerical simulation of the uniaxial tensile test under the two working conditions of 313 K and 353 K is carried out, and the stress parameters are compared with the experimental results from the literature [33], as shown in Figure 5. In the simulations, the strain rate is constant at 0.5% s^{-1} during uniaxial tensile testing. The two-dimensional geometry of the sample used is illustrated in Figure 5a, and the three-dimensional model is given in Figure 5b. The stress at the tapered collar obtained by the numerical simulation are compared with the experimental results. The mechanical properties of the 63Sn–37Pb solder are shown in Table 3.

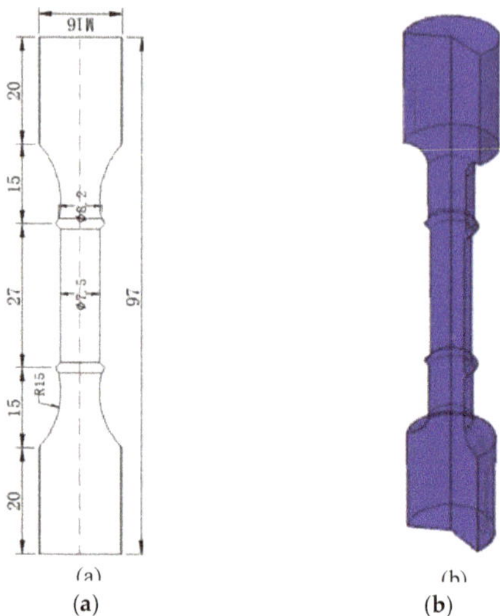

(a)   (b)

Figure 5. Uniaxial tensile test finite element numerical calculation model: (**a**) two-dimensional diagram and (**b**) three-dimensional diagram.

Table 3. Performance parameters of 63Sn–37Pb tin-lead solder at a strain rate of 0.5% s^{-1}.

Temperature (K)	Elastic Modulus (MPa)
313	41350
353	38540
398	34568

Figure 6 shows the good agreement between the numerical simulation and experimental results for the stresses at 313 K and 353 K, which indicates that our model is able to sufficiently describe the mechanical characteristics of the solder at various working temperatures.

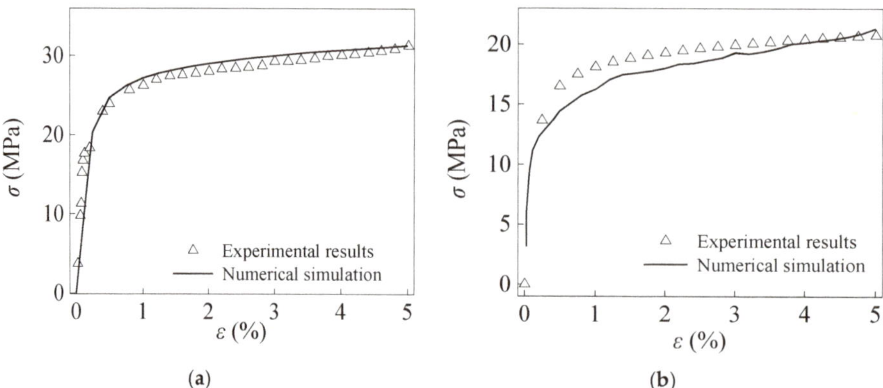

(a)   (b)

Figure 6. Comparison of the numerical results and experimental data of the 63Sn–37Pb solder: (**a**) operating temperature of 313 K and (**b**) operating temperature of 353 K.

3. Results and Discussion

The thermal fatigue life distributions in the flip chip component under typical working conditions is obtained using the Darveaux energy method, where the maximum power density of the silicon chip is $P_{max} = 5 \times 10^7$ W·m^{-3} and the minimum power density is $P_{min} = 1 \times 10^7$ W·m^{-3}. The conversion time of the chip power is $t_{trans} = 600$ s. As illustrated in Figure 7, the solder joint with the shortest thermal fatigue life in the flip chip component appears at the outer corner point, which experience 19,055 ($10^{4.28}$) cycles. Since the malfunction of one solder joint in the flip chip component can lead to the failure of the whole flip chip component, the solder joint with the shortest thermal fatigue life in the flip chip component is regarded as the key solder joint. The thermal fatigue life of this key solder joint is regarded as the final life of the flip chip component under typical conditions. In this paper, the thermal fatigue lives of the flip chip components under various working conditions are investigated, and the dependence of the thermal fatigue life on the maximum power density of the device P_{max} and the power conversion time t_{trans}, ambient temperature T_0, ambient pressure p_0, and t is analyzed.

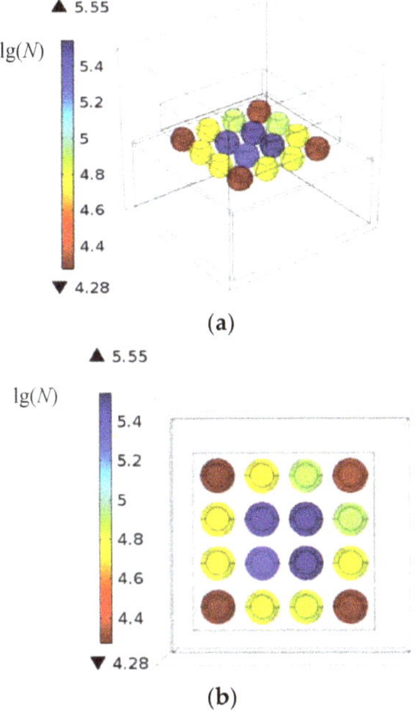

(a)

(b)

Figure 7. Thermal fatigue life distributions in the flip chip component under typical operating conditions: (**a**) three-dimensional view and (**b**) top view.

The temperature and stress distribution of the flip chip component is shown in Figure 8. The highest temperature appears at the silicon chip and the solder joint during the high power dissipation period. The stress at the connections between the solder joint and the silicon chip is the maximum. The stress inside the solder joint is much smaller than the stress at the connections.

To describe the mechanism that lead to thermal failure of the electronic devices in detail, the evolution of the temperature and the stress data at the connections between the key solder joint and the silicon chip are shown in Figure 9, under the same typical condition in Figure 3. Once the power dissipation transfers from high value to low value or from low value to high value, the variation rate of the temperature at the connection is very high and the stress at the connection suddenly increases. When the temperature of the solder joint is nearly stable, the stress at the connection decreases evidently.

The variation of the stress can be attributed to the variation rate of the temperature field, which obviously depends on the maximum heating power and the power conversion time.

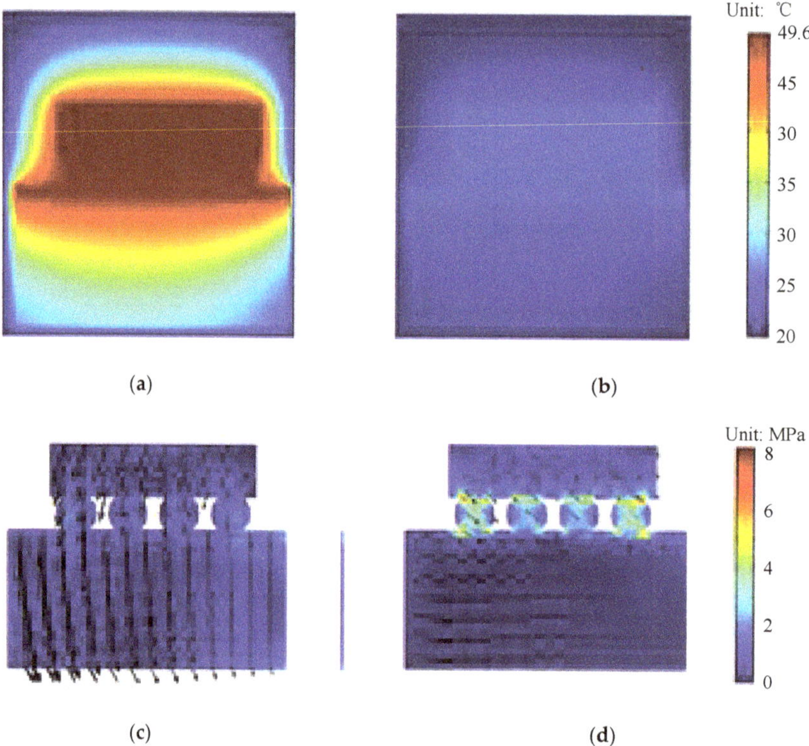

Figure 8. Temperature and stress distribution of the flip chip component at the vertical section of key solder point: (**a**) temperature distribution in the high power dissipation period (t = 4.11 h); (**b**) temperature distribution in the low power dissipation period (t = 6.11 h); (**c**) stress distribution in the high power dissipation period (t = 4.11 h); and (**d**) stress distribution in the low power dissipation period (t = 6.11 h).

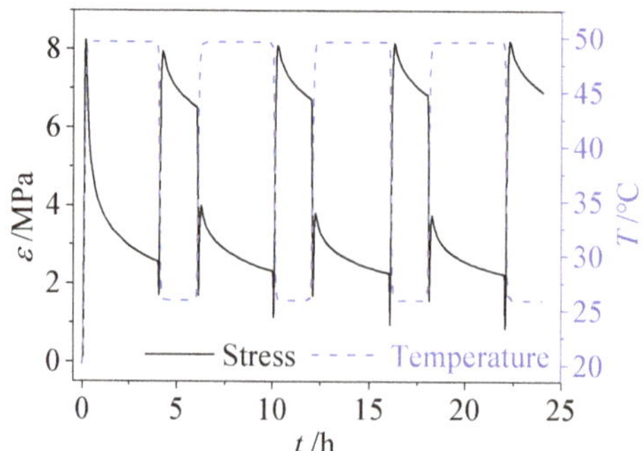

Figure 9. Temperature and the stress data at the connections between the key solder joint and the silicon chip. ($P_{max} = 5 \times 10^7$ W·m^{-3}, $P_{min} = 1 \times 10^7$ W·m^{-3}, and t_{trans} = 600 s).

3.1. Effects of the Maximum Power Density

The maximum power density P_{max} plays a key role in the thermal fatigue lives. To study the effect of P_{max}, the parameters of the operating conditions are set as follows:

$$\left\{ \begin{array}{l} t_{trans} = 600 \text{ s} \\ p_0 = 1 \text{ bar} \\ T_0 = 20\,^{\circ}\text{C} \end{array} \right. . \tag{8}$$

The other parameters are set to be the same as those under typical operating conditions.

As shown in Figure 10a, when P_{max} is less than 50 MW·m^{-3}, the thermal fatigue lives decreases rapidly with P_{max}. When P_{max} is greater than 50 MW·m^{-3}, the rate of decrease in the thermal fatigue lives in slows down as P_{max} increases. In particular, the thermal fatigue lives are reduced to 1/10,730 when P_{max} increases from 20 MW·m^{-3} to 100 MW·m^{-3} indicating that high heat load is not beneficial for the safe operation. Figure 10b shows that the temperatures of the silicon and solder joint increase with P_{max}.

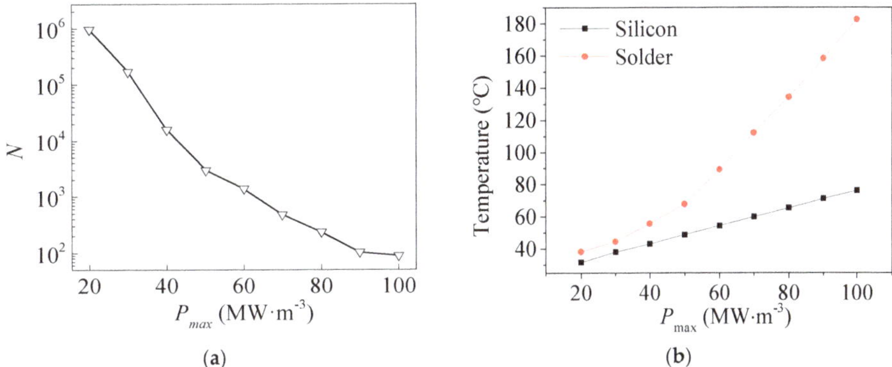

(a)

(b)

Figure 10. Effects of the maximum power density on (**a**) thermal fatigue lives of the key solder joints and (**b**) silicon temperatures and maximum of solder temperatures.

3.2. Effects of the Power Conversion Time

The dependence of thermal fatigue lives on the power conversion time t_{trans} is analyzed. The parameters of the operating conditions are set as follows:

$$\left\{ \begin{array}{l} P_{max} = 50 \text{ MW·m}^{-3} \\ p_0 = 1 \text{ bar} \\ T_0 = 20\,^{\circ}\text{C} \end{array} \right. . \tag{9}$$

The other parameters are set to be the same as those under typical operating conditions.

As shown in Figure 11, as t_{trans} increases, the thermal fatigue lives increase gradually. When t_{trans} is less than 500 s, the thermal fatigue lives are particularly short, which indicates that the thermal shock generated by the fast conversion from high power to low power bring forth disadvantage in the safe operation of electronic device. Therefore, the conversion rate from high power to low power needs to be carefully controlled.

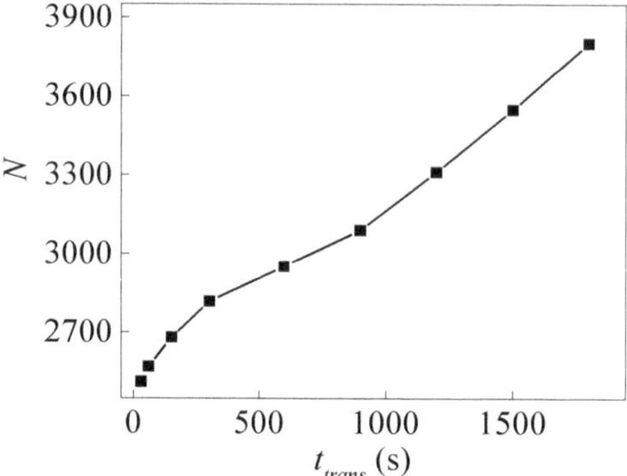

Figure 11. Effects of the power conversion time on the thermal fatigue lives of the key solder joints.

3.3. Effects of the Ambient Temperature

The temperature difference between flip chip component and the ambient air determines whether the heat can be brought away from the flip chip component. The variation of thermal fatigue lives of the key solder joints versus the ambient air temperature is plotted in Figure 12. The parameters of the operating conditions are set as follows:

$$\begin{cases} P_{max} = 50 \text{ MW·m}^{-3} \\ p_0 = 1 \text{ bar} \\ t_{trans} = 600 \text{ s} \end{cases} \qquad (10)$$

The other parameters are set to be the same as those under typical operating conditions.

As seen from Figure 12, when the ambient temperature is less than −20 °C, the thermal fatigue lives remain at high values because of the large temperature difference between the key solder joint and ambient environment. These results indicate that the good heat dissipation effect caused by the low ambient temperature can improve the thermal fatigue lives. When the ambient temperature rises and is greater than −20 °C, the heat dissipation is reduced, which leads to a drastic decrement in the thermal fatigue lives. When T_0 rises from −20 °C to 20 °C, the thermal fatigue lives are reduced to approximately 1/107. These results show that the working lives of electronic devices decrease with increasing ambient temperature under normal working conditions.

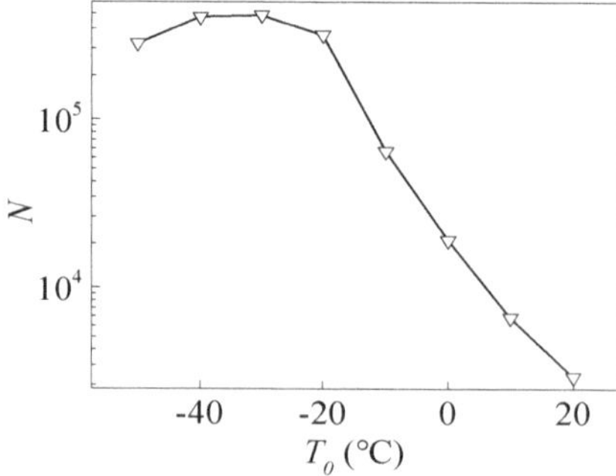

Figure 12. Effects of the ambient temperature on the thermal fatigue lives of the key solder joints.

3.4. Effects of the Ambient Pressure

The ambient pressure p_0 also have significance during the heat transfer between the flip chip component and the ambient environment as illustrated in Figure 13. The parameters of the operating conditions are set as follows:

$$\left\{ \begin{array}{l} P_{max} = 50 \text{ MW·m}^{-3} \\ T_0 = 20\,^{\circ}\text{C} \\ t_{trans} = 600 \text{ s} \end{array} \right. . \tag{11}$$

The other parameters are set to be the same as those under typical operating conditions.

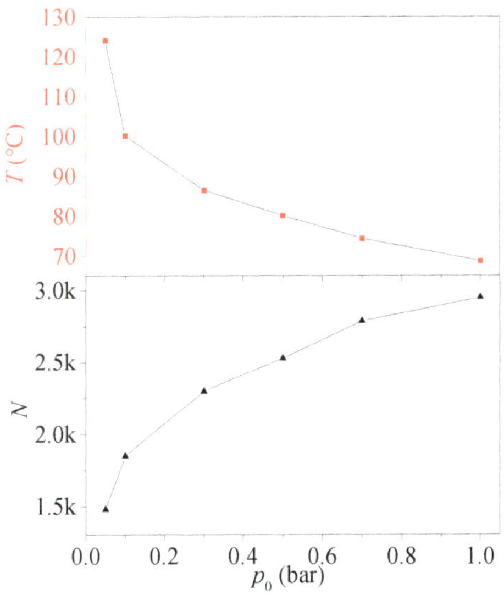

Figure 13. Effects of the ambient pressure on the thermal fatigue lives of the key solder joints and the maximum temperature of the key solder joints at the time $t = 7$ min.

In Figure 13, as p_0 decreases, the thermal fatigue lives gradually decrease. As p_0 decreases from 1 bar to 0.1 bar, the thermal fatigue lives decrease by approximately 37.4%. The intermediate temperature of the solder joint at the time $t = 7$ min increases as the ambient pressure drops. These results indicate that under natural convection conditions, the heat dissipation of an electronic device is weakened by decreases in the ambient air pressure, which leads to quick increases in the temperature and larger stress variation of the solder joint. Therefore, the thermal fatigue life of the key solder joint decreases. For the ambient pressure $p_0 = 0.05$ bar, which is nearly a vacuum environment, the natural convection is extremely weak at this ambient pressure, the heat dissipation of the electronic devices mainly depends on the heat conduction and the heat radiation. So the temperature of the electronic device is relatively high and the thermal fatigue life of the key solder joint is relatively short.

4. Conclusions

In this paper, based on the Darveaux energy method, a theoretical model for BGA products under cyclic thermal loading is developed in an effort to predict the thermal fatigue lives of solder joints in flip chip component. Based on the simulation, the effects of power load factors and environmental factors on thermal fatigue life are analyzed. The main conclusions drawn from the results are as follows:

1. Under the condition of cyclic thermal load, the location of the solder joint with the shortest life in a flip chip component is at the outer corner point in the array. The final life of the flip chip component under typical conditions is equal to the thermal fatigue life of this key solder joint.

2. For the properties of the thermal load, an increase in the power density or a decrease in the power conversion time will result in a short thermal fatigue life of the key solder joint in the flip chip component. Moreover, the thermal shock generated by the fast conversion from high power to low power will have a great disadvantage in the safe operation of the device.

3. When the ambient temperature is lower than $-20\ ^{\circ}$C, the thermal fatigue life of the key solder joint is notably sensitive to the variation in ambient temperature. As a result of the relatively large temperature difference between the ambient environment and the key solder joint, the thermal fatigue life of the key solder joint is high which is beneficial for the reliability of the device. When the ambient temperature rises and is greater than $-20\ ^{\circ}$C, the heat dissipation from the flip chip component to the environment is reduced, which leads to a rapid decrease in the thermal fatigue lives of the key solder joints.

4. Under the condition of natural convection, the heat dissipation of an electronic device is weakened with decreases in the ambient air pressure, which leads to increases in the temperature and failure potential of an electronic device.

Author Contributions: W.Y. provided the guidance and supervision. L.W. and X.H. implemented the main research, discussed the results, and wrote the paper. C.S. and F.Y. collected the data. All authors read and approved the final manuscript.

Funding: The research is financially supported by National Natural Science Foundation of China (Nos. 51706194 and 51876184) and Natural Science Foundation of Yangzhou City (No.YZ2017103).

Conflicts of Interest: The authors declare no conflict of interest.

References

1. Sun, L.; Wu, G.; Xue, Y.; Shen, J.; Li, D.; Lee, K.Y. Coordinated control strategies for fuel cell power plant in a microgrid. *IEEE Trans. Energy Convers.* **2018**, *33*, 1–9. [CrossRef]

2. Zhang, C.; Chen, Y.; Wu, R.; Shi, M. Flow boiling in constructal tree-shaped minichannel network. *Int. J. Heat Mass Transf.* **2011**, *54*, 202–209. [CrossRef]

3. Deng, Z.; Liu, X.; Zhang, C.; Huang, Y.; Chen, Y. Melting behaviors of pcm in porous metal foam characterized by fractal geometry. *Int. J. Heat Mass Transf.* **2017**, *113*, 1031–1042. [CrossRef]

4. Chen, Y.P.; Deng, Z.L. Hydrodynamics of a droplet passing through a microfluidic t-junction. *J. Fluid Mech.* **2017**, *819*, 401–434. [CrossRef]

5. Sun, L.; Jin, Y.; Pan, L.; Shen, J.; Lee, K.Y. Efficiency analysis and control of a grid-connected pem fuel cell in distributed generation. *Energ. Convers. Manag.* **2019**, *195*, 587–596. [CrossRef]

6. Graebner, J.; Jin, S.; Kammlott, G.; Bacon, B.; Seibles, L.; Banholzer, W. Anisotropic thermal conductivity in chemical vapor deposition diamond. *J. Appl. Phys.* **1992**, *71*, 5353–5356. [CrossRef]

7. Chen, Y.; Liu, X.; Shi, M. Hydrodynamics of double emulsion droplet in shear flow. *Appl. Phys. Lett.* **2013**, *102*, 051609. [CrossRef]

8. Ghovanloo, M.; Atluri, S. A wide-band power-efficient inductive wireless link for implantable microelectronic devices using multiple carriers. *IEEE Trans. Circuits Syst. I Regul. Pap.* **2007**, *54*, 2211–2221. [CrossRef]

9. Ashraf, M.W.; Tayyaba, S.; Afzulpurkar, N. Micro electromechanical systems (mems) based microfluidic devices for biomedical applications. *Int. J. Mol. Sci.* **2011**, *12*, 3648–3704. [CrossRef] [PubMed]

10. Lupinacci, A.; Shapiro, A.; Suh, J.; Minor, A. A study of solder alloy ductility for cryogenic applications. In Proceedings of the 2013 IEEE International Symposium on Advanced Packaging Materials (APM), Irvine, CA, USA, 27 February–1 March 2013; pp. 82–88.

11. Liu, X.; Chen, Y.; Shi, M. Dynamic performance analysis on start-up of closed-loop pulsating heat pipes (clphps). *Int. J. Therm. Sci.* **2013**, *65*, 224–233. [CrossRef]

12. Shapiro, A.A.; Tudryn, C.; Schatzel, D.; Tseng, S. Electronic packaging materials for extreme, low temperature, fatigue environments. *IEEE Trans. Adv. Packag.* **2010**, *33*, 408–420. [CrossRef]

13. Niessner, M.; Schuetz, G.; Birzer, C.; Preu, H.; Weiss, L. Accurate prediction of snagcu solder joint fatigue of qfp packages for thermal cycling. In Proceedings of the 2014 15th International Conference on Thermal, Mechanical and Multi-Physics Simulation and Experiments in Microelectronics and Microsystems (Eurosime), Ghent, Belgium, 7–9 April 2014; pp. 1–6.

14. Li, J.; Karppinen, J.; Laurila, T.; Kivilahti, J.K. Reliability of lead-free solder interconnections in thermal and power cycling tests. *IEEE Trans. Compon. Packag. Technol.* **2009**, *32*, 302–308.

15. Wang, J.; Sun, L.; Zou, M.; Gao, W.; Liu, C.; Shang, L.; Gu, Z.; Zhao, Y. Bioinspired shape-memory graphene film with tunable wettability. *Sci. Adv.* **2017**, *3*, e1700004. [CrossRef] [PubMed]

16. Wang, J.; Gao, W.; Zhang, H.; Zou, M.H.; Chen, Y.P.; Zhao, Y.J. Programmable wettability on photocontrolled graphene film. *Sci. Adv.* **2018**, *4*, eaat7392. [CrossRef] [PubMed]

17. Cheng, H.-C.; Cheng, H.-K.; Lu, S.-T.; Juang, J.-Y.; Chen, W.-H. Drop impact reliability analysis of 3-d chip-on-chip packaging: Numerical modeling and experimental validation. *IEEE Trans. Device Mater. Reliab.* **2014**, *14*, 499–511. [CrossRef]

18. Jen, Y.-M.; Wu, Y.-L.; Fang, C.-K. Impact of the number of chips on the reliability of the solder balls for wire-bonded stacked-chip ball grid array packages. *Microelectron. Reliab.* **2006**, *46*, 386–399. [CrossRef]

19. Perkins, A.; Sitaraman, S.K. Universal fatigue life prediction equation for ceramic ball grid array (cbga) packages. *Microelectron. Reliab.* **2007**, *47*, 2260–2274. [CrossRef]

20. Wu, K.-C.; Lin, S.-Y.; Hung, T.-Y.; Chiang, K.-N. Reliability assessment of packaging solder joints under different thermal cycle loading rates. *IEEE Trans. Device Mater. Reliab.* **2015**, *15*, 437–442. [CrossRef]

21. Jacques, S.; Caldeira, A.; Batut, N.; Schellmanns, A.; Leroy, R.; Gonthier, L. Lifetime prediction modeling of non-insulated to-220ab packages with lead-based solder joints during power cycling. *Microelectron. Reliab.* **2012**, *52*, 212–216. [CrossRef]

22. Che, F.; Pang, J.H. Fatigue reliability analysis of sn–ag–cu solder joints subject to thermal cycling. *IEEE Trans. Device Mater. Reliab.* **2013**, *13*, 36–49. [CrossRef]

23. Chen, C.; Suhling, J.C.; Lall, P. Improved submodeling finite element simulation strategies for bga packages subjected to thermal cycling. In Proceedings of the 2018 17th IEEE Intersociety Conference on Thermal and Thermomechanical Phenomena in Electronic Systems (ITherm), San Diego, CA, USA, 29 May–1 June 2018; pp. 1146–1154.

24. Jiang, L.; Zhu, W.; He, H. Comparison of darveaux model and coffin-manson model for fatigue life prediction of bga solder joints. In Proceedings of the 2017 18th International Conference on Electronic Packaging Technology (ICEPT), Harbin, China, 16–19 August 2017; pp. 1474–1477.

25. Ghaffarian, R. Accelerated thermal cycling and failure mechanisms for bga and csp assemblies. *J. Electron. Packag.* **2000**, *122*, 335–340. [CrossRef]

26. Evans, J.W. *A Guide to Lead-Free Solders: Physical Metallurgy and Reliability*; Springer Science & Business Media: Berlin, Germany, 2007.

27. Mattila, T.; Xu, H.; Ratia, O.; Paulasto-Kröckel, M. Effects of thermal cycling parameters on lifetimes and failure mechanism of solder interconnections. In Proceedings of the 2010 Proceedings 60th Electronic Components and Technology Conference (ECTC), Las Vegas, NV, USA, 1–4 June 2010; pp. 581–590.

28. Zhang, C.B.; Chen, Y.P.; Shi, M.H. Effects of roughness elements on laminar flow and heat transfer in microchannels. *Chem. Eng. Process.* **2010**, *49*, 1188–1192. [CrossRef]

29. Zhang, C.B.; Deng, Z.L.; Chen, Y.P. Temperature jump at rough gas-solid interface in couette flow with a rough surface described by cantor fractal. *Int. J. Heat Mass Transf.* **2014**, *70*, 322–329. [CrossRef]

30. Ghaffarian, R. Ccga packages for space applications. *Microelectron. Reliab.* **2006**, *46*, 2006–2024. [CrossRef]

31. Ramesham, R. Reliability of sn/pb and lead-free (snagcu) solders of surface mounted miniaturized passive components for extreme temperature (−185 °C to +125 °C) space missions. In Proceedings of the International Society for Optics and Photonics Reliability, Packaging, Testing, and Characterization of MEMS/MOEMS and Nanodevices X, San Francisco, CA, USA, 11 February 2011; p. 79280F.

32. Ramesham, R. Halt to qualify electronic packages: A proof of concept. In Proceedings of the International Society for Optics and Photonics Reliability, Packaging, Testing, and Characterization of MOEMS/MEMS, Nanodevices, and Nanomaterials XIII, San Francisco, CA, USA, 3–4 February 2014; p. 89750J.

33. Zhang, L. Fatigue Life of Bga Composite Solder Joint under Thermal Cycle Loading. Master's Thesis, Harbin Institute of Technology, Harbin, China, 2011.

Article

Development of Engine Efficiency Characteristic in Dynamic Working States

Piotr Bera

Department of Machine Design and Technology, Faculty of Mechanical Engineering and Robotics, AGH University of Science and Technology, al. Mickiewicza 30, 30-059 Krakow, Poland; pbera@agh.edu.pl

Received: 4 July 2019; Accepted: 26 July 2019; Published: 28 July 2019

Abstract: The objective of this paper is to present a new approach to the problem of combustion engine efficiency characteristic development in dynamic working states. The artificial neural network (ANN) method was used to build a mathematical model of the engine comprising the following parameters: Engine speed, angular acceleration, engine torque, torque change intensity, and fuel mass flow, measured on a test bed on a spark ignition engine in static and dynamic working states. A detailed analysis of ANN design, data preparation, the training method, and the ANN model accuracy are described. The paper presents conducted calculations that clearly show the suitability of the approach in every aspect. Then, a simplified ANN was created, which allows a two dimensional characteristic in dynamic states, including 4 variables, to be determined.

Keywords: combustion engine efficiency; dynamic states; artificial neural network

---

1. Introduction

The overall efficiency of the internal combustion engine is one of its most important operational parameters. It directly influences fuel consumption, which is an extremely important factor both for drivers and car manufacturers who have to meet the strict EU regulations [1].

The overall engine efficiency is defined as the quotient of the mechanical power on the flywheel to the power of the fuel injected into the cylinders. It takes into account all losses related to both thermodynamic processes, internal friction mainly among moving parts in the crank and piston system, as well as losses in the alternator drive, the coolant pump, and other accessories.

Currently, the efficiency of internal combustion engines is up to 40% for spark ignition (SI) naturally aspirated engines [2] and up to 36% for SI turbocharged engines [3]. Over the last 20 years, efficiency has grown by about 10%, due to numerous factors. The main ones are as follows: The use of fuel injection systems, modern lubricants, coatings on pistons and cylinders [4], optimization of the valve timing [5], the use of the Atkinson cycle [2], improved fuels, increased compression ratio and many others.

A well-known method for presenting the efficiency of engines is with the specific fuel consumption, which has been used for many years [6]. It represents isolines of specific fuel consumption as a function of engine rotational speed and torque. It illustrates the efficiency of the engine in the whole working range and makes it possible to compare different constructions. An important limitation of this characteristic is the possibility of applying it only to steady states of engine operation, where the rotational speed and torque are constant and the throttle position is fixed. Therefore, its use in simulations of vehicle fuel consumption is dedicated to significant errors, because vehicle engines work dominantly in dynamic working states [7], where energy accumulation in rotating masses, mixture enrichment at throttle opening, and fuel cut off during deceleration significantly influence engine efficiency. Despite this big disadvantage, this characteristic is often used in simulations of vehicle fuel consumption in different driving cycles. However, it can only be used with sufficient accuracy in comparative simulations, e.g., concerning gear ratio selection [8], allowing differences between different constructions to be shown, rather than nominal values of fuel consumption to be calculated.

A technical problem, where many factors that influence other factors can often be solved with the ANN method, which is capable of approximating non-linear relationships based on a data set from performed measurements. The applications of the ANN method for solving engine performance problems are numerous. The authors of Reference [9] have developed dynamic models of the engine, where speed, throttle angle, and angular acceleration are considered as ANN inputs and engine torque and fuel consumption are the outputs. In References [7,10], the author undertook another attempt. The inertia-related engine features are included in the ANN model enabling fuel flow calculation in dynamic states. In contrast with Reference [9], this model includes not only engine speed, torque, and angular acceleration, but also torque increase over time, which is an important factor including air-fuel mixture enrichment. This method is very precise, but no 2D or 3D characteristic was presented, thus a simple and quick assessment of the engine dynamic parameters is impossible in this case. Computational simulations must be performed to obtain engine efficiency for each particular engine working state. In Reference [11], the ANN method was used to predict the exhaust emissions and performance (e.g., brake specific fuel consumption) of a compression ignition engine based on engine load, engine speed, and the percentage of biodiesel fuel derived from waste cooking oil in diesel fuel. Although this article provides relationships between engine speed, load, and performance, it refers only to static working states. Very similar problems and approaches are presented in References [12,13], however, in these cases the fuel blends are as follows: Diethyl ether diesel fuel mixtures and Jatropha biodiesel, respectively. Artificial intelligence can also be used in more sophisticated problems concerning the efficiency of a combustion engine working in a hybrid drivetrain [14]. The literature shows that, apart from the ANN method, other approaches are also possible when solving the problems of engine efficiency (specific fuel consumption). In Reference [15], the authors' approach was to analytically calculate energy expenditure and fuel consumption, taking into account the instantaneous specific fuel consumption, approximated by two generalized single dimension polynomial functions. The authors of Reference [16] developed a dynamic model based on the static characteristic complemented by a factor including vehicle speed and acceleration. This model provides a closer characteristic to real engine behaviour, however, such an attempt also includes vehicle features such as gear ratios. The authors of Reference [17] used the mean value model representing the engine. A non-linear model includes air flow and engine inertia, however fuel-air mixture composition is assumed constant and values of some efficiencies are assumed, not measured. The authors of Reference [18] invented the new transient fuel consumption model whose basic structure is steady-state estimation plus transient correction, including engine speed and torque change resulting from the instantaneous speed and acceleration of the vehicle in specific driving cycles.

The need to reduce research costs and shorten the time of prototype preparation leads to the use of computer simulations to the widest possible extent. This, in turn, involves the need to develop more and more accurate mathematical models that reflect the physical construction as closely as possible. The above examples present different approaches to the problem of combustion engine efficiency and the use of different calculation methods. Unfortunately, the development of a graphic representation of combustion engine efficiency in dynamic states is complex, because it requires a multidimensional relationship to be developed. Taking the above into consideration, the author used the ANN to develop a novel method that allows graphical representations to be created with the highest possible match with measurement data from dynamic engine working states.

The rest of the manuscript is organized as follows. Section 2 describes the basics of engine efficiency, including the well-known specific fuel consumption characteristic as well as the problem of engine inertia influencing efficiency in dynamic working states. Section 3 provides information concerning engine tests and the measurement system in static and dynamic states. All the aspects of the ANN (creating mathematical dependency between engine speed, torque, and its efficiency) concerning its architecture, data scaling, and training method, are precisely described. Section 4 presents a novel approach to the problem of drawing engine efficiency in dynamic working states and, finally, shows two dimensional characteristics, allowing engine efficiency to be quickly and precisely calculated. Section 5 presents the simulation results, which prove the correctness of the adopted approach.

2. Problem Formulation

Engine efficiency can be measured for every point in the whole field of work, however, only the maximum values of individual engines are usually given. The efficiency of the engine in the entire work area shows the specific fuel consumption, where, in the coordinate system, torque M vs. the rotational speed n, the contours of specific fuel consumption are presented. Such a characteristic is presented in Figure 1.

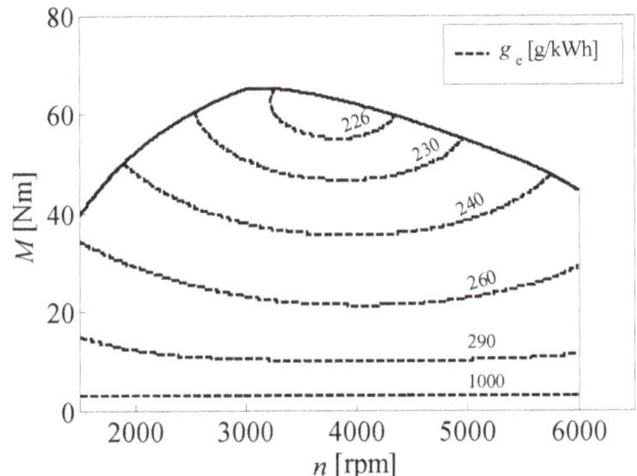

Figure 1. Engine specific fuel consumption characteristic.

The graph in Figure 1 is created based on a method presented in Reference [8], where the algorithm of drawing such a characteristic, based on only 4 operating points, is described in detail. This characteristic can be directly converted to a characteristic of engine efficiency:

$$\eta = \frac{N}{G \cdot W_d} = \frac{3600}{g_e \cdot W_d},$$

(1)

where N is the engine power in kW, G is fuel mass flow in g/s, and W_d is the heat value of the fuel in MJ/kg.

As mentioned in the Introduction, these characteristics are based on measurements in static states. In dynamic states, the efficiency of the engine can vary significantly. This results from the following factors: Enrichment of the mixture during load increase (in relation to static states), delay of the system response to the control signal, and accumulation of combustion energy in moving parts of the engine (Figure 2). Although this energy can be partially recovered during deceleration with engine braking, it is usually lost in the form of pumping losses when the throttle is closed and the engine speed is reduced.

It is therefore important to determine the characteristics that, as well as the engine speed n and torque M, also take into account the angular acceleration and the increase in the torque over time, which significantly affect efficiency, as shown in Figure 3.

Such an attempt is strongly desired, because car engines work for a predominant percentage of time in dynamic working conditions, which results from the specificity of road traffic, especially in the area of large urban agglomerations, the impact of road resistance, wind resistance, or the driver's own operation of the vehicle. It turns out that, in such working conditions, the efficiency of the engine is even smaller. In the WLTP homologation test (Worldwide harmonized light duty vehicle test procedure), the phases in which the engine operates with a constant load are only under 17% of the entire test time. This clearly shows that the use of fuel consumption characteristics in dynamic work states is insufficiently accurate.

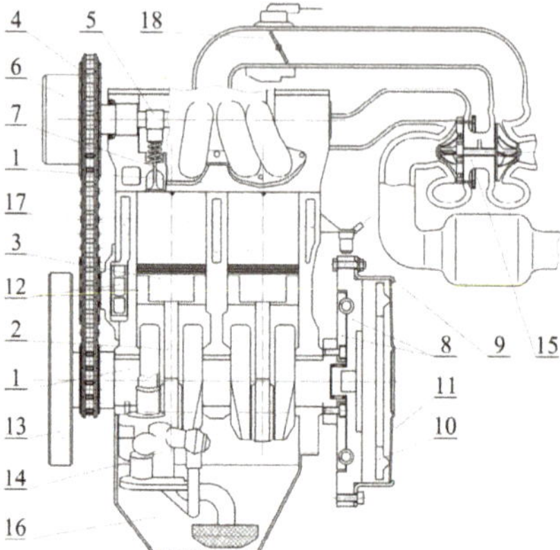

Figure 2. Main parts of the combustion engine influencing its overall inertia: 1—crankshaft, 2—connecting rod, 3—piston, 4—valve-train chain, 5—camshaft, 6—variable valve timing, 7—valve, 8—dual mass fly-wheel, 9—clutch cover, 10—pressure plate, 11—diaphragm spring, 12—coolant pump, 13—accessories pulley, 14—oil pump, 15—turbocharger, 16—oil, 17—coolant, and 18—throttle valve.

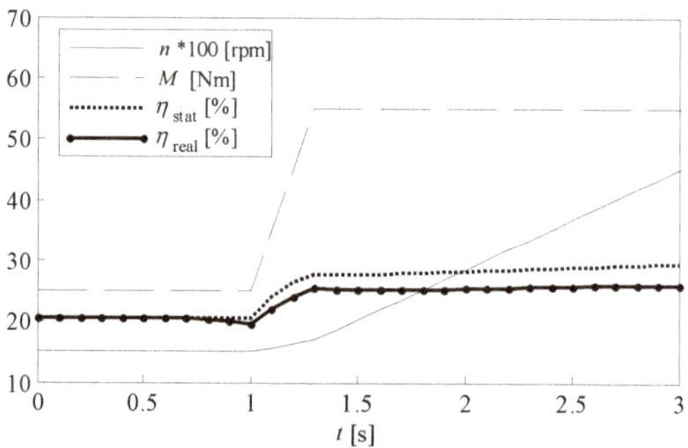

Figure 3. Engine efficiency in dynamic states.

3. Methodology

3.1. Measurements in Dynamic Working States

The presented engine efficiency characteristic, using the ANN, is determined based on measurement data from the engine test bench, which allows measurements of the fuel consumption in dynamic operating states to be performed. This is necessary to ensure an appropriate training set for the ANN. Such a set must contain measurement data from all the possible operating states of the engine, covering the entire engine work field, in this case n = 1000–5500 rpm and M = 0–65 Nm. The measurement methodology is described in detail in Reference [7]. Measurements were performed on an SI engine with a displacement of 899 cm^3, a nominal power of 29 kW at 5500 rpm, and a nominal torque of 65 Nm at 3000 rpm. Resistant torque was generated by a hydraulically controlled friction brake enabling smooth and rapid load changes (Figure 4). Two parameters, the brake load and throttle opening angle (in the full range 0–90°), were controlled so all the engine working states could be realized. Measurements were performed at a nominal engine temperature. No warm-up period was included. The engine speed was measured with the use of a Tacho GT 3.10 L/405 tachometer. To match the measurement range of the data acquisition device (0–10 V), an additional tooth belt drive, with a ratio of 2.909, was used. The fuel flow was measured indirectly. A Bosch 0 280 217 123 air mass flow meter measured the airflow and an LSU-4.1 Innovate Motorsport LC-1 wideband oxygen sensor (measurement range of $\lambda = 0.5 \div 1.523$), with a controller, constantly measured the air-fuel ratio. The brake caliper was pivoted and the reaction force was generated by a compression spring, whose shortening was measured by a potentiometer, and a lever system. Such a system measured the moment of resistance, taking into account both the contact pressure between the pads and the disc, as well as the friction coefficient between these elements. The moment of inertia of all elements is known and is taken into account at the stage of building the training set for the ANN. All 4 sensors generated analog voltage signals and were connected to a National Instruments NI USB-6009 data acquisition device. LabVIEW software was used to collect the data on the computer.

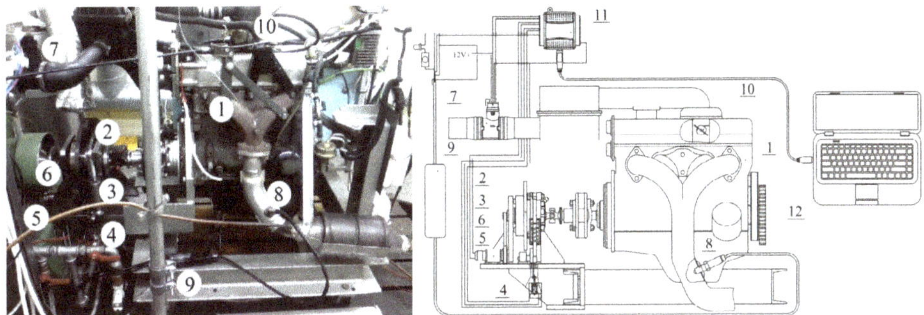

Figure 4. Test stand used for measurements in dynamic states: 1—engine, 2—friction brake, 3—compression spring, 4—potentiometer, 5—tachometer, 6—tooth belt gear, 7—mass air flow meter, 8—wideband oxygen sensor, 9—oxygen sensor controller, 10—throttle valve, 11—NI USB-6009, 12—computer.

All the possible engine states that might occur during normal engine operation in a vehicle must be performed during measurements. All of them are presented in Figure 5.

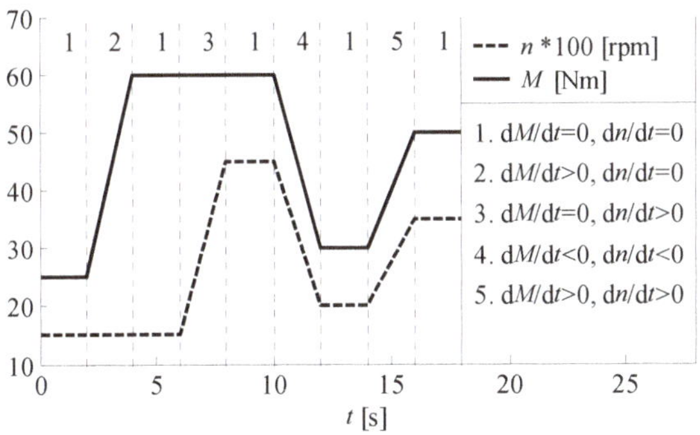

Figure 5. Possible engine working states.

Figure 5 represents all the situations that might occur during driving. The first one is driving at constant speed on a flat road, the second one is sudden throttle opening—a demand for acceleration, the third one is acceleration with constant torque, the fourth case is engine braking, and the fifth one is slow acceleration with torque and speed increase.

3.2. Data Analysis with the Use of ANN

To use the ANN for measurement data approximation, the ANN architecture, data scaling, transfer functions, and training methods must be defined. The aim is to calculate engine efficiency (resulting from fuel mass flow), which is a function of 4 parameters, as follows: Engine speed n, torque M and angular acceleration ε_n, and torque increase over time ε_M. Thus, the considered ANN must have 4 inputs and one output. Only one hidden layer can be used because, with a proper number of hidden neurons, such a network is capable of approximating non-linear relationships with high accuracy [19]. The non-linear function in the hidden layer was set to logsig, which is one of the transfer functions available. It ensures similar results to other functions (e.g., tansig), however, it can be calculated quickly, thus accelerating calculations [19]. Data scaling for all inputs and the output is proportional, because an increase of every single input results in an increase of the output. The ANN used for calculations is presented in Figure 6.

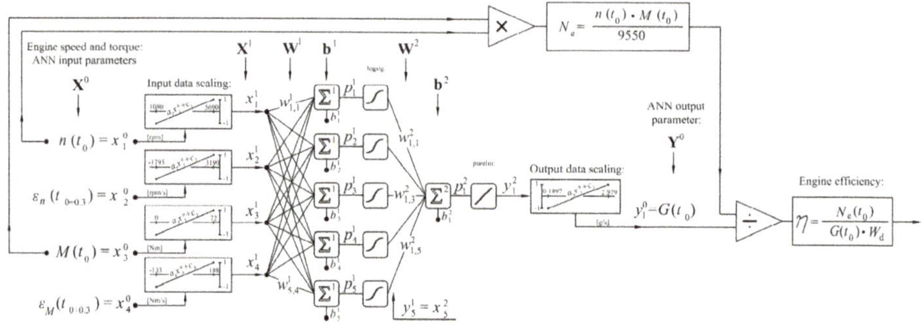

Figure 6. ANN used for computing engine efficiency in dynamic states.

The number of hidden neurons depends on many factors, e.g., the data number (in this case, the training set contained nearly 13,800 data parts from measurements in dynamic and static states), the homogeneity of measurement data, and the method of data scaling, and thus cannot be set up front. The author verified mean squared error (*mse*) for subsequent numbers of hidden neurons, as presented in Figure 7.

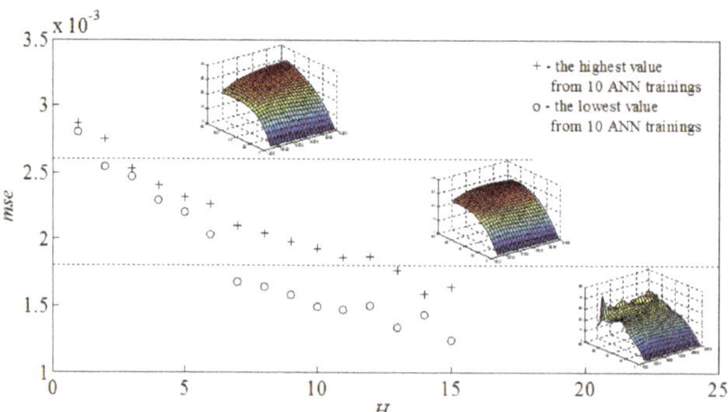

Figure 7. Mean squared error vs. hidden neurons number H.

The *mse* decreases with increasing numbers of hidden neurons H, however, the higher the H the higher the possibility of data overfitting ($mse < 0.00200$), resulting in discontinuities on the graph (Figure 7), which are created based on the ANN. Taking these two factors into consideration, it turned out that $H = 5$ is optimal, because it ensures high accuracy and does not lead to data overfitting.

The ANN approximates measurement data, thus setting the dependency between engine speed torque and fuel mass flow. The input data vector $\mathbf{X}^0$ are engine parameters given in their basic units, as follows: Engine speed n in rpm, engine angular acceleration ε_n rpm/s, engine torque M in Nm, engine torque increase ε_M in Nm/s, and the fuel flow G in g/s. The parameters ε_n and ε_M are considered in the time interval 0.3 s [7]. The input vector,

$$\mathbf{X}^0 = \left[x_1^0,\, x_2^0,\, x_3^0,\, x_4^0\right]^{\mathrm{T}} = \left[n(t_0),\, \varepsilon_n(t_{0 \div 0.3}),\, M(t_0),\, \varepsilon_M(t_{0 \div 0.3})\right]^{\mathrm{T}}, \tag{2}$$

must be scaled to the range $(-1;\,1)$ resulting in the vector $\mathbf{X}^1$:

$$\mathbf{X}^1 = \left[x_1^1,\, x_2^1,\, x_3^1,\, x_4^1\right]^{\mathrm{T}}. \tag{3}$$

To ensure proper ANN training, data scaling is conducted according to the following formulas (with regard to Figure 6):

$$\left.\begin{aligned}
x_1^1 &= 0.0004338 \cdot x_1^0 - 1.4686 \\
x_2^1 &= 0.0004012 \cdot x_2^0 - 0.2800 \\
x_3^1 &= 0.0277800 \cdot x_3^0 - 1.0000 \\
x_4^1 &= 0.0062300 \cdot x_4^0 - 0.1713
\end{aligned}\right\}, \tag{4}$$

The ANN output $\mathbf{Y}^0$, which is the fuel mass flow G.

$$\mathbf{Y}^0 = y_1^0 = G(t_0), \tag{5}$$

was also scaled to the range $(-1;\,1)$ according to the following formula:

$$y_1^2 = 0.717 \cdot y_1^0 - 1.136. \tag{6}$$

The hidden layer weights matrix is:

$$\mathbf{W}^1 = \begin{bmatrix} w_{1,1}^1 & \cdots & \\ \cdots\, w_{h,in}^1 & \cdots & \\ & \cdots & w_{H,IN}^1 \end{bmatrix}, \tag{7}$$

and hidden layer biases are:

$$\mathbf{b}^1 = [b_1^1 \dots b_h^1 \dots b_H^1]^{\mathrm{T}}, \tag{8}$$

where h ($h = 1 \dots H$) is the number of a hidden neuron, and in is the number of the input ($in = 1 \dots IN = 4$). In a summing member, the input, multiplied by its weights, is summed with bias:

$$p_h^1 = \sum_{in=1}^{IN} {}^1 w_{h,in}^1 \cdot x_{in}^1 + b_h^1. \tag{9}$$

The sum $p_h{}^1$ is then calculated with the use of the non-linear logsig transfer function f^1:

$$y_h^1 = f^1(p_h^1) = \frac{1}{1 + e^{-p_h^1}}, \tag{10}$$

and multiplied by the output layer weights:

$$\mathbf{W}^2 = [w_{1,1}^2 \cdots w_{1,h}^2 \cdots w_{1,H}^2]^{\mathrm{T}}. \tag{11}$$

The sum of $p_1{}^2$ is converted with the use of the simple linear transfer function f^2:

$$y_1^2 = f^2(p_1^2) = p_1^2. \tag{12}$$

The output layer bias matrix is:

$$\mathbf{b}^2 = [b_1^2].$$ (13)

The whole training set was divided as follows: 60% training set, 20% validation set, and 20% test set. The ANN was trained with the Lavenberg–Marquardt algorithm [19,20]. The training process took 49 epochs, which resulted in the mean squared error of 0.00224. Its graph during the learning process is presented in Figure 8.

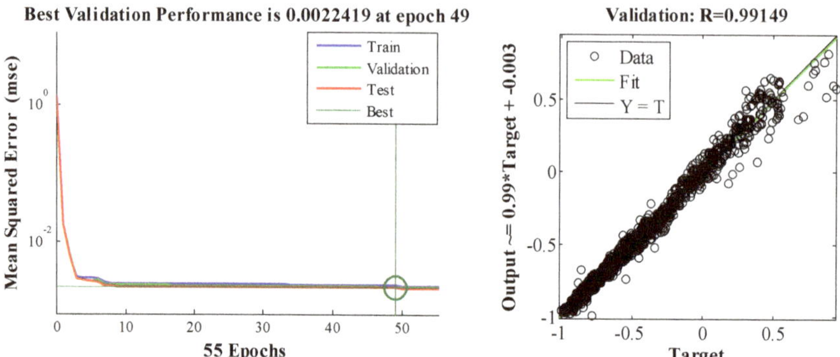

Figure 8. Mean squared error vs. training epochs and regression plot.

Achieving the expected value of *mse* is equivalent to the completion of the ANN training. The matrices $\mathbf{W}^1$, $\mathbf{b}^1$, $\mathbf{W}^2$, $\mathbf{b}^2$, which represent the result of the network learning and represent the ANN engine dynamic model, are the following:

$$\mathbf{W}^1 = \begin{bmatrix} -2.9957 & -2.9347 & -0.4181 & 1.5487 \\ -0.6117 & 0.8342 & 0.6878 & -0.8130 \\ -1.9564 & -0.9297 & -0.8992 & -2.3007 \\ 3.3275 & -1.5137 & -0.4126 & -2.1666 \\ -1.3134 & -2.1830 & -2.5913 & -4.4844 \end{bmatrix},$$ (14)

$$\mathbf{b}^1 = [3.9769 \; 2.2505 - 0.5819 \; 3.5082 \; 3.1910]^T,$$ (15)

$$\mathbf{W}^2 = [-1.4632 \; 2.2144 - 1.2476 \; 0.7342 - 1.9261]^T,$$ (16)

$$\mathbf{b}^2 = [b_1^2] = 0.9945.$$ (17)

After ANN training, engine torque is re-scaled from the $(-1; 1)$ range to the basic range $(0.1897 \div 2.979 \text{ g/s})$, based on the following relationship:

$$\mathbf{Y}^0 = G(t_0) = \frac{y_1^2 + 1.136}{0.717}.$$ (18)

By simulation of such an ANN, a static efficiency characteristic can be obtained. It is shown in Figure 9.

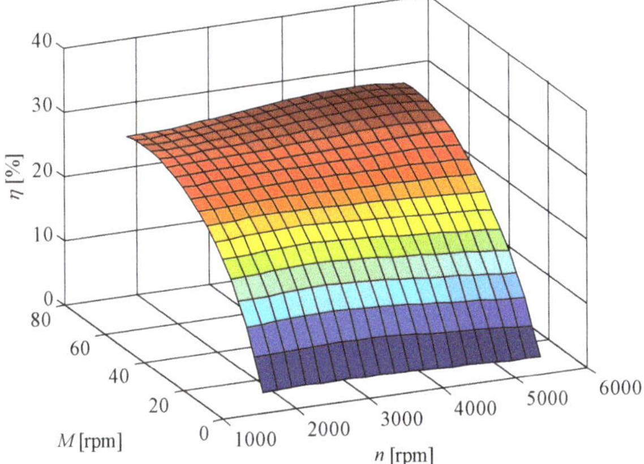

Figure 9. Engine efficiency characteristic in static states.

The characteristic, which is entangled in weights, biases (Equations (14)–(17)), transfer functions (Equations (10) and (12)), and scaling factors (Equation (4) and (6)), can be used in different simulations for precise calculations of engine efficiency. However, the engine efficiency cannot be assessed in a straightforward manner, as in the case of graphic representation. As a consequence of this, the author's method for graphical representation of such a characteristic is presented in the next chapter.

4. Development of the Graphic Engine Efficiency Characteristic

The ANN presented in Section 3 can be used in computer simulations, because it ensures high accuracy. However, it does not allow a simple 2D or 3D characteristic to be obtained. Thus, a modified ANN, with a new architecture, will be used in this chapter to develop such a characteristic. First of all, an ANN with only one hidden neuron will be used. Secondly, it will differ from the previous one, because, in this case, engine efficiency is directly the ANN output. The logsig shape of the transfer function f^1 will be the base for the characteristic. All of the 4 variables must be presented simultaneously, because they influence each other. The architecture used is presented in Figure 10.

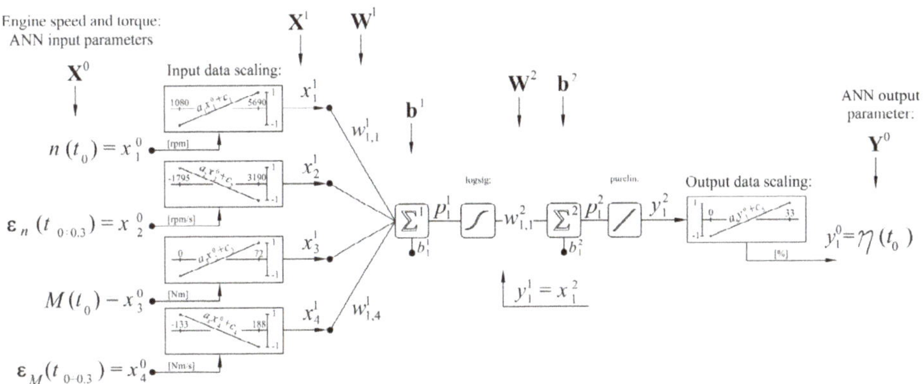

Figure 10. ANN used for engine efficiency graph development.

The fact that the output is engine efficiency means that the 2nd and 4th inputs must be scaled inversely proportionally (different than in the first ANN with 5 hidden neurons (Figure 6)), because both high angular acceleration and increased torque reduce engine efficiency. The input values remain the same:

$$\mathbf{X}^0 = \left[x_1^0,\, x_2^0,\, x_3^0,\, x_4^0\right]^{\mathrm{T}} = \left[n(t_0),\, \varepsilon_n(t_{0\div 0.3}),\, M(t_0),\, \varepsilon_M(t_{0\div 0.3})\right]^{\mathrm{T}}. \tag{19}$$

They are scaled to the range $(-1;\,1)$, thus creating the vector $\mathbf{X}^1$:

$$\mathbf{X}^1 = \left[x_1^1,\, x_2^1,\, x_3^1,\, x_4^1\right]^{\mathrm{T}}. \tag{20}$$

Data scaling is the following:

$$\left.\begin{aligned}
x_1^1 &= 0.0004338 \cdot x_1^0 - 1.4686 \\
x_2^1 &= -0.0004012 \cdot x_2^0 + 0.2800 \\
x_3^1 &= 0.0277800 \cdot x_3^0 - 1.0000 \\
x_4^1 &= -0.0062300 \cdot x_4^0 + 0.1712
\end{aligned}\right\}. \tag{21}$$

The ANN output $\mathbf{Y}^0$ is the engine efficiency η [%]:

$$\mathbf{Y}^0 = y_1^0 = \eta(t_0). \tag{22}$$

It was also scaled to the range $(-1;\,1)$, according to the following formula:

$$y_1^2 = 0.0606 \cdot y_1^0 - 1. \tag{23}$$

The training process took 18 epochs this time, which resulted in the mean squared error of 0.00854. Its graph during the learning process is presented in Figure 11.

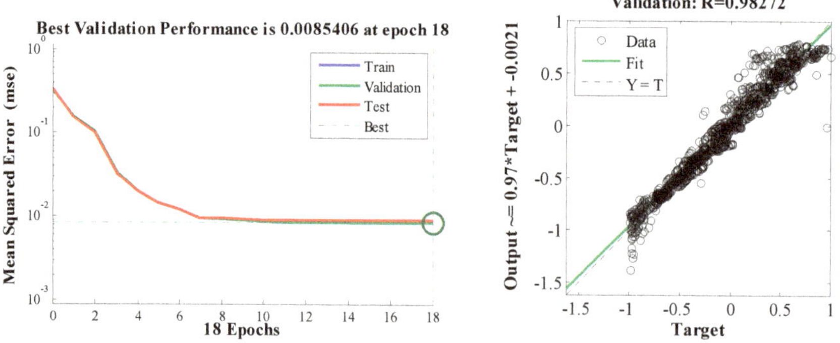

Figure 11. Mean squared error vs. training epochs and regression plot.

The matrices $\mathbf{W}^1$, $\mathbf{b}^1$, $\mathbf{W}^2$, $\mathbf{b}^2$ are now the following:

$$\mathbf{W}^1 = \left[-0.0463\ \ 0.2889\ \ 1.6809\ \ 0.2402\right], \tag{24}$$

$$\mathbf{b}^1 = \left[3.2176\right], \tag{25}$$

$$\mathbf{W}^2 = \left[11.2440\right], \tag{26}$$

$$\mathbf{b}^2 = \left[-10.3915\right]. \tag{27}$$

After ANN training, engine torque is re-scaled from the $(-1; 1)$ range to the basic range $(0–33\%)$ based on the following relationship:

$$\mathbf{Y}^0 = \eta(t_0) = \frac{y_1^2 + 1}{0.0606}. \tag{28}$$

Then, it must be verified if only one hidden neuron is capable of approximating the whole data set, because it is essential to create a characteristic with reasonable accuracy. Based on the two neural networks above, static characteristics are created. The result obtained with the use of the ANN with one hidden neuron differs no more than 5% for 95% of the engine working field from the ANN with 5 neurons. Thus, the accuracy is reasonable and the ANN with 1 neuron can be used further for 2D characteristic development.

To obtain a simple and direct relationship between the engine parameters and its efficiency scaling coefficients (Equation (21)), its input weights are multiplied and simplified, according to the following equations:

$$\Sigma^1 = \left(a_1 \cdot x_1^0 + c_1\right) \cdot w_{1,1}^1 + \left(a_2 \cdot x_2^0 + c_2\right) \cdot w_{1,2}^1 + \left(a_3 \cdot x_3^0 + c_3\right) \cdot w_{1,3}^1 + \left(a_4 \cdot x_4^0 + c_4\right) \cdot w_{1,4}^1 + b_1^1. \tag{29}$$

After transformation:

$$\Sigma^1 = \left(a_1 \cdot x_1^0 + c_1\right) \cdot w_{1,1}^1 + \left(a_2 \cdot x_2^0 + c_2\right) \cdot w_{1,2}^1 + \left(a_3 \cdot x_3^0 + c_3\right) \cdot w_{1,3}^1 + \left(a_4 \cdot x_4^0 + c_4\right) \cdot w_{1,4}^1 + b_1^1, \tag{30}$$

$$\Sigma^1 = A_1 \cdot x_1^0 + A_2 \cdot x_2^0 + A_3 \cdot x_3^0 + A_4 \cdot x_4^0 + S. \tag{31}$$

The final coefficients are the following: $A_1 = -0.0000200$, $A_2 = -0.0001159$, $A_3 = 0.0467000$, $A_4 = -0.0015000$, $S = 1.7265$.

Taking into account (Equation (2)), this results in the final dependency, which can be directly moved further on the graph:

$$\Sigma^1 = A_1 \cdot n(t_0) \, [\text{rpm}] + A_2 \cdot \varepsilon_n(t_{0 \div 0.3}) \, [\text{rpm/s}] + A_3 \cdot M(t_0) \, [\text{Nm}] + A_4 \cdot \varepsilon_M(t_{0 \div 0.3}) \, [\text{Nm/s}] + S. \tag{32}$$

Finally, the logsig function is converted by the output layer weight W^2, and then by rescaling the output data. This is presented in Figure 12.

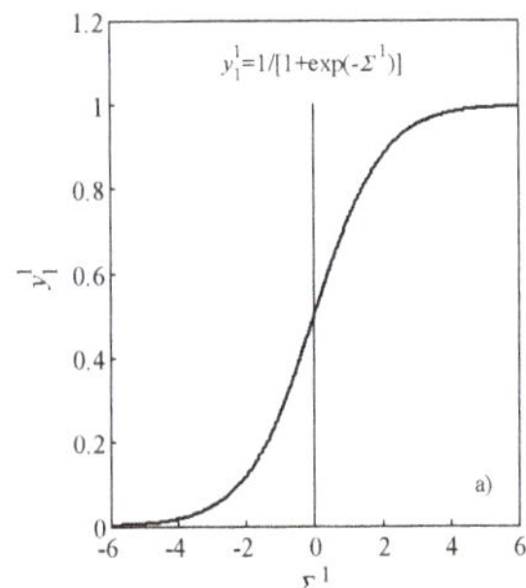
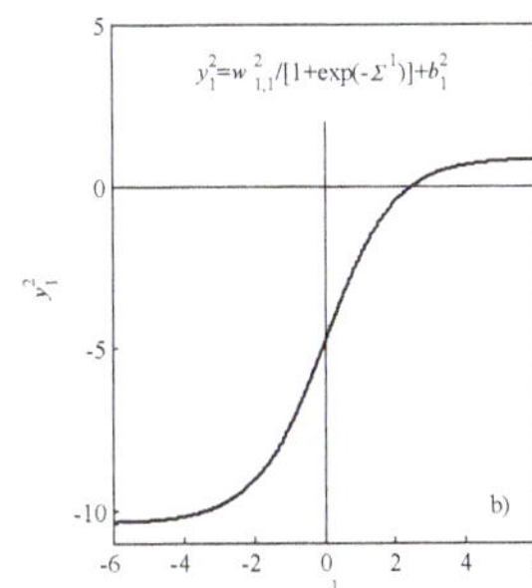
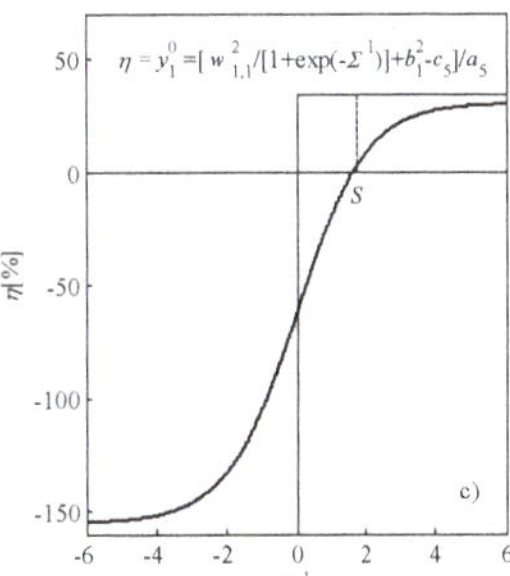

Figure 12. Operations on the logsig function inside the AN.

To create a graphic representation, the coefficients are scaled to specific lengths in a graph. This allows a simple and quick assessment of engine efficiency. The final characteristic is presented in Figure 13.

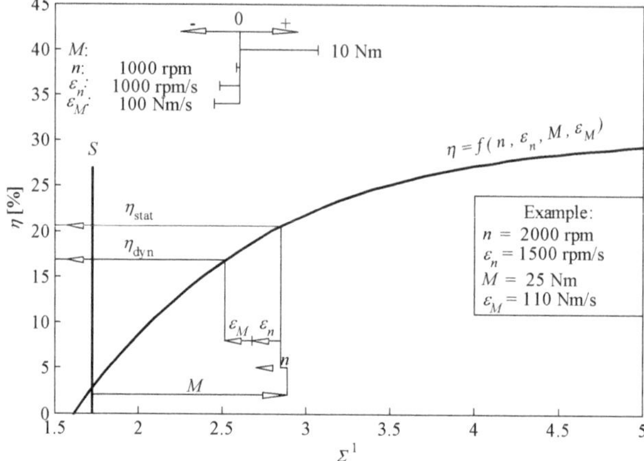

Figure 13. A new engine efficiency characteristic.

This characteristic allows a simple assessment of engine efficiency in the whole working range. The length of the measurement sections clearly shows the influence of individual components on the overall efficiency of the engine and, in particular, how it decreases where angular acceleration or torque increase occur. The main advantage of the characteristic is that it is a continuous and differentiable function, which allows the efficiency to be calculated at any working point.

5. Simulation Results and Comparison of Dynamic and Static Characteristic

The ANN network with 5 hidden neurons can be used, as mentioned, in simulations of vehicle fuel consumption in WLTP homologation tests, for example. Based on vehicle speed and vehicle parameters (wheel radius, gearbox ratios, and vehicle weight), one can calculate engine speed and torque, and the ANN will calculate engine efficiency at every moment. A sample course of engine parameters is presented in Figure 14, which allows the newly developed characteristic η_{dyn} to be compared with the static characteristic η_{stat}.

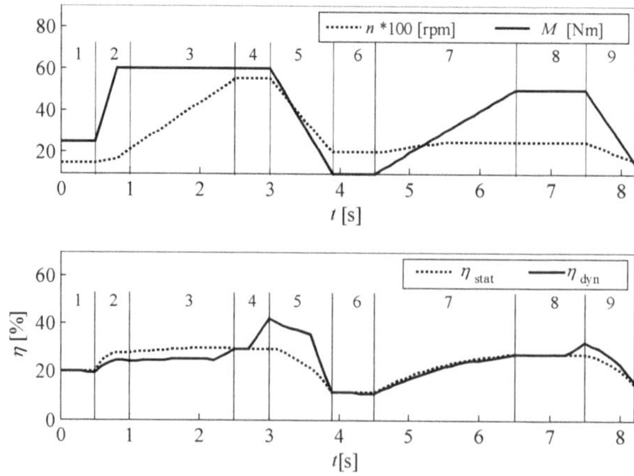

Figure 14. Engine efficiency in different working states based on the ANN characteristic (η_{dyn}) and static characteristic (η_{stat}).

It can be clearly seen that the engine efficiency drops by 10% during sudden throttle opening (2). In phase (3) when the torque is constant, the increase in engine speed decreases the efficiency by up to 15%. An unusual situation occurs in phase (4–5). Sudden throttle closing results in air-fuel mixture depletion, however, the engine generates torque (due to the speed decrease). This leads to the situation that the engine efficiency reaches up to 42%, whereas its maximum efficiency in steady states is 33%. Phase (7) represents slow engine load and speed increase. In this situation, the difference between characteristics is only 4%. Phases (1,4,6,8) represent static states, so $\eta_{dyn} = \eta_{stat}$. This proves that static characteristics (specific fuel consumption characteristic) can only be used in quasi-static conditions. Taking into consideration the accuracy requirements of modern simulation programs, as well as the wide range of engine operation, this is not enough. Only the newly developed characteristic can calculate engine efficiency in any working state with high accuracy.

6. Summary and Conclusions

The article presents a detailed algorithm for developing the combustion engine efficiency characteristic, both in a full version with 5 hidden neurons, dedicated for computer simulations, and a simplified one with only one hidden neuron, which visually describes the engine properties with the accuracy of 5% with regard to the detailed characteristic. Both methods describe every step exactly, including the coefficients of data scaling and weights and biases of both networks, which can be recreated by any researcher. ANN design and the training method prove the correctness of the presented attempt. The *mse* and regression plots are correct and $R > 0.98$, which also means that the network architecture and scaling methods were assumed correctly. However, one should remember that the specific values of weights and lengths reflect the specificity of the tested engine. With different propulsion, the weight and length coefficients will be different; however, the methodology remains the same. The example presented in Section 5 shows that the discrepancy between the static characteristic and the new characteristic can reach up to 15% in the case of high angular acceleration and sudden throttle opening. Scaling lengths (Figure 13) clearly show the influence of each component (n, ε_n, M, and ε_M) on the overall efficiency in the whole working range and allow a quick and precise assessment of the engine properties.

Funding: This research was funded by subvention 16.16.130.942.

Conflicts of Interest: The author declares no conflict of interest.

References

1. Worldwide Emissions Standards. Passenger Cars and Light Duty. 2016. Available online: www.delphi.com/gdi (accessed on 10 June 2019).
2. Hyundai Ioniq. Available online: https://www.hyundai.pl/ (accessed on 1 June 2019).
3. Toyota 1.2 Turbo D-4T 116 KM. Available online: https://www.toyota.pl/innovation/ (accessed on 1 June 2019).
4. Wong, V.W.; Tung, S.C. Overview of automotive engine friction and reduction trends. *Friction* **2016**, *4*, 1–28. [CrossRef]
5. Sabaruddin, A.A.; Wiriadidjaja, S.; Rafie, A.S.M.; Romli, F.I.; Djojodihardjo, H. Engine optimization by using variable valve timing system at low engine revolution. *ARPN J. Eng. Appl. Sci.* **2015**, *10*, 9730–9735.
6. Goering, C.E.; Cho, I. Engine model for mapping bsfc contours. *Math. Comput. Model.* **1988**, *11*, 514–518. [CrossRef]
7. Bera, P. Fuel consumption analysis in dynamic states of the engine with use of artificial neural network. *Combust. Eng.* **2013**, *155*, 16–25.
8. Bera, P. A design method of selecting gear ratios in manual transmissions of modern passenger cars. *Mech. Mach. Theory* **2019**, *132*, 133–153. [CrossRef]
9. Yin, X.; Ge, A. A dynamic model of engine using neural network description. In Proceedings of the IEEE International Vehicle Electronics Conference (IVEC), Tottori, Japan, 25–28 September 2001. [CrossRef]
10. Bera, P. Method for Preparation of the Dynamic Characteristics of Fuel. Patent PL 223780 B1, 15 March 2016.

11. Jaliliantabar, F.; Ghobadian, B.; Najafi, G.; Yusaf, T. Artificial neural network modeling and sensitivity analysis of performance and emissions in a compression ignition engine using biodiesel fuel. *Energies* **2018**, *11*, 2410. [CrossRef]

12. Uslu, S.; Celik, M.B. Prediction of engine emissions and performance with artificial neural networks in a single cylinder diesel engine using diethyl ether. *Eng. Sci. Technol. Int. J.* **2018**, *21*, 1194–1201. [CrossRef]

13. Kawade, G.H.; Satpute, S.T.; Parane, K.A. Optimization of CI engine performance parameters for jatropha biodiesel blending fuel by using ANN software. *Int. Res. J. Eng. Technol.* **2015**, *2*, 1151–1156.

14. Liu, X.; Qinm, D.; Wang, S. Minimum energy management strategy of equivalent fuel consumption of hybrid electric vehicle based on improved global optimization equivalent factor. *Energies* **2019**, *12*, 2076. [CrossRef]

15. Ben-Chaim, M.; Shmerling, E.; Kuperman, A. Analytic modeling of vehicle fuel consumption. *Energies* **2013**, *6*, 117–127. [CrossRef]

16. Zhou, M.; Jin, H. Development of a transient fuel consumption model. *Transport. Res. Part D* **2017**, *51*, 82–93. [CrossRef]

17. Bastida, H.; Ugalde-Loo, C.E.; Abeysekera, M. Dynamic modelling and control of a reciprocating engine. *Energy Procedia* **2017**, *142*, 1282–1287. [CrossRef]

18. Guang, H.; Jin, H. Fuel consumption model optimization based on transient correction. *Energy* **2019**, *169*, 508–514. [CrossRef]

19. Beale, M.H.; Hagan, M.T.; Demuth, H.B. Neural Network ToolboxTM 6. User's Guide. Available online: www.mathworks.com (accessed on 4 March 2019).

20. Bera, P. The use of artificial neural networks trained in supervised mode to the analysis of measurement data of combustion engines and automotive vehicles. In *Combustion Engines and Ecology*; Mitianiec, W., Ed.; Cracow University of Technology: Cracow, Poland, 2014; ISBN 978-83-7242-763-2.

Article

Temperature Dependent Parameter Estimation of Electrical Vehicle Batteries

Anna I. Pózna [1,*]**, Katalin M. Hangos** [1,2] **and Attila Magyar** [1]

[1] Department of Electrical Engineering and Information Systems, University of Pannonia, Egyetem Street 10, H-8200 Veszprém, Hungary; hangos.katalin@virt.uni-pannon.hu (K.M.H.); magyar.attila@virt.uni-pannon.hu (A.M.)

[2] Institute for Computer Science and Control, Systems and Control Laboratory, Kende Street 13-17, H-1111 Budapest, Hungary

[*] Correspondence: pozna.anna@virt.uni-pannon.hu

Received: 10 September 2019; Accepted: 27 September 2019; Published: 30 September 2019

Abstract: Parameter estimation of electrical vehicle batteries in the presence of temperature effect is addressed in this work. A simple parametric temperature dependent battery model is used for this purpose where the temperature dependence is described by static relationships. A two-step method is used that includes a parameter estimation step of the key parameters at different temperatures followed by a static optimization step that determines the temperature coefficients of the corresponding parameters. It was found that the temperature dependent parameter characteristics can be reliably estimated from charging profiles only. The proposed method can be used as a computationally effective way of determining the key battery parameters at a given temperature from their actual estimated values and from their previously determined static temperature dependence. The proposed parameter estimation method was verified by simulation experiments on a more complex battery model that also describes the detailed dynamic thermal behavior of the battery.

Keywords: dynamic modeling; thermal management; parameter estimation; energy storage operation and planning; electric and solar vehicles

---

1. Introduction

Lithium-ion batteries are popular energy sources of our everyday life because of their high energy density, low self-discharge and lightweight. Portable electronic devices (mobile phones and laptops), home electronics, electronic tools and electric vehicles (EVs) all run on some type of lithium-ion battery. In applications such as electrical vehicles, batteries are connected in parallel and series to meet the power needs. The optimal performance and safe operation of the set of battery cells are managed by the battery management system (BMS). Another essential role of the BMS is the state of charge (SOC) and state of health (SOH) estimation. The former quantity informs the driver on the remaining charge of the battery bank (i.e., the remaining mileage that can be traveled with the electrical vehicle), while the latter shows the the ratio between the capacity of a new battery in relation to the actual capacity of the battery. Similar to any other battery, the performance of the lithium-ion battery is not constant but slowly degrades during the operation and strongly depends on the ambient temperature. The battery health conditions cannot be measured directly therefore it should be estimated based on measurable quantities.

Thermal modeling and the analysis of lithium-ion batteries under different temperatures has been addressed by several authors. The thermal modeling of batteries as well as the modeling without temperature dependency can be classified based on the scientific background (e.g., equivalent circuit models and electrochemical models). The review by Madani *et al.* [1] gives a thorough analysis of not only the different electrochemical models but also the parameter identification methods.

In such applications where the computational complexity (i.e., time) is crucial, e.g., in BMSs (battery management systems), *equivalent circuit models* are widely used [2]. The authors of [3] addressed the study of open circuit voltage-state of charge (OCV-SOC) characterization under the influence of different temperatures. The results show that the OCV-SOC characteristics curve highly depends on the temperature. An online estimation method for model parameters and SOC is proposed in [4], for applications in EVs under various temperatures. Their model is based on the RC circuit equivalent of the investigated battery. In [5], a design of experiment approach is used for the development of the electro-thermal model of electric vehicle batteries. The basis of their work is also an equivalent circuit model of the battery. The authors of [6] investigated the influence of thermal effect on the performance of their dual Kalman filter based (state and parameter estimation) method.

Another class of battery models is the *electrochemical models* where the chemical reactions and mechanisms taking place in the battery serves as a basis for the modeling equations. An electro-thermal model is developed and validated experimentally in [7], where electronic conduction, heat transfer, energy balance and electrochemical mechanisms are included in the model. A computationally more efficient electrochemical lithium-ion battery model is proposed in [8]. The simplified single particle model is compared with more complicated electrochemical models as well as experimental data. Hosseinzadeh *et al.* [9] gave a systematic approach for the development of thermal electrochemical models of large lithium-ion batteries for EV applications. An *et al.* [10] addressed the problem of non-uniformity of heat generation and electrochemical reaction increase with the discharge rate in an electrochemical-thermal coupled lithium-ion battery model.

Pure thermal models are also present in the literature; the authors of [11] developed a lumped parameter thermal model of the widely used $LiFePO_4$ lithium-ion battery. Using thermal measurements and the model, they determined the heat transfer coefficient and the heat capacity of the examined battery.

Due to the above mentioned thermal effects taking place in lithium-ion batteries, the previously mentioned roles of BMSs are usually extended with *thermal management*. The most frequently used thermal management solutions of lithium-ion batteries (used either in HEVs or in EVs) are reviewed in [12]. Temperature dependence of the key battery parameters and variables motivated the authors of [13] to develop a two stage battery capacity estimation method. In the first stage, battery core temperature is estimated and, afterwards, SOC and capacity are estimated by a sliding model observer.

Due to the nonlinear nature of parametric lithium-ion battery models and the fact that parameters might also depend on time and/or external variables, the computational complexity of battery parameter estimation can be demanding. Wang *et al.* [14] overcame this problem by a parallel Java algorithm implemented on GPU (CUDA) architecture. The authors of [15] developed and compared three different solutions for the internal resistance estimation of lithium-ion batteries (direct resistance estimation, Extended Kalman Filter (EKF), and recursive least squares) and concluded that EKF approach performed the best in terms of computational efficiency.

In our previous work [16], we proposed a parameter estimation for lithium ion batteries based on their first-order equivalent circuit model. The aim of the present paper is to generalize that work to the case when the temperature dependence of the parameters are also taken into account in such a way that a computationally effective way of temperature dependent parameters important from the viewpoint of applicability (e.g., actual capacity) could be given for implementation in future BMSs.

Therefore, the method presented in this paper proposes temperature dependent static characteristics for the battery parameters using a very simple dynamic model. The obtained characteristics can be used for determining the estimated parameter values for given thermal circumstances. Since the characteristics are determined for the parameters independently as static functional relationships, they can be calculated effectively in a real-time environment, e.g., a battery management system.

The paper is organized as follows. Section 2 introduces the parametric lithium-ion battery model that is used as the basis of our further steps. In Section 3.3, the parameter sensitivity analysis of the model is performed together with the discussion of its results. Section 3 is the main contribution of this paper and proposes our novel parameter estimation method for lithium-ion batteries. In Section 4, the proposed parameter estimation method is analyzed by simulation experiments, and the paper is closed by concluding remarks.

2. Parametric Battery Model

The parametric lithium-ion battery model is an important basis of the methods proposed in the sequel, thus it is presented here. This is a modified version of that used in [16].

2.1. Modeling Assumptions

The following assumptions were made for the battery model [17] with temperature dependency:

- The capacity of the battery does not change with the amplitude of the current (no Peukert effect).
- The self-discharge of the battery is not represented.
- The battery has no memory effect (no ageing is assumed).
- The voltage and the current can be influenced.
- The capacity depends on the ambient temperature.
- The constant potential, the polarization coefficient, the polarization resistance and the internal resistance depend on the internal (cell) temperature of the battery.

2.2. Temperature Dependent Battery Model

From the potential modeling methodologies, the equivalent electrical circuit type was selected to create the basic battery model. The selected model is originally developed in [17]; in our previous work, we described that model without temperature effect [18].

The structure of the model can be seen in Figure 1.

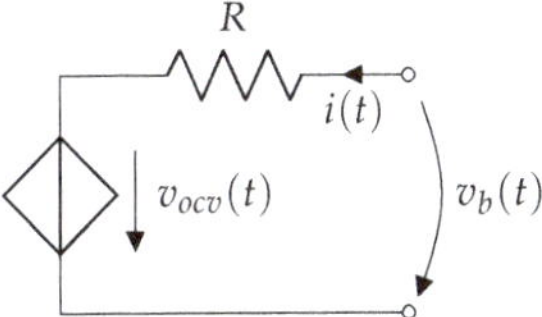

Figure 1. Equivalent electrical circuit model of the battery. Voltage $v_{ocv}(t)$ of the controlled voltage source is different in the case of charge and discharge.

The input of the model is the battery current (i) and the output is the battery voltage (v_b). The open circuit voltage (v_{ocv}) is represented by a controlled voltage source, and it is different during charging and discharging. The model was extended with temperature effects as it can be found in the Matlab Simulink Battery block (Simscape/Electrical/Specialized Power Systems/Electric Drives/Extra Sources). The difference with respect to the basic model is that some of the parameters depend on the ambient or cell temperature. As a result, the temperature dependent state space model of the battery is obtained in the form of Equations (1)–(6) as follows [17].

State equations:

$$\frac{d}{dt}q(t) = \frac{1}{3600}i(t) \tag{1}$$

$$\frac{d}{dt}i^*(t) = -\frac{1}{\tau}i^*(t) + \frac{1}{\tau}i(t) \tag{2}$$

The state variables have the following meaning:

- q is the actual extracted capacity of the battery. The initial values of q are $q(t_0) = 0$, if the battery is fully charged and $q(t_0) = Q$, if the battery is fully discharged.
- i^* is the polarization current. It can be computed by applying a low-pass filter to the battery current i, where τ is the time constant of the filter (see Equation (2)).

Output equations:

- Charge model

$$
\begin{aligned}
v_{ocv}^{ch}(t, T, T_a) =& E_0(T) - K_1(T)\frac{Q(T_a)}{q(t) + 0.1Q(T_a)}i^*(t) \\
& - K_2(T)\frac{Q(T_a)}{Q(T_a) - q(t)}q(t) + A\exp(-Bq(t)) - Cq(t)
\end{aligned}
\tag{3}
$$

$$v_b^{ch}(t, T) = v_{ocv}^{ch}(t, T, T_a) - R(T)i(t) \tag{4}$$

- Discharge model

$$
\begin{aligned}
v_{ocv}^{dch}(t, T, T_a) =& E_0(T) - K_1(T)\frac{Q(T_a)}{Q(T_a) - q(t)}i^*(t) \\
& - K_2(T)\frac{Q(T_a)}{Q(T_a) - q(t)}q(t) + A\exp(-Bq(t)) - Cq(t)
\end{aligned}
\tag{5}
$$

$$v_b^{dch}(t, T) = v_{ocv}^{dch}(t, T, T_a) - R(T)i(t) \tag{6}$$

The *output* of the model is the battery voltage v_b^X that is composed of the open circuit voltage (v_{ocv}^X) and the voltage drop across the internal resistance ($R(T)i(t)$). The open circuit voltage is the voltage of the battery when no external load is connected to it. $X = \{ch, dch\}$ denotes the charge/discharge mode of the battery.

The variables of the model with their meaning and units can be seen in Table 1.

The indirect temperature dependency of the model defined by Equations (1)–(6) is realized through a static temperature dependence of the model parameters. The *temperature dependency of the parameters* can be described with the following equations [19]:

- The change of polarization coefficient, polarization resistance and internal resistance with the battery temperature T can be derived from the Arrhenius law:

$$K_1(T) = K_1|_{T_{ref}} \exp\left(\alpha_1\left(\frac{1}{T} - \frac{1}{T_{ref}}\right)\right) \tag{7}$$

$$K_2(T) = K_2|_{T_{ref}} \exp\left(\alpha_2\left(\frac{1}{T} - \frac{1}{T_{ref}}\right)\right) \tag{8}$$

$$R(T) = R|_{T_{ref}} \exp\left(\beta\left(\frac{1}{T} - \frac{1}{T_{ref}}\right)\right) \tag{9}$$

- The temperature dependency of the capacity and the constant potential can be written in the following form:

$$Q(T_a) = Q|_{T_{ref}} + \frac{\Delta Q}{\Delta T}(T_a - T_{ref}) \tag{10}$$

$$E_0(T) = E_0|_{T_{ref}} + \frac{\partial E}{\partial T}(T - T_{ref}) \tag{11}$$

The parameters of the temperature dependent battery model with their meaning and nominal values can be found in Table 1. Our examined battery is a Samsung INR18650-20Q type battery (Cheonan, Korea) with 2000 mAh nominal capacity and 3.6 V nominal voltage. The nominal parameters of the battery were extracted from the datasheet and the Matlab Simulink model [20].

Table 1. Variables and parameters of the examined Samsung INR18650-20Q Li-ion battery.

Name	Type	Meaning	Unit	Value
i	input variable	battery current	A	-
i^*	state variable	polarization current	A	-
q	state variable	extracted capacity	Ah	-
t	independent variable	time	s	-
v_{ocv}	variable	open circuit voltage	V	-
v_b	output variable	battery voltage	V	-
T	external variable	battery cell temperature	K	-
T_a	external variable	ambient temperature	K	-
T_{ref}	parameter	nominal ambient temperature	K	298.15
τ	parameter	time constant of the filter	s	0.003
E_0	parameter	constant potential of the electrodes	V	-
$E_0|_{T_{ref}}$	parameter	constant potential of the electrodes at nominal ambient temperature	V	3.9388
$\partial E/\partial T$	parameter	reversible voltage temperature coefficient	V/K	0.002
R	parameter	internal resistance	Ω	-
$R|_{T_{ref}}$	parameter	internal resistance at nominal ambient temperature	Ω	0.005
β	parameter	Arrhenius rate constant for the internal resistance	K	3839.8
K_1	parameter	polarization constant	V/Ah	-
$K_1|_{T_{ref}}$	parameter	polarization constant at nominal ambient temperature	V/Ah	0.0018
α_1	parameter	Arrhenius rate constant for the polarization coefficient	K	8415.3
K_2	parameter	polarisation resistance	Ω	-
$K_2|_{T_{ref}}$	parameter	polarization resistance at nominal ambient temperature	Ω	0.0018
α_2	parameter	Arrhenius rate constant for the polarization resistance	K	8415.3
Q	parameter	battery capacity	Ah	-
$Q|_{T_{ref}}$	parameter	battery capacity at nominal ambient temperature	Ah	2.0
$\Delta Q/\Delta T$	parameter	maximum capacity temperature coefficient	Ah/K	0.016
A	parameter	exponential voltage	V	0.1589
B	parameter	exponential capacity	$(Ah)^{-1}$	15.0
C	parameter	nominal discharge curve slope	V/Ah	0.2362

Remark on the battery cell temperature:

To obtain a simple model for parameter estimation, we have omitted the energy balance and considered the battery cell temperature T as an *external variable* that does not change too much during a charge or discharge operation.

3. Parameter Estimation Methodology

Based on the modeling and analysis results of Sections 2 and 3.3, the parameter estimation method is presented in this section.

3.1. Input Signal

The pseudo-random binary sequence (PRBS) is chosen as the input signal for the parameter estimation. It is a widely used signal in the field of parameter estimation [21] because it is easy to generate and provides sufficient excitation. The PRBS has only two values in between the signal changes randomly. The two parameters of the PRBS are the range (the upper and lower level of the signal) and the frequency of the change that should be chosen considering the system dynamics. In our case, the clock frequency of the PRBS was chosen to be five times the time constant of the system; the latter can be approximately determined from a step input to the system (see in Section 3.3).

An other important factor of our parameter estimation method is the ambient temperature. At each experiment, the ambient temperature is chosen to be constant, thus we hold the ambient temperature constant during an experiment.

The minimum and maximum battery temperatures of the experiments should be chosen according to the recommended operating temperatures of the examined battery. Then, this range is evenly divided to get the list of ambient temperatures at which the experiments should be carried out.

3.2. Simulation Setup

The parameter estimation methods were implemented and tested by simulation experiments in Matlab. To simulate the heat dissipation of the battery during charge/discharge, the battery model in Simulink/Simscape/Electrical/Specialized Power Systems/Electric Drives/Extra Sources (an *extended model*) was used. *This model contains additional energy balance equations* that describe the temperature effects of the battery [22]. This means that the cell temperature and the heating/cooling effects of the battery (including self-heating) during the operation can be simulated. It is important to note that the model that we used for parameter estimation (Equations (1)–(11)) is much simpler, as it does not contain the internal energy balance equation. The advantage of the Simulink model is that *the battery cell temperature can be directly extracted from the model*, which can be used as measurement data for the cell temperature.

The simulated battery was a Samsung INR18650Q-20Q battery with 2000 mAh capacity. The nominal parameters of the battery can be seen in Table 1. The operating temperatures of the battery from the datasheet are 0–50 °C for charge and -20–75 °C for discharge. Based on these values, we decided to simulate the battery in the range 0–50 °C. The charge and the discharge of the battery was simulated at 11 different ambient temperatures with PRBS input signal of 1–99% state of charge. The simulation setup in case of charge and discharge can be seen below.

Simulation setup for charge:

- PRBS input: $I_{min} = -2A$, $I_{max} = -0.5A$, $T_s = 160s$
- initial values: $q(t_0) = 0.99Q$, $i^*(t_0) = 0$, $T = T_a$
- ambient temperatures: $T_a = 0, 5, 10, 15, 20, 25, 30, 35, 40, 45, 50$ °C
- stopping criterion: $q(t) = 0$;

Simulation setup for discharge:

- PRBS input: $I_{min} = 0.5A$, $I_{max} = 2A$, $T_s = 160s$
- initial values: $q(t_0) = 0.01Q$, $i^*(t_0) = 0$, $T = T_a$
- ambient temperatures: $T_a = 0, 5, 10, 15, 20, 25, 30, 35, 40, 45, 50\ ^\circ C$
- stopping criterion: $q(t) = 0.99Q$;

All simulations were performed on a PC (Intel i5 CPU with 4GB RAM) (Intel, Santa Clara, CA, USA).

3.3. Parameter Sensitivity Analysis

As a first step of the parameter estimation, the parameter sensitivity of the charge and discharge model of the battery was analyzed. It is important to note that the temperature has an indirect effect on the model output through the parameters directly depend on the temperature. We used our previously described method [16] for the sensitivity analysis, i.e., the parameter values were changed one by one with $\pm 10\%$ with respect to the nominal values, then the difference between the nominal and the perturbed model was evaluated using a quadratic loss function:

$$W_s(\tilde{\theta}) = \frac{1}{N} \sum_{k=1}^{N} \frac{1}{2} \left(v_b(\theta;k) - v_b(\tilde{\theta};k) \right)^2 \tag{12}$$

where θ denotes the parameter vector, and $\tilde{\theta}$ is the perturbed parameter vector. First, the step response of the model was simulated to get the time constant of the system (τ_s). The sample time of the PRBS signal (T_s) was chosen to be $T_s = \tau_s/5$. The sensitivity analysis was repeated at six different temperatures: $0\ ^\circ C$, $10\ ^\circ C$, $20\ ^\circ C$, $30\ ^\circ C$, $40\ ^\circ C$ and $50\ ^\circ C$. The battery was charged/discharged between 0% and 100% state of charge with PRBS current input (amplitude: charge $\{-2\ A, -0.5\ A\}$ and discharge $\{0.5\ A, 2\ A\}$; and sample time: 160 s). Both the charge and the discharge models were analyzed. The nominal model was the charge/discharge model at the nominal ambient temperature $T_{ref} = 25\ ^\circ C$.

The models were simulated in Matlab using the model equations (Equations (1)–(6)). At each temperature, the nominal parameters were perturbed one-by-one and the value of the loss function was computed. The result of the sensitivity analysis of the charge and the discharge model can be seen in Tables 2 and 3. The graphical representation of the results is depicted in Figure 2.

Table 2. Values of the loss function in case of the parameter sensitivity analysis of the charge model.

Parameter	Change	0 °C	10 °C	20 °C	30 °C	40 °C	50 °C
E_0	-10%	0.1100	0.0710	0.0728	0.0837	0.0922	0.0999
	$+10\%$	0.1342	0.0939	0.0830	0.0721	0.0652	0.0592
K_1	-10%	0.0437	0.0047	0.0003	0.0003	0.0011	0.0020
	$+10\%$	0.0455	0.0051	0.0004	0.0003	0.0011	0.0020
K_2	-10%	0.0365	0.0041	0.0003	0.0003	0.0011	0.0020
	$+10\%$	0.0537	0.0059	0.0004	0.0003	0.0011	0.0020
Q	-10%	0.0376	0.0069	0.0028	0.0013	0.0005	0.0007
	$+10\%$	0.0562	0.0054	0.0016	0.0025	0.0039	0.0055
R	-10%	0.0446	0.0049	0.0003	0.0003	0.0011	0.0020
	$+10\%$	0.0446	0.0049	0.0004	0.0003	0.0011	0.0020

Table 3. Values of the loss function in case of the parameter sensitivity analysis of the discharge model.

Parameter	Change	0 °C	10 °C	20 °C	30 °C	40 °C	50 °C
E_0	-10%	0.3795	0.1374	0.0912	0.0687	0.0591	0.0517
	$+10\%$	0.1305	0.0581	0.0680	0.0886	0.1013	0.1119
K_1	-10%	0.1641	0.0184	0.0018	0.0011	0.0026	0.0042
	$+10\%$	0.1913	0.0220	0.0022	0.0011	0.0026	0.0042
K_2	-10%	0.1578	0.0182	0.0017	0.0011	0.0026	0.0042
	$+10\%$	0.1982	0.0223	0.0023	0.0010	0.0026	0.0042
Q	-10%	0.1362	0.0408	0.0020	0.0002	0.0023	0.0042
	$+10\%$	0.1852	0.0346	0.0004	0.0015	0.0027	0.0042
R	-10%	0.1769	0.0200	0.0200	0.0011	0.0026	0.0042
	$+10\%$	0.1780	0.0203	0.0020	0.0011	0.0026	0.0042

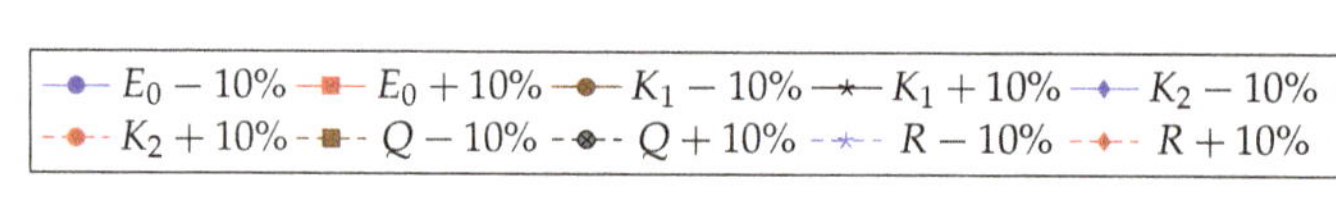

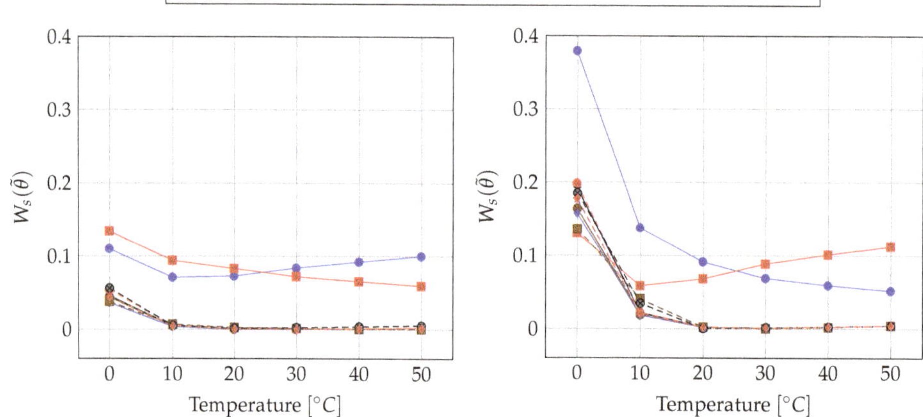

(a) Sensitivity of the charge model   (b) Sensitivity of the discharge model

Figure 2. Results of the parameter sensitivity analysis of the charge and the discharge model.

It can be seen that the discharge model is a bit more sensitive to the change of the parameters as the magnitude of the error is greater in that case. Both the charge and the discharge models have similar characteristics with respect to the parameter sensitivity:

- The models are highly sensitive to the constant potential E_0.
- The models are less sensitive to K_1, K_2 and Q.
- The rate of sensitivity is similar in the case of K_1, K_2 and Q.
- The sensitivity of the models increases as the temperature decreases.
- At ambient temperatures greater than the nominal temperature, the effect of changing the parameters is really small (except for E_0), especially in case of the discharge model.
- The change of the internal resistance R at different temperatures has no effect on the models, as the errors related to the $\pm 10\%$ change are the same. In these cases, only the temperature affects the models.

Based on these statements, *the parameters E_0, K_1, K_2 and Q will be estimated while R is fixed to its nominal value.*

3.4. Methods for Parameter Estimation

Our parameter estimation method consists of two steps. At first the battery is charged or discharged at different constant ambient temperatures. At each temperature, the parameters E_0, K_1, K_2 and Q of the battery are estimated. In the second step, the temperature coefficients of these parameters are estimated.

3.4.1. Estimation of the Battery Parameters

The first step of our method is the estimation of the battery parameters at different constant ambient temperatures to see how these parameters change with that temperature. The inputs of the parameter estimation are the battery current and voltage at different temperatures during a full charge or discharge process. The result of the estimation is a set of battery parameters at different temperatures.

It can be seen from Equations (1)–(6) that the battery model has a nonlinear output equation and four parameters to be estimated as we fixed the internal resistance R to its nominal value. Therefore, a suitable nonlinear parameter estimation method should be chosen. In our work, the nonlinear least-squares method is chosen. Nonlinear parameter estimation problems are usually solved as nonlinear optimization problems where a suitable cost function should be minimized. In our case, the cost function is the sum of squared deviation between the model and the measurement data at every time instance (see Equation (13) below).

$$W(\theta) = \sum_{k=1}^{N} \left(\hat{v}_b^X(k) - v_b^X(\theta;k) \right)^2 \tag{13}$$

$$X \in \{ch; dch\}$$

where $\hat{v}_b^X(k) = \hat{v}_b^X(k\,T_s)$ is the measured value of the battery voltage at the kth sample, $v_b^X(\theta;k)$ is the output of the model (Equation (4) or Equation (6)) with the parameter vector $\theta = [E_0, K_1, K_2, Q]$, and N is the total number of samples.

As all of the parameters to be estimated have physical meaning, the range and scale of the parameter values are usually known in advance. Therefore, upper and lower bounds for the parameters can be defined that is useful to limit the searching space of the optimization. As a result, a constrained nonlinear optimization problem should be solved. From the potential algorithms, the trust region reflective algorithm [23] is chosen in our work.

3.4.2. Estimation of the Temperature Dependency of the Parameters

The second step of the parameter estimation method is the estimation of the reference values and the temperature dependency coefficients of the parameters. The inputs of this parameter estimation problem are the estimated parameters at different temperatures from the previous step (Section 3.4.1). It can be seen from the temperature dependent battery model that the battery parameters can be divided into two groups based on the type of their temperature dependency:

- Parameters with linear temperature dependency: E_0, Q.
- Parameters with nonlinear (exponential) temperature dependency: K_1, K_2.

Moreover, it can be seen from Equations (7)–(11) that some of the parameters (Q) depend on the ambient temperature and others (E_0, K_1, K_2) depend on the battery cell temperature. The problem is that the cell temperature usually cannot be measured. To overcome this, we made the *following additional assumptions:*

- The cell temperature does not change a lot during charge/discharge (maximum $\pm 2\,^{\circ}\text{C}$).
- The cell temperature is substituted by the average surface temperature during charge/discharge.
- Initially, the cell temperature and the ambient temperature are equal.
- The surface temperature of the battery is measured.

With the above assumptions, the temperature coefficients of the parameters can be estimated. The *coefficients to be estimated* are:

- $E_0|_{T_{ref}}$ and $\partial E/\partial T$ for the temperature dependency of E_0
- $Q|_{T_{ref}}$ and $\Delta Q/\Delta T$ for the temperature dependency of Q
- $K_1|_{T_{ref}}$ and α_1 for the temperature dependency of K_1
- $K_2|_{T_{ref}}$ and α_2 for the temperature dependency of K_2

The coefficients of $E_0(T)$ and $Q(T_a)$ can be estimated with the simple linear least squares method because Equations (11) and (10) are linear.

The coefficients of $K_1(T)$ and $K_2(T)$ can also be estimated by the least squares method by transforming the equations and their dependent variables.

4. Simulation Results

In this section, the results of the simulation based experiments are introduced and analyzed. In Section 4.1, the results of the estimation of the battery parameters at 11 different temperatures are presented. Then, in Section 4.2, the results of the estimation of the temperature dependency of the battery parameters are described.

4.1. Estimated Battery Parameters

The battery parameters at different temperatures were estimated using the *lsqnonlin* function from Matlab Optimization Toolbox [24]. This function needs at least two input arguments: a function to minimize and the vector of initial parameter values. Additional input arguments such as lower and upper bounds of the parameters and other options can be also defined. In our case, the following bounds were defined for the parameters: $0 \leq E_0 \leq 5$, $0 \leq Q \leq 3$, $0 \leq K_1 \leq 0.1$, $0 \leq K_2 \leq 0.1$.

The function to minimize is the cost function in Equation (13) and the parameters to be estimated are $\theta = [E_0, Q, K_1, K_2]^T$. The initial values of the parameters were set to the nominal parameter values (see in Table 1).

The results of the parameter estimation can be seen in Tables 4 and 5. It can be noticed in the second row of Table 4 that above 35 °C the battery reached its maximum capacity during charge.

Table 4. Estimated battery parameters at different temperatures during charge.

T_a [°C]	0	5	10	15	20	25	30	35	40	45	50
E_0 [V]	3.9175	3.9154	3.9190	3.9259	3.9343	3.9436	3.9532	3.9631	3.9651	3.9783	3.9893
Q [Ah]	1.6001	1.6800	1.7599	1.8399	1.9201	2.0004	2.0811	2.1623	2.1576	2.1579	2.1582
K_1 [V / Ah]	0.0169	0.0099	0.0059	0.0036	0.0023	0.0015	0.0010	0.0007	0.0012	0.0008	0.0007
K_2 [Ω]	0.0246	0.0140	0.0082	0.0049	0.0030	0.0019	0.0012	0.0008	0.0000	0.0000	0.0000

Table 5. Estimated battery parameters at different temperatures during discharge.

$T_a\,[°C]$	0	5	10	15	20	25	30	35	40	45	50
$E_0\,[V]$	3.8877	3.8980	3.9083	3.9185	3.9286	3.9388	3.9490	3.9591	3.9693	3.9795	3.9884
$Q\,[Ah]$	1.6010	1.6801	1.7599	1.8393	1.9188	1.9980	2.0764	2.1540	2.2300	2.3035	2.1583
$K_1\,[V/Ah]$	0.0239	0.0138	0.0081	0.0048	0.0029	0.0018	0.0011	0.0007	0.0004	0.0003	0.0000
$K_2\,[\Omega]$	0.0243	0.0139	0.0081	0.0048	0.0029	0.0018	0.0011	0.0007	0.0005	0.0003	0.0000

It can be seen from the estimated values that they are in good agreement with the nominal parameters of the investigated battery type, and coincide well with the parameters in the detailed dynamic battery model in Simulink/Simscape/Electrical/Specialized Power Systems/Electric Drives/Extra Sources.

The accuracy of the parameter estimation can be characterized by the covariance matrix of the estimation. In our results, the elements of the covariance matrices are really small (with orders between 10^{-8} and 10^{-12}) in both charge and discharge cases. This means that the parameter estimation is very accurate. Note that the experimental data were obtained *from the simulation of the model equations of the extended model with energy balance equation* and not from real measurements, therefore no external noise or modeling errors are included.

The confidence region of the estimated parameters can be approximated by the $1.05 \cdot W_{min}$ contour line of the cost function W. To analyze and illustrate the confidence regions, we analyzed the parameters in pairs. We fixed two parameters and computed the value of the cost function when changing the other two parameter values around their estimated value. The two parameters pairs were chosen as E_0, Q and K_1, K_2. Some examples of the confidence regions in the case of charge/discharge at different temperatures are illustrated on Figures 3 and 4. The order of magnitude on the x and y axes are the same in Figures 3a–d and 4a–c respectively. In Figure 4d, the axes are magnified for better visibility. Comparing the confidence regions at different temperatures and operating modes, the following conclusions can be drawn:

- In the case of both charge and discharge, the confidence of Q increases while E_0 decreases as the temperature rises (see Figure 3a–d).
- E_0 and Q are uncorrelated because the axes of the ellipse are almost parallel with the x and y axes.
- In the case of charge, the confidence region of K_1, K_2 becomes smaller as the temperature rises (see Figure 4a,b).
- A linear relationship between K_1 and K_2 can be assumed in the case of discharge (see Figure 4c,d).

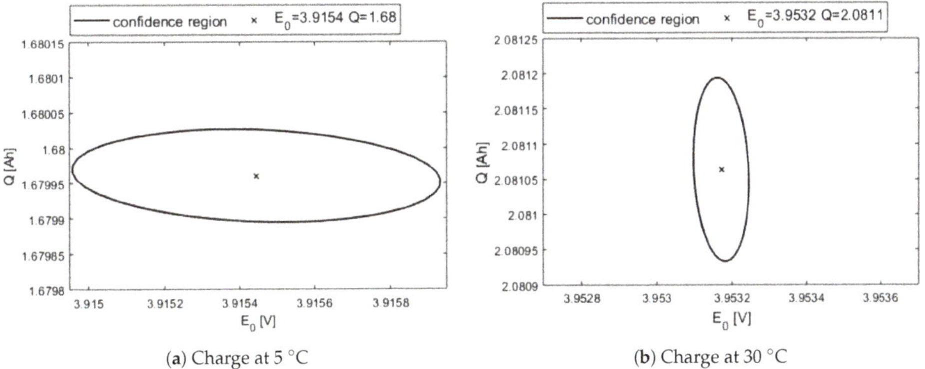

(a) Charge at 5 °C

(b) Charge at 30 °C

Figure 3. *Cont.*

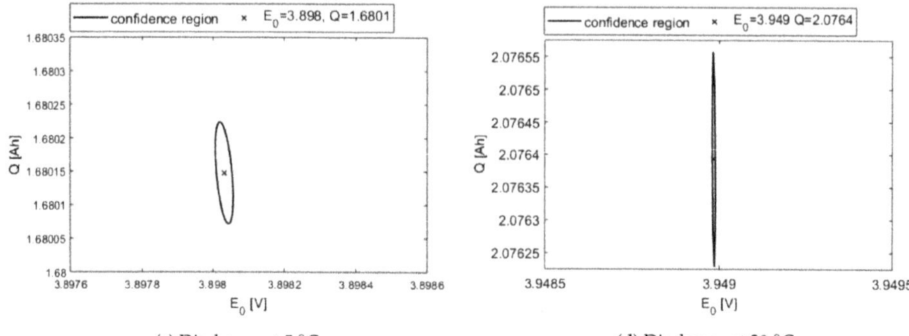

(**c**) Discharge at 5 °C

(**d**) Discharge at 30 °C

Figure 3. Confidence regions of the estimated parameters E_0, Q during charge/discharge at different temperatures. x axis, E_0; y axis, Q; x axis range, 1×10^{-3}; y axis range, 3.5×10^{-4}; $-$, confidence region; $\times$, estimated parameter value.

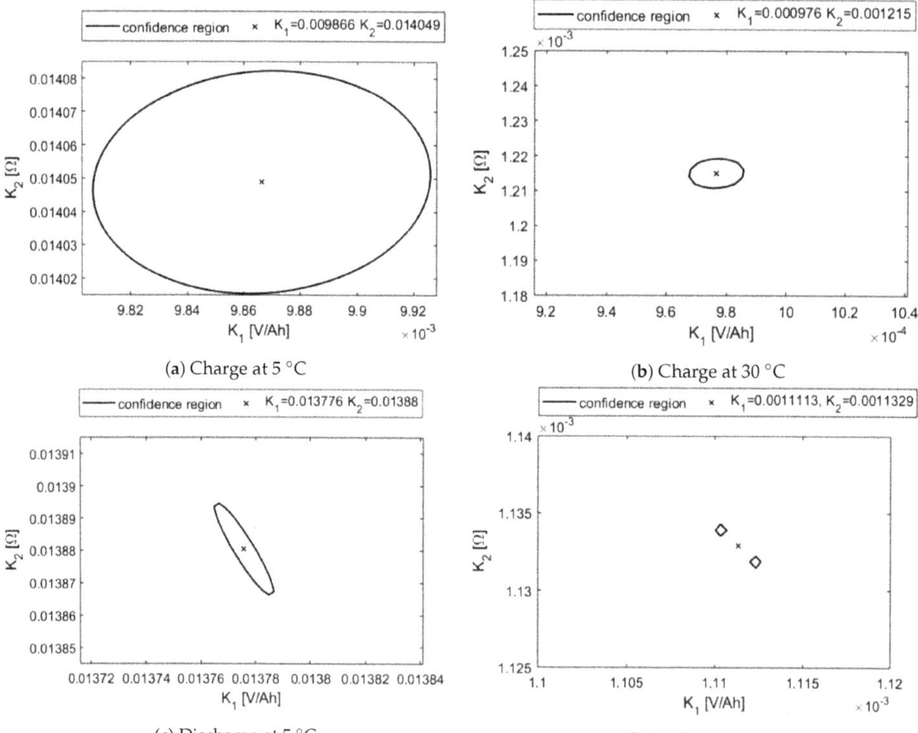

(**a**) Charge at 5 °C

(**b**) Charge at 30 °C

(**c**) Discharge at 5 °C

(**d**) Discharge at 30 °C

Figure 4. Confidence regions of the estimated parameters K_1, K_2 during charge/discharge at different temperatures. x axis, K_1; y axis, K_2; x axis range, 1.25×10^{-4}; y axis range, 7×10^{-5}; $-$, confidence region; $\times$, estimated parameter value.

Looking at Figures 5 and 6, it is apparent that the estimated values of E_0 and Q are more uncertain in the case of charge. This phenomenon can be explained by the confidence regions depicted in Figure 3. It can be seen that the confidence region is wider in the case of charge, hence the uncertainty of the parameters are greater. It can be also noticed that the shape of the confidence region changes with temperature. At low temperatures, the confidence of the estimated Q is smaller than the confidence of E_0. On the contrary, at high temperatures, the confidence of Q becomes greater while the confidence of E_0 decreases. That is why we can better estimate Q at low temperatures and E_0 at high temperatures. Additionally, the estimates are results of nonlinear optimization, which is affected by the initial values, stopping criteria, and the shape of the cost function.

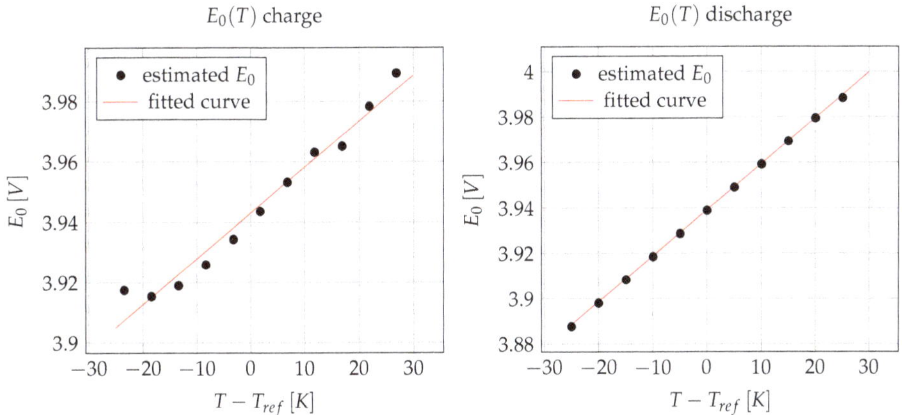

(a) The fitted thermal characteristics of parameter $E_0(T)$ from the charge data. $r^2 = 0.9691$ (b) The fitted thermal characteristics of parameter E_0 from the discharge data. $r^2 = 0.9999$

Figure 5. Estimation of the temperature dependency of E_0.

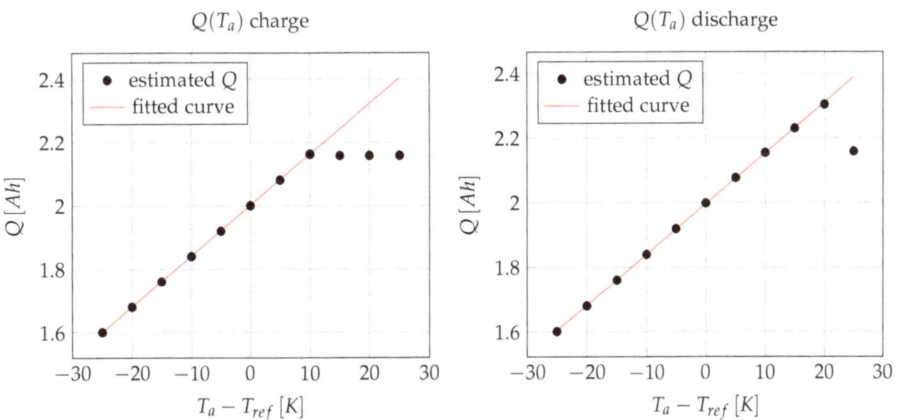

(a) The fitted thermal characteristics of parameter Q from the charge data. $r^2 = 1$ (b) The fitted thermal characteristics of parameter Q from the discharge data. $r^2 = 0.9999$

Figure 6. Estimation of the temperature dependency of Q.

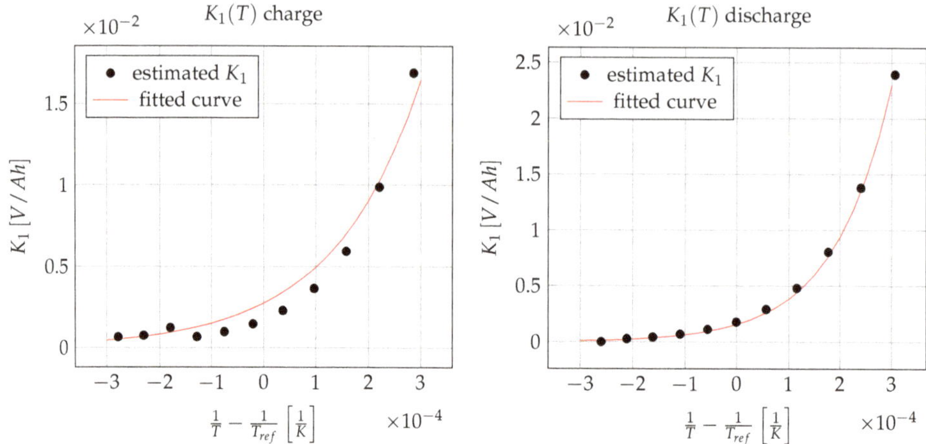

(**a**) The fitted thermal characteristics of parameter $K_1(T)$ from the charge data. $r^2 = 0.9656$
(**b**) The fitted thermal characteristics of parameter $K_1(T)$ from the discharge data. $r^2 = 0.9995$

Figure 7. Estimation of the temperature dependency of K_1.

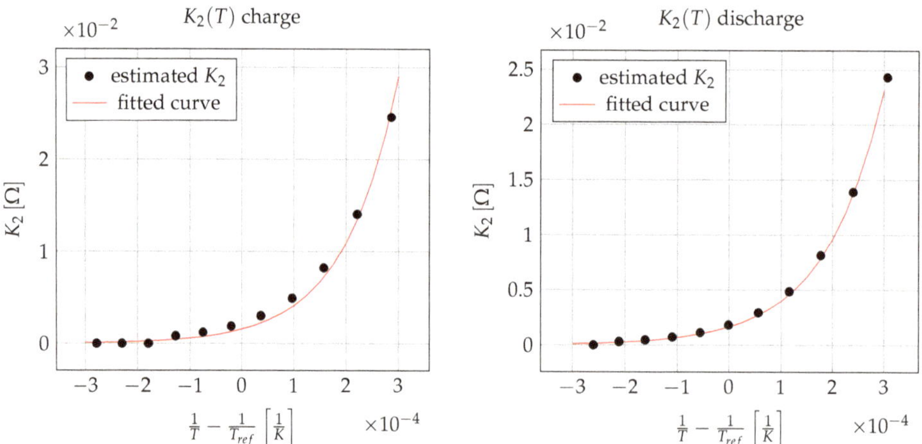

(**a**) The fitted thermal characteristics of parameter $K_2(T)$ from the charge data. $r^2 = 0.9925$
(**b**) The fitted thermal characteristics of parameter $K_2(T)$ from the discharge data. $r^2 = 0.9988$

Figure 8. Estimation of the temperature dependency of K_2.

4.2. Estimated Temperature Dependent Parameters

Having estimated the battery parameters at different ambient temperatures, the temperature dependency of the parameters was estimated with the help of the Matlab Curve Fitting Toolbox [25]. Each parameter has two coefficients that describe the temperature dependency: the parameter value at the reference temperature and the temperature coefficient. The independent variables of the four different parameter estimation tasks are the following:

- $T - T_{ref}$, in case of $E_0(T)$;
- $T_a - T_{ref}$, in case of $Q(T_a)$;
- $\frac{1}{T} - \frac{1}{T_{ref}}$ in case of $K_1(T)$ and $K_2(T)$.

As mentioned in Section 3.4.2, the cell temperature T was substituted by the average surface temperature of the battery. The dependent variables are the estimated parameter values of the previous step that can be seen in Tables 4 and 5.

The coefficients of the temperature dependency were estimated during both charge and discharge. The results of the estimation can be seen in Tables 6 and 7. The 95% confidence bounds shows the uncertainty of the estimated coefficients.

It can be seen that the estimated temperature dependency of E_0 and Q is close to the nominal values in both charge and discharge cases. The estimation of $Q|_{T_{ref}}$ and $\Delta Q/\Delta T$ is better in the case of charge because the differences between the nominal and estimated parameter are smaller. However, the estimation of the other parameters is better in the case of discharge.

The fitted curves of the temperature dependency can be seen in Figures 5–8 with red line. The goodness of fit was characterized by the r^2 value that is computed by:

$$r^2 = 1 - \frac{\sum_i (\hat{y}_i - y_i)^2}{\sum_i (\hat{y}_i - \bar{y})^2}$$

where $\hat{y}$ is the measured data, y is the model predicted value, and $\bar{y}$ is the mean of the measured data. The results can be seen in Table 8.

It can be seen that the curve fitting is a bit more accurate in case of discharge, except Q.

Table 6. Estimated parameters of the temperature dependency of the battery parameters during charge.

Parameter	Nominal Value	Estimated Value	95% Confidence Bounds	Unit
$E_0\|_{T_{ref}}$	3.9388	3.943	(3.94, 3.946)	V
$\partial E/\partial T$	2.0×10^{-3}	1.518×10^{-3}	$(1.314 \times 10^{-3}, 1.723 \times 10^{-3})$	V/K
$Q\|_{T_{ref}}$	2.0	2.001	(2.0, 2.001)	Ah
$\Delta Q/\Delta T$	1.6×10^{-2}	1.605×10^{-2}	$(1.601 \times 10^{-2}, 1.610 \times 10^{-2})$	Ah/K
$K_1\|_{T_{ref}}$	1.8×10^{-3}	2.735×10^{-3}	$(1.866 \times 10^{-3}, 3.604 \times 10^{-3})$	V/Ah
α_1	8415	5989	(4684, 7294)	K
$K_2\|_{T_{ref}}$	1.8×10^{-3}	1.545×10^{-3}	$(1.866 \times 10^{-3}, 1.987 \times 10^{-3})$	Ω
α_2	8415	9785	(8706, 10,860)	K

Table 7. Estimated parameters of the temperature dependency of the battery parameters during discharge.

Parameter	Nominal Value	Estimated Value	95% Confidence Bounds	Unit
$E_0\|_{T_{ref}}$	3.9388	3.939	(3.938, 3.939)	V
$\partial E/\partial T$	2.0×10^{-3}	2.025×10^{-3}	$(2.009 \times 10^{-3}, 2.041 \times 10^{-3})$	V/K
$Q\|_{T_{ref}}$	2.0	1.995	(1.993, 1.997)	Ah
$\Delta Q/\Delta T$	1.6×10^{-2}	1.568×10^{-2}	$(1.554 \times 10^{-2}, 1.581 \times 10^{-2})$	Ah/K
$K_1\|_{T_{ref}}$	1.8×10^{-3}	1.588×10^{-3}	$(1.418 \times 10^{-3}, 1.757 \times 10^{-3})$	V/Ah
α_1	8415	8908	(8528, 9289)	K
$K_2\|_{T_{ref}}$	1.8×10^{-3}	1.661×10^{-3}	$(1.542 \times 10^{-3}, 1.781 \times 10^{-3})$	Ω
α_2	8415	8793	(8538, 9048)	K

Table 8. Goodness of curve fitting characterized by the r^2 value.

	E_0	Q	K_1	K_2
charge	0.9691	1	0.9656	0.9925
discharge	0.9999	0.9999	0.9995	0.9988

5. Conclusions and Future Work

An optimization based lithium-ion battery parameter estimation method is proposed in this paper that is capable of describing the thermal behavior of batteries. The basis of the method is a nonlinear charge and discharge model that describes the temperature dependency as a parametric function of temperature as an external variable.

Parameter sensitivity analysis was carried out on the model to find the parameters to be estimated, which are the electrode potential, the battery capacity, and two polarization constants. The internal resistance was found to be non-sensitive to the model output, thus it was not estimated.

The proposed parameter estimation method contains two steps. At first, the parameters E_0, Q, K_1, K_2 are estimated from measured data of charging/discharging at different constant ambient temperatures. In the second step, the temperature coefficients of these parameters are estimated.

The proposed parameter estimation method was verified by a set of simulation experiments on an electro-thermal battery model capable of describing the energy balance (i.e., the thermal behavior) of the battery. The temperature dependent parameter characteristics obtained generated by the proposed method can be used as a computationally effective way of determining the key battery parameters at a given temperature. The *novelty of the method* is that the temperature dependent parameter characteristics can be estimated from charging profiles by the proposed method can be used as a computationally effective way of determining the key battery parameters at a given temperature. The proposed parameter estimation method was verified by simulation experiments on a more complex battery model that also describes the thermal behavior of the battery.

Further research directions include the use of this parameter estimation method for determining the state of health of the battery, and to estimate the temperature dependent state of charge during its life cycle. This is possible through a suitable experiment policy that estimates the battery capacity from well chosen charging operations under different thermal conditions. Therefore, extensive climate chamber experiments will be performed to validate the results of the present work.

Author Contributions: A.I.P. implemented the parameter estimation method, implemented simulation experiments, and wrote the main part of the manuscript. K.M.H. and A.M. designed the parameter estimation method and wrote other parts of the manuscript.

Funding: This research was funded by the Hungarian National Research, Development and Innovation Office (NKFIH) through grant Nos. K115694 and 120422. A. Magyar was supported by the János Bolyai Research Scholarship of the Hungarian Academy of Sciences. We acknowledge the financial support of Széchenyi 2020 under the EFOP-3.6.1-16-2016-00015.

Conflicts of Interest: The authors declare no conflicts of interest.

References

1. Madani, S.S.; Schaltz, E.; Kær, S.K. A Review of Different Electric Equivalent Circuit Models and Parameter Identification Methods of Lithium-Ion Batteries. *ECS Trans.* **2018**, *87*, 23–37. [CrossRef]
2. Panchal, S.; Mathew, M.; Fraser, R.; Fowler, M. Electrochemical thermal modeling and experimental measurements of 18650 cylindrical lithium-ion battery during discharge cycle for an EV. *Appl. Therm. Eng.* **2018**, *135*, 123–132. [CrossRef]
3. Zhang, R.; Xia, B.; Li, B.; Cao, L.; Lai, Y.; Zheng, W.; Wang, H.; Wang, W.; Wang, M. A Study on the Open Circuit Voltage and State of Charge Characterization of High Capacity Lithium-Ion Battery Under Different Temperature. *Energies* **2018**, *11*, 2408. [CrossRef]
4. Feng, F.; Lu, R.; Wei, G.; Zhu, C. Online estimation of model parameters and state of charge of LiFePO4 batteries using a novel open-circuit voltage at various ambient temperatures. *Energies* **2015**, *8*, 2950–2976. [CrossRef]
5. Mathew, M.; Mastali, M.; Catton, J.; Samadani, E.; Janhunen, S.; Fowler, M. Development of an electro-thermal model for electric vehicles using a design of experiments approach. *Batteries* **2018**, *4*, 29. [CrossRef]
6. Campestrini, C.; Walder, G.; Jossen, A.; Lienkamp, M. Temperature influences on state and parameter estimation based on a Dual Kalman Filter. In Proceedings of the CoFAT, Munich, Germany, 6–8 March 2014.
7. Ye, Y.; Shi, Y.; Cai, N.; Lee, J.; He, X. Electro-thermal modeling and experimental validation for lithium ion battery. *J. Power Sources* **2012**, *199*, 227–238. [CrossRef]
8. Astaneh, M.; Dufo-Lopez, R.; Roshandel, R.; Golzar, F.; Bernal-Agustín, J.L. A computationally efficient Li-ion electrochemical battery model for long-term analysis of stand-alone renewable energy systems. *J. Energy Storage* **2018**, *17*, 93–101. [CrossRef]
9. Hosseinzadeh, E.; Genieser, R.; Worwood, D.; Barai, A.; Marco, J.; Jennings, P. A systematic approach for electrochemical-thermal modelling of a large format lithium-ion battery for electric vehicle application. *J. Power Sources* **2018**, *382*, 77–94. [CrossRef]
10. An, Z.; Jia, L.; Wei, L.; Dang, C.; Peng, Q. Investigation on lithium-ion battery electrochemical and thermal characteristic based on electrochemical-thermal coupled model. *Appl. Therm. Eng.* **2018**, *137*, 792–807. [CrossRef]
11. Forgez, C.; Do, D.V.; Friedrich, G.; Morcrette, M.; Delacourt, C. Thermal modeling of a cylindrical LiFePO4/graphite lithium-ion battery. *J. Power Sources* **2010**, *195*, 2961–2968. [CrossRef]
12. Wang, Q.; Jiang, B.; Li, B.; Yan, Y. A critical review of thermal management models and solutions of lithium-ion batteries for the development of pure electric vehicles. *Renew. Sustain. Energy Rev.* **2016**, *64*, 106–128. [CrossRef]
13. Zhang, D.; Dey, S.; Perez, H.E.; Moura, S.J. Real-time capacity estimation of lithium-ion batteries utilizing thermal dynamics. *IEEE Trans. Control Syst. Technol.* **2019**. [CrossRef]
14. Wang, L.; Zhang, Z.; Huang, C.; Tsui, K.L. A GPU-accelerated parallel Jaya algorithm for efficiently estimating Li-ion battery model parameters. *Appl. Soft Comput.* **2018**, *65*, 12–20. [CrossRef]
15. Mathew, M.; Janhunen, S.; Rashid, M.; Long, F.; Fowler, M. Comparative analysis of lithium-ion battery resistance estimation techniques for battery management systems. *Energies* **2018**, *11*, 1490. [CrossRef]
16. Pózna, A.I.; Magyar, A.; Hangos, K.M. Model identification and parameter estimation of lithium ion batteries for diagnostic purposes. In Proceedings of the 2017 International Symposium on Power Electronics (Ee 2017), Novi Sad, Serbia, 19–21 October 2017; pp. 1–6. doi:10.1109/PEE.2017.8171673. [CrossRef]
17. Tremblay, O.; Dessaint, L.A.; Dekkiche, A.I. A generic battery model for the dynamic simulation of hybrid electric vehicles. In Proceedings of the 2007 IEEE Vehicle Power and Propulsion Conference, Arlington, TX, USA, 9–12 September 2007; pp. 284–289.
18. Pózna, A.I.; Hangos, K.M.; Magyar, A. Design of Experiments for Battery Aging Estimation. *IFAC-PapersOnLine* **2018**, *51*, 386–391. [CrossRef]
19. Tremblay, O.; Dessaint, L.A. Experimental validation of a battery dynamic model for EV applications. *World Electr. Veh. J.* **2009**, *3*, 289–298. [CrossRef]
20. The Mathworks, Inc. *Simulink Version: 9.3*; The Mathworks, Inc.: Natick, MA, USA, 2019.
21. Ljung, L. *System Identification: Theory for the User*, 2nd ed.; Prentice Hall Inc.: Upper Saddle River, NJ, USA, 1999.

22. Saw, L.; Somasundaram, K.; Ye, Y.; Tay, A. Electro-thermal analysis of Lithium Iron Phosphate battery for electric vehicles. *J. Power Sources* **2014**, *249*, 231–238. [CrossRef]
23. Byrd, R.H.; Schnabel, R.B.; Shultz, G.A. A trust region algorithm for nonlinearly constrained optimization. *SIAM J. Numer. Anal.* **1987**, *24*, 1152–1170. [CrossRef]
24. The Mathworks, Inc. *MATLAB Optimization Toolbox Version: 9.6.0,1047502 (R2019a)*; The Mathworks, Inc.: Natick, MA, USA, 2019.
25. The Mathworks, Inc. *MATLAB Curve Fitting Toolbox Version: 9.6.0,1047502 (R2019a)*; The Mathworks, Inc.: Natick, MA, USA, 2019.

Article

Stacked Auto-Encoder Modeling of an Ultra-Supercritical Boiler-Turbine System

Hao Zhang [1], Xiangjie Liu [1], Xiaobing Kong [1],* and Kwang Y. Lee [2]

[1] The State Key Laboratory of Alternate Electrical Power System with Renewable Energy Sources, North China Electric Power University, Beijing 102206, China; zhanghaooa@163.com (H.Z.); liuxj@ncepu.edu.cn (X.L.)

[2] The Department of Electrical and Computer Engineering, Baylor University, Waco, TX 76798-7356, USA; Kwang_Y_Lee@baylor.edu

* Correspondence: kongxiaobing@ncepu.edu.cn; Tel.: +86-10-6177-2103

Received: 10 September 2019; Accepted: 17 October 2019; Published: 23 October 2019

Abstract: The ultra-supercritical (USC) coal-fired boiler-turbine unit has been widely used in modern power plants due to its high efficiency and low emissions. Since it is a typical multivariable system with large inertia, severe nonlinearity, and strong coupling, building an accurate model of the system using traditional identification methods are almost impossible. In this paper, a deep neural network framework using stacked auto-encoders (SAEs) is presented as an effective way to model the USC unit. In the training process of SAE, maximum correntropy is chosen as the loss function, since it can effectively alleviate the influence of the outliers existing in USC unit data. The SAE model is trained and validated using the real-time measurement data generated in the USC unit, and then compared with the traditional multilayer perceptron network. The results show that SAE has superiority both in forecasting the dynamic behavior as well as eliminating the influence of outliers. Therefore, it can be applicable for the simulation analysis of a 1000 MW USC unit.

Keywords: ultra-supercritical unit; deep neural network; stacked auto-encoder; maximum correntropy

1. Introduction

With the fast development of China's economy in the 21st century, the demand for electricity is growing rapidly. Although the installed capacity of renewable energy, such as wind power and solar power, have increased in recent years, coal-fired power generation still accounts for a large proportion of the power generation. In China, the coal-fired installed capacity reached 921.2 GW by the end of 2017, accounting for nearly 72% of the total electricity generation [1]. In the process of coal burning, many air pollutants may be released, e.g., sulfur dioxide (SO_2), nitrogen oxides (NO_X), and carbon dioxide (CO_2), which are extremely dangerous to the global climate [2]. In this case, the Chinese government pledges to reduce the CO_2 emissions per unit of GDP by 60–65% in 2030 compared to 2005 levels [3]. To meet the above requirement, it is an inevitable trend to develop coal-fired power generation technology with large-capacity, low-pollution, and high-efficiency.

At present, most power plant designers are attempting to improve the boiler-turbine efficiency by increasing steam parameters [4]. Thus, ultra-supercritical (USC) coal-fired power plants operating at higher temperature and pressure levels have been gaining increasing attention worldwide. Theoretically, every 20 °C rise in the main steam temperature can result in an approximately 1% increase in efficiency [5]. The cycling heat efficiency of the USC units is up to 49%, which is approximately 10% higher than that of subcritical units. Meanwhile, the release of CO_2 and SO_2 can be reduced by 145 g/kWh and 0.4 g/kWh, respectively [6]. In the past decades, the USC power plants have been greatly promoted in China, with more than one hundred 1000 MW USC units put into operation by the end of 2017.

While the USC units enjoy higher efficiency and lower emissions, they are also more complicated than subcritical units. For instance, there is no obvious boundary between water and steam under the once-through operation, resulting in the strong coupling effect among boiler parameters. In addition, the load-cycling operation of USC units lead to the operating point changing in a wide range, making the nonlinearity of the plant variables even more serious. Due to the high-complexity as well as the nonlinearity, multi-variable, strong-coupling characteristics, the modeling and control of USC units faces greater challenges.

The modeling of power plants can be categorized into two groups: first-principle modeling [7–9] and experimental modeling [10]. A classical nonlinear dynamic model derived from first-principles for a natural circulation 160 MW drum-boiler was presented in [11], which was developed on the basis of several fundamental physical laws. Due to its clear physical structure, this model has been widely used for controller design [12–14]. In [15], a mathematical simulation model was developed to study the stability of a steam boiler drum subjected to all of the possible initial operating conditions, including both stable and unstable. Papers [16,17] present the static and dynamic mathematic model of a supercritical power plant and its application to improve the load changes and start-up processes. In [18], three different flexible dynamic models of the same single-pressure combined-cycle power plant have been successfully developed, and based on these models, an evaluation of the drum lifetime reduction was performed. However, owing to the complexity of the USC unit, it is hardly possible to build an accurate first-principle model, and experimental modeling offers a good framework. In [6], the dynamic model of a 1000 MW power plant was established by combining the experimental modeling approach and the first-principle modeling approach, which can be feasible and applicable for simulation analysis and testing control algorithms. Based on this model, a sliding mode predictive controller was proposed in [19] to achieve excellent load tracking ability under wide-range operation. In [20], this model was further improved with added closed-loop validations and more reasonable structure.

In 1995, Irwin originally developed a feedforward neural network (NN) to model a 200 MW oil-fired and drum-type turbo-generator unit [21]. Due to its practicability and flexibility, NNs become useful tools for power plant modeling [22,23]. In [24], an effective NN modeling method for a steam boiler was proposed: this model maps the influence of flue gas losses and energy losses due to unburned combustibles on the main operational parameters of the boiler. In [25], two separate NN models were developed for the boiler and the steam turbine, which are eventually integrated into a single NN model representing a 210 MW real power plant. Subsequently, some other methods were also introduced in NN, such as fuzzy logic. In [26], Liu et al. firstly presented a model of a USC unit using a fuzzy neural network, the results showing that the model's built-in fuzzy neural network had satisfactory accuracy and performance. In [27], a fuzzy model of the USC unit was firstly developed, and then based on the model, an extended state observer-based model predictive control was proposed. In [28], an improved Takagi-Sugeno fuzzy framework was applied to the modeling of a 1000 MW USC unit, the parameters were identified by a k-means++ algorithm and an improved stochastic gradient algorithm.

During the past decades, computer technology has been widely used in USC power plants. The supervisory information system, which provides comprehensive optimization for the plant's real-time production, collecting all the process data and storing the data in the historical database. These massive datasets are of great value since they can reflect the actual operational condition of the USC unit and embody the unit's complex physical and chemical characteristics. The big data generated in the USC power plant are generally characterized by massiveness, multi-source, heterogeneity, and high-dimension. Developing an advanced modeling technology based on big data is of great significance to the USC unit. Traditional NNs with shallow architecture have low efficiency in digging and extracting effective information from big data, since they often suffer from uncontrolled convergence speed and local optima. Meanwhile, it is more difficult to optimize the parameters of NNs as the number of hidden layers and the training sample size increase.

The deep neural network (DNN), proposed by Hinton et al. in 2006 [29], provides an effective tool to deal with the big data modeling problem. In the DNN, layer-by-layer unsupervised learning is performed for pre-training before the subsequent supervised fine-tuning [30]. The lower layers represent the low-level features from inputs while the upper layers extract the high-level features that explain the input samples. Through layer-wise-greedy-learning, DNNs can effectively extract the compact, hierarchical, and inherently abstract features in the original data and, thus, are able to achieve high-performance modeling with big data. As one of the commonly used DNN architectures, a stacked auto-encoder (SAE) is constructed by stacking several shallow auto-encoders (AE) [31], which learns features by first encoding the input data and then reconstructing it. Due to its remarkable representation ability, SAE has been successfully applied in fault diagnosis [32], electricity price forecasting [33], and wind speed forecasting [34].

During the training procedure of SAE, the mean square error (MSE) has been widely used as the loss function, owing to its simplicity and efficiency. SAE under MSE usually performs well when the training data are not disturbed by outliers. However, in practical application, the dataset obtained from a USC power plant will inevitably contain outliers due to various reasons, which makes the performance of the SAE deteriorate rapidly. Therefore, it is quite important to develop a new loss function. Unlike MSE, maximum correntropy (MC) [35] is a Gaussian-like weighting function, it is a local criterion of similarity and thus can be very useful for cases when the measurement data contains outliers. Since it could attenuate the large error terms effectively, the outliers would have a less impact.

Accordingly, the main contributions of this paper are summarized as follows:

(1) In order to establish an accurate USC unit model using generated big data, SAE is adopted as the DNN model structure in this paper. The SAE model can generalize very well and yield better performance when compared to conventional shallow architectures. The SAE model is concise and suitable for big data analysis.

(2) In order to reduce the bad influence of outliers on the modeling, a loss function using MC is developed in this paper.

The rest of the paper is organized as follows: Section 2 presents a brief description for the USC unit. Section 3 proposes the USC power plant modeling using SAE. The simulation results are given in Section 4. Finally, conclusions are drawn in Section 5.

2. The Ultra-Supercritical Coal-Fired Boiler-Turbine Unit

2.1. Brief Description of USC Unit

The power plant considered in this paper is a pulverized coal firing, once-through steam-boiler generation unit with a power rate of 1000 MW. The maximum steam consumption of the power plant is 2980 T/h with a superheated steam pressure and temperature of 26.15 MPa and 605 °C, respectively. Figure 1 shows the simplified diagram of the USC boiler-turbine unit. The boiler mainly includes the economizer, the waterwall, the separator, the superheater, and the reheater. The tandem compound triple turbine consists of a high-pressure (HP) turbine, an intermediate-pressure (IP) turbine, and a low-pressure (LP) turbine.

As shown in Figure 1, the pulverizing system transforms the raw coal into the pulverized coal so that it can fully burn in the furnace. The feedwater is first warmed by the economizer, and further heated in the waterwall which surrounds the furnace vertically and spirally. Eventually, it turns into steam with high temperature and pressure. There is a separator on top of the furnace, and the steam passes through the separator and superheats in the superheater. The superheater consists of four parts: primary, division, platen, and finish. The turbine governor valve controls the quantity of superheated steam delivered to the HP turbine. The extraction steam from the HP turbine goes to the reheater. The reheated steam is used to drive the IP/LP turbine.

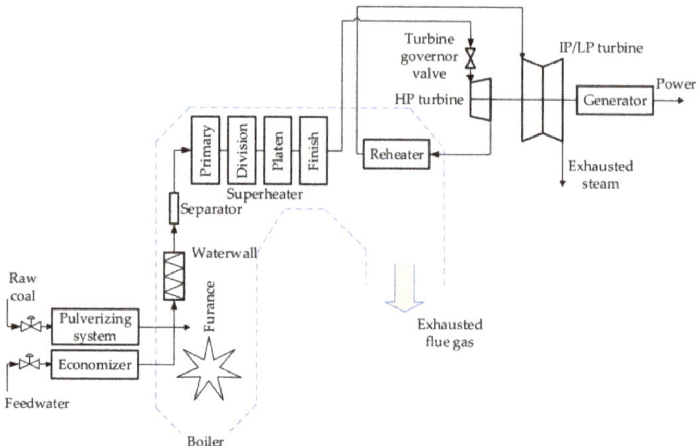

Figure 1. The layout of the 1000 MW boiler-turbine unit.

2.2. Determination of Input-Output Variables

In the once-through operation of the USC unit, there is no obvious boundary between water and steam. The feedwater is continuously heated, evaporated, and superheated from the inlet of the economizer. Without the buffering of the steam drum, the USC unit will suffer greater disturbances than a subcritical unit. This leads to the strong non-linearity and coupling of the USC unit, which can be seen as a complicated system with multiple inputs and multiple outputs.

In order to reduce the impact of external disturbances and simplify the model structure of the USC unit, the following assumptions are made:

1. The fuel flow and the forced draft volume are balanced to ensure the combustion stability.
2. The ratio between the forced draft volume and the induced draft volume remains constant, to ensure that the pressure in the furnace is stable.
3. The control of the main steam temperature is relatively independent.

If the above assumptions are satisfied, the USC unit can be depicted as a three-input, three-output nonlinear system, as shown in Figure 2. The inputs u_1, u_2, u_3 are the fuel flow rate, the turbine governor valve opening, and the feedwater flow rate, respectively. The outputs y_1, y_2, y_3 are the electric power, the main steam pressure, and the separator outlet steam temperature, respectively. The direct correlation property between water and steam causes the strong coupling between the inputs and the outputs.

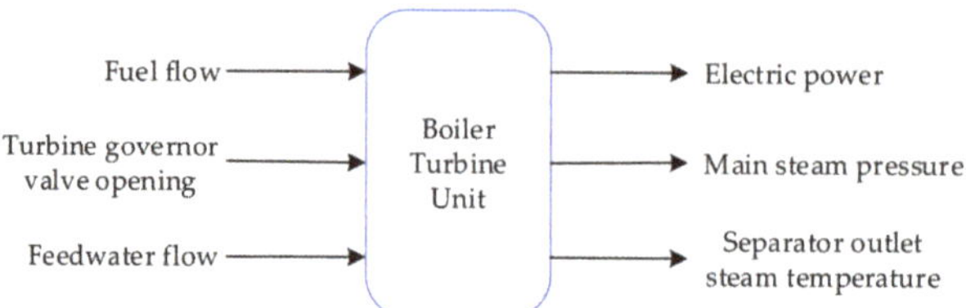

Figure 2. The three-input, three-output system of the USC unit.

3. Stacked Auto-Encoder

The SAE adopts a multi-layer structure, which is hierarchically stacked by a series of AEs, as shown in Figure 3. Denote the kth hidden layer to be h^k, and then the AE associated with h^{k-1} and $h^k (k = 1, 2, \cdots, l)$ is indicated as AE^k.

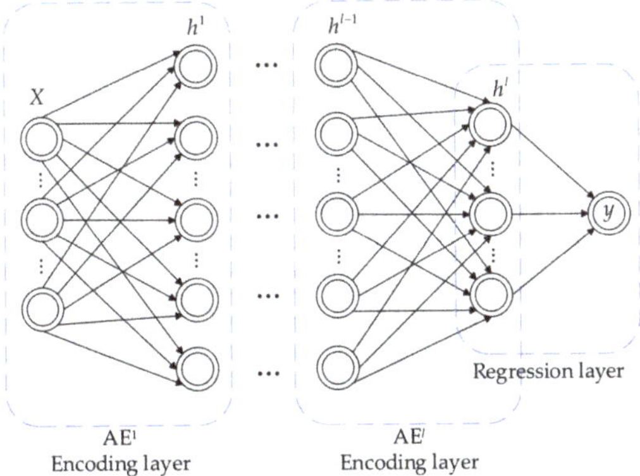

Figure 3. The architecture of the SAE.

3.1. Auto-Encoder

The AE is a one-hidden-layer feedforward NN with an encoder and a decoder. The encoder converts the input data from a high-dimensional representation into a low-dimensional abstract representation. Then the decoder reconstructs the input data from the corresponding codes. The main purpose of the AE is to learn an approximation in the hidden layer so that the input data can be perfectly reconstructed in the output layer. The structure of AE^k is shown in Figure 4.

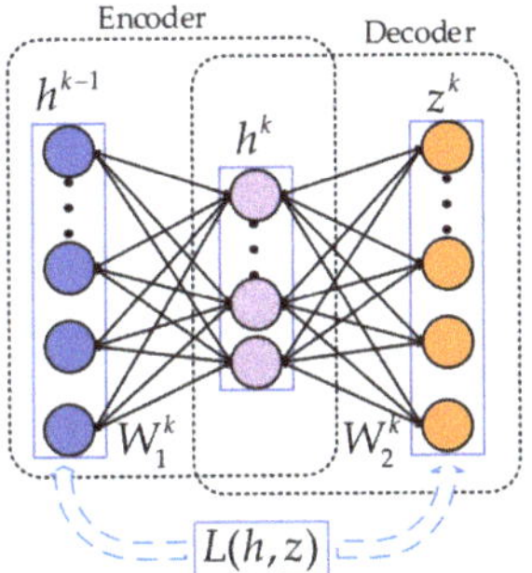

Figure 4. The kth auto-encoder.

Given the input h^{k-1} of AE^k, the hidden representation h^k can be obtained through the encoder based on Equation (1), and then maps back to a reconstructed vector z by the decoder as in Equation (2):

$$h^k = f(W_1^k h^{k-1} + b_1^k),\tag{1}$$

$$z^k = g(W_2^k h^k + b_2^k),\tag{2}$$

where the function $f(x) = g(x) = 1/(1 + e^{-x})$, W_2^k and b_1^k represent the weight matrix and bias term of the encoder, and W_2^k and b_2^k represent the weight matrix and bias term of the decoder, respectively. The parameter set in AE^k is $\theta^k = \{W_1^k, b_1^k, W_2^k, b_2^k\}$.

The parameter set θ^k can be optimized by minimizing the reconstruction error:

$$J_{AE}(\theta^k) = \frac{1}{m}\sum_{i=1}^{m} L(h_i^{k-1}, z_i^k),\tag{3}$$

where m is the sample size and L is the mean square error (MSE) expressed as:

$$L_{AE-MSE}\left(h_i^{k-1}, z_i^k\right) = \frac{1}{2}\left\|h_i^{k-1} - z_i^k\right\|^2, \tag{4}$$

3.2. New Loss Function Design Using Maximum Correntropy

Usually large amounts of operating data are captured continuously by the online plant data acquisition system in the USC power plant. Before using these data for network training, data preprocessing is required, since they will always contain some outliers. The generation of outliers may include faulty sensors, human errors, errors in data capturing system, etc. However, it is very difficult to remove all outliers manually since the sample size is too large.

During the training procedure of AE^k $(k = 1, 2, \cdots, l)$, the MSE is used as the loss function, owing to its simplicity and efficiency. The AE under MSE usually performs well when the training data are not disturbed by outliers. However, when the outliers are mixed within the training data, the performance of the AE under MSE may deteriorate greatly.

Notice that the MSE function is a quadratic function in the joint space with a valley along the $h^{k-1} = z^k$ line. The quadratic term has the net effect of amplifying the contribution of samples which are far away from the $h^{k-1} = z^k$ line, so that the outliers would have a great impact on the normal training of the model.

Unlike MSE, MC [35] uses a Gaussian-like weighting function so that it is a local criterion of similarity and, thus, can be very useful for cases when the measurement data contains large outliers. Since it could attenuate the large error terms effectively, the outliers would have less of an impact. The MC function to be maximized is expressed as:

$$L_{AE-MC}\left(h_i^{k-1}, z_i^k\right) = \frac{1}{d}\sum_{i=1}^{d} K_\sigma\left(h_i^{k-1}, z_i^k\right), \tag{5}$$

where d is the number of the output units, and $K_\sigma(\cdot, \cdot)$ is the Gauss kernel, which is defined as:

$$K_\sigma(a, b) = \frac{1}{\sqrt{2\pi}\sigma}\exp\left(-\frac{(a-b)^2}{2\sigma^2}\right), \tag{6}$$

where σ is the kernel size.

Owing to the effectiveness of MC function, it is chosen as the loss function of each AE in this paper.

3.3. SAE Model Structure and Learning Algorithm

The SAE model can be established by stacking several AEs. Figure 5 shows the structure of the SAE used for USC modeling. Due to the inertia and delay of the system, the historical data of $\{u_1(k), u_2(k), u_3(k), y_1(k-1), y_2(k-1), y_3(k-1)\}$ in the last two steps are also adopted as the inputs of the model. Thus, the total number of inputs and outputs are 18 and three, respectively. In this model, multiple AEs are used to obtain the intrinsic features from the original USC data, while the regression layer is responsible for outputting the expected normal behaviors of the system along the time axis.

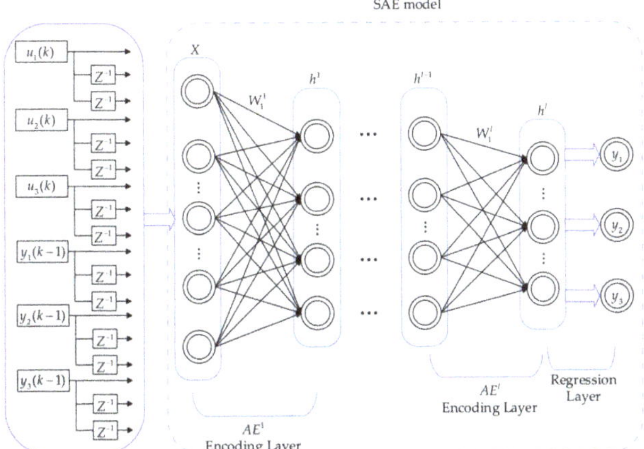

Figure 5. Structure of the SAE used for USC modeling.

The training of the SAE includes two steps: an unsupervised layer-wise pre-training step and a supervised fine-tuning step, as shown in Figure 6. In Figure 6a, with the original training data, the AE at the bottom layer is first trained by minimizing the reconstruction error in Equation (3) using the gradient descent method. Then, the generated hidden representation can be used as the input for training the higher-level AE. In this way, multiple AEs can be stacked hierarchically. After the layer-wise pre-training, all the obtained hidden layers are stacked, and the regression layer can be added on top of the SAE to generate the final outputs, as shown in Figure 6b. The parameters of the whole SAE network can be fine-tuned in a supervised way using the gradient descent method.

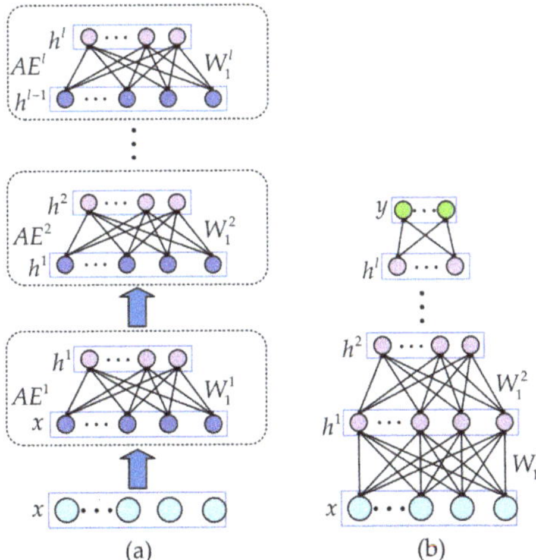

Figure 6. (**a**) Unsupervised layer-wise pre-training of the SAE. (**b**) Supervised fine-tuning of the SAE.

4. USC Unit Modeling

4.1. Experimental Settings

Training of the SAE with a dataset including all possible variations in the range of working conditions is very crucial. The dataset used for training was carefully selected from the very large amount of data logged in the historical database, during which the working condition varies frequently.

Twenty thousand sets of continuous I/O data with 1 s sampling were selected for training, with load changing conditions ranging from 550 MW to 1000 MW. Another 3000 sets of I/O data were selected for validation.

Within the datasets, there exist outliers that need to be removed in advance. In practice, the outliers can be identified in different ways. Usually, the data points which deviate substantially from the general trend of variations of its neighboring points can be considered as outliers. Additionally, the outliers can be found by checking the relationship between trends of data for highly correlated parameters. For example, the increase of fuel flow must correspond to the effect on that of the electric power, with a regular correspondence. By using these method, the detected outliers are listed in Table 1. As for the identified outliers in the dataset, they are replaced by the data from their neighboring points. These datasets used for training and validation, after preprocessing the outliers, are shown in Figure 7. All the preprocessed data are normalized in the range of [0,1] before establishing the SAE model.

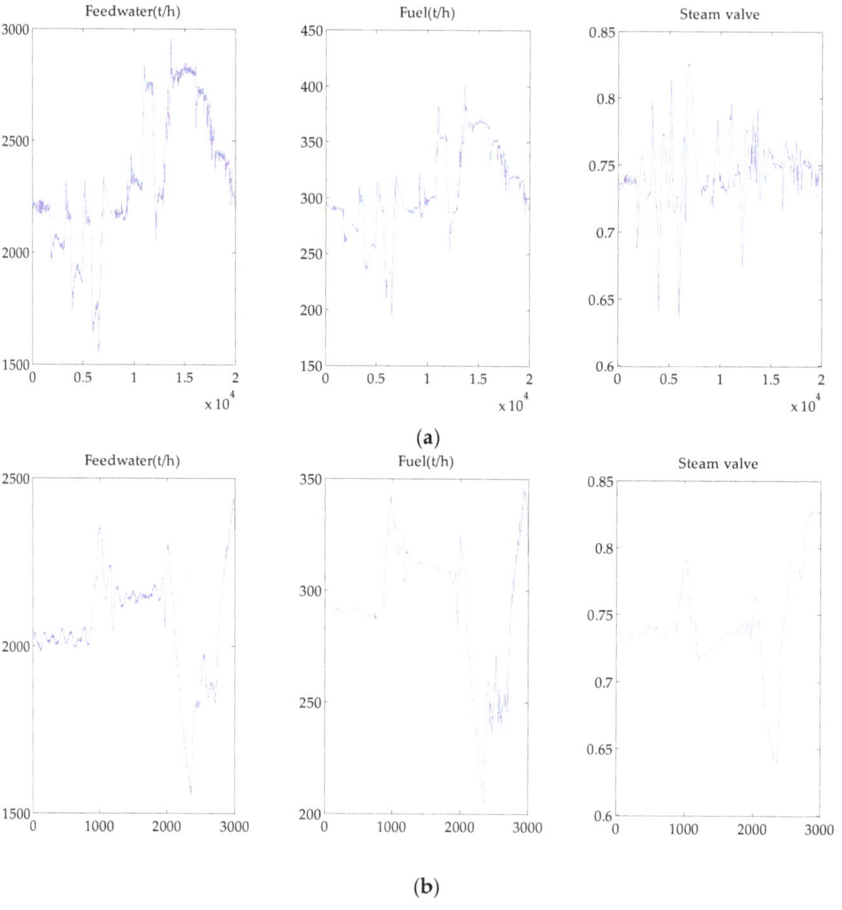

Figure 7. The datasets used for training (**a**) and validating (**b**).

Table 1. The number of outliers in training and validating sample sets.

Parameter	Training Sample Set	Validating Sample Set
u_1	36/20,000	14/3000
u_2	24/20,000	8/3000
u_3	34/20,000	11/3000
y_1	28/20,000	14/3000
y_2	23/20,000	12/3000
y_3	29/20,000	9/3000

The root mean squared error (*RMSE*) in Equation (7) is employed as the evaluation metric:

$$RMSE = \sqrt{\frac{\sum_{k=1}^{K} \left(y_k^* - y_k\right)^2}{K}}, \tag{7}$$

where y^* is the model output and y is the plant output, K is the total data number. Notice that y^* and y are normalized values. There are several parameters that have to be defined for the SAE model, such as the nodes in each layer, the number of AEs, learning rate, momentum, etc. These parameters are determined through cross-validation only on the training set. The initial weights and biases of each AE were chosen to be small random values sampled from a zero-mean Gauss distribution with a standard deviation of 0.01. The maximum number of epochs is set to 100 and the fine-tuning stage terminates when the variation in *RMSE* of the validation set is less than 10^{-3}. This criterion will reduce the model complexity and, thus, result in a better generalization by avoiding overfitting.

In order to determine the optimal structure of SAE model, i.e., the number of AEs and hidden layer units in each AE, experiments were repeatedly done by choosing the number of AEs ranging from 1 to 10, while the number of units in hidden layers from $\varphi = [18, 17, \cdots, 5, 4]$. The optimal structure is found from different configurations considering the *RMSE* value.

The relationship between the number of AEs and the *RMSE* of the learning network is shown in Figure 8. The network is unable to generalize well when the number of AEs is too small because of the insufficient number of tunable parameters in the model. The performance of the network gradually improves as the number of AEs increases, especially when the number reaches 8. However, when the number of AEs increase further, the improvement seems to be very little, as using more AEs will lead to more complex structures that are prone to overfitting. Moreover, the vanishing gradient problem also imposes negative impacts on the fine-tuning of the SAE when the number of AEs increases. Therefore, the network structure of SAE is set to be eight hidden layers.

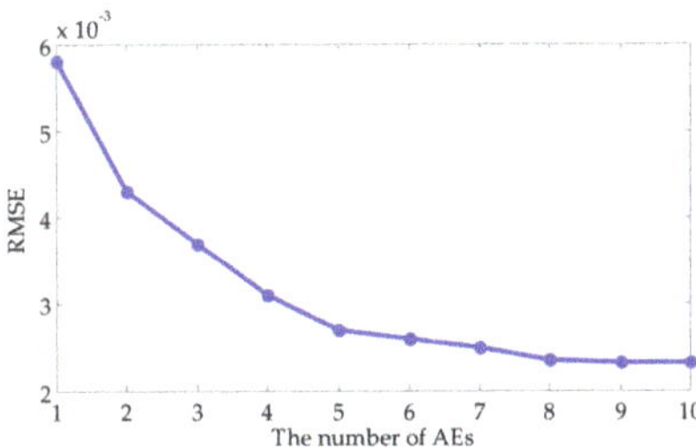

Figure 8. The relationship between the number of AEs and the *RMSE* value.

4.2. The Modeling Results

Figure 9 shows the modeling results with the 20,000 sets of training data. Then, this training model was validated by the 3000 sets of validating data, as shown in Figure 10. When adopting very different sets of operating data, the SAE model is still able to achieve good performance. From both the training and validating, it is clearly seen that the SAE model can predict the USC dynamic accurately over a wide range of loads.

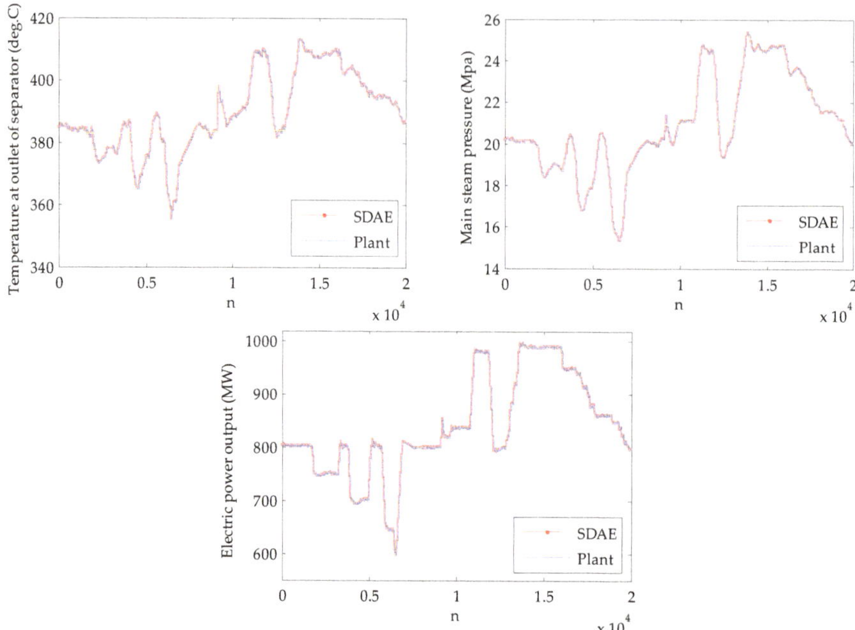

Figure 9. A comparison of the boiler system and the SAE model (training).

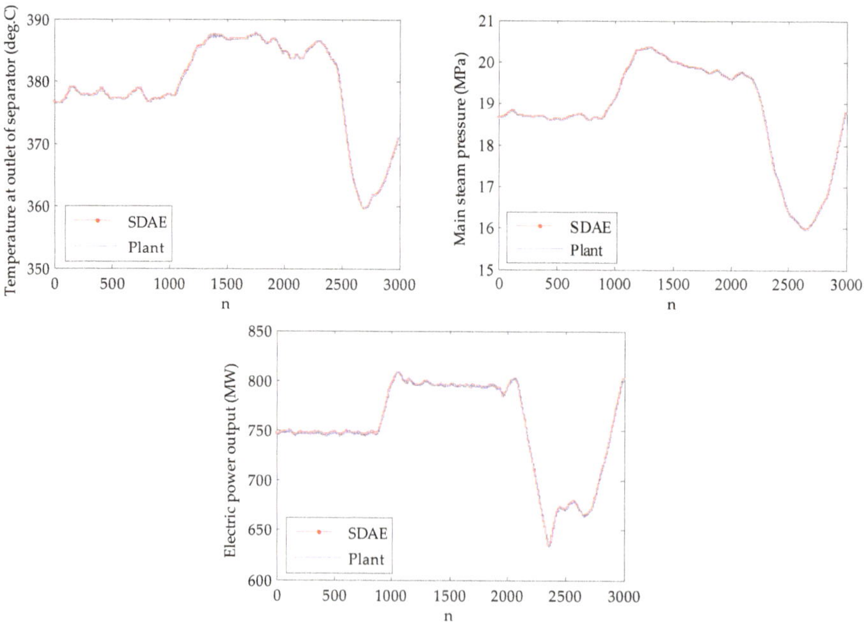

Figure 10. A comparison of the boiler system and the SAE model (validating).

This SAE model was then compared with a general multi-layer perceptron (MLP) network adopting the structure in [26], under the same I/O data. Table 2 lists the comparing results. It is found that the performance of the MLP network is lower than that of the SAE network, as it is a shallow architecture which often suffers from uncontrolled convergence speed and local optimality, especially when the training sample size grows too large.

Table 2. The root mean square errors of three adopted models.

		Temperature	Pressure	Power
MLP	Training	0.0039	0.0022	0.0034
	Validating	0.0076	0.0065	0.0072
SAE	Training	0.0016	0.0007	0.0015
	Validating	0.0031	0.0019	0.0034

4.3. The Modeling Using Maximum Correntropy

As listed in Table 1, there exists outliers in both the training and validating sample sets. In order to disclose the essential influence of these outliers on the modeling effect, the simulation is repeated without the preprocessing process. It can be found that the modeling performance soon deteriorates, as shown by the green line in Figures 11 and 12.

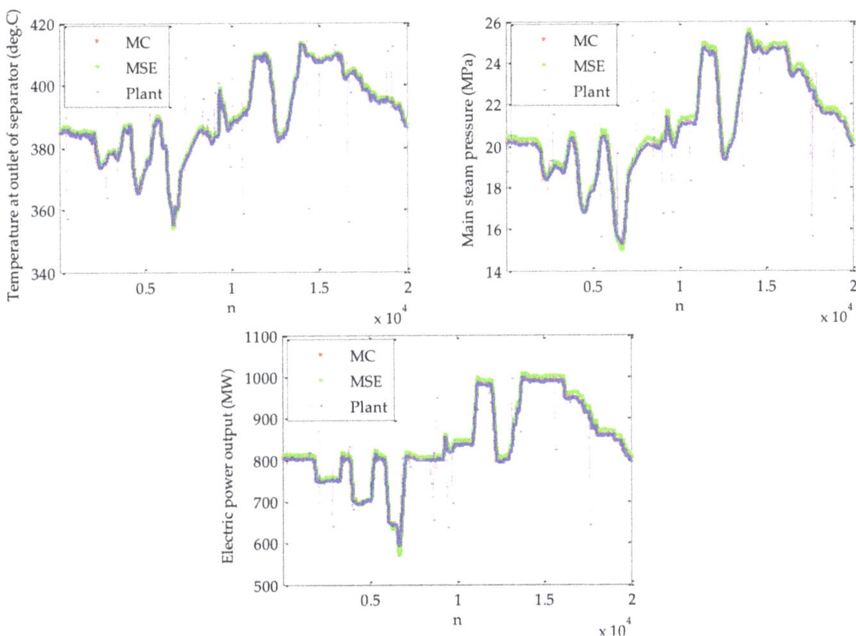

Figure 11. A comparison of SAE models using different loss functions without the preprocessing process (training).

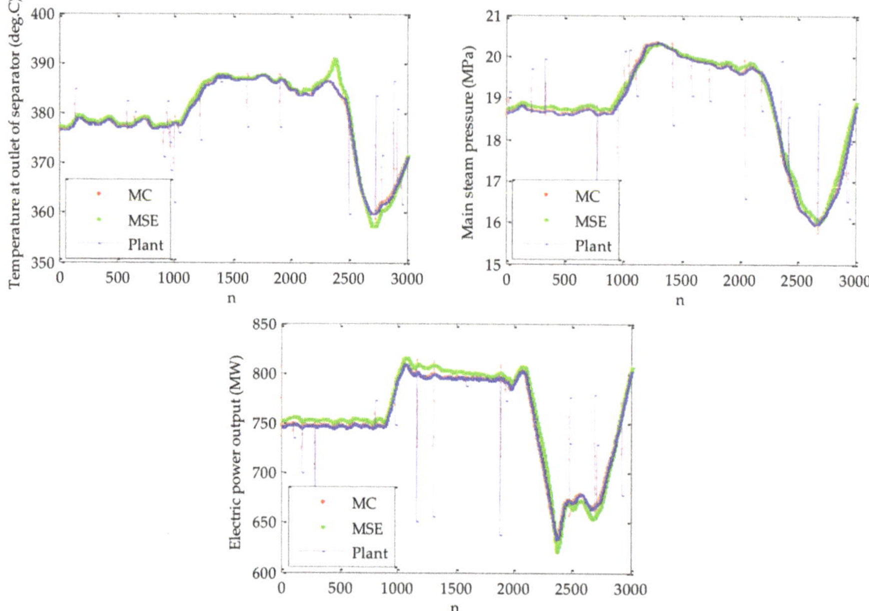

Figure 12. A comparison of SAE models using different loss functions without the preprocessing process (validating).

The simulation using the MC loss function is shown by the red line of Figures 11 and 12. Table 3 lists the *RMSE* values under these two methods, which clearly shows the advantage of the SAE model incorporating the MC function in alleviating the influence of outliers.

Table 3. The root mean square errors of SAE without the preprocessing process.

	RMSE	Temperature	Pressure	Power
MSE	Training	0.0231	0.0301	0.0307
	Validating	0.0529	0.0476	0.0559
MC	Training	0.0185	0.0209	0.0169
	Validating	0.0217	0.0275	0.0248

5. Conclusions

For the modeling of a 1000 MW USC coal-fired boiler-turbine unit, a DNN framework using an SAE was proposed in this paper. Real-time measurement big data generated from a wide range of operating points were used for the network training and validating. To evaluate the effectiveness of the proposed model, a comparative analysis of the SAE and MLP network was constructed. From the results, the following conclusions can be summarized.

(1) Compared with the shallow-layer NN, the DNN architecture adopting the SAE model trained by an unsupervised greedy layer-by-layer pre-training and a supervised fine-tuning is very efficient to deal with the big data modeling problem since it can effectively extract the compact, hierarchical, and inherent abstract features in the original USC unit data through the layer-wise-greedy-learning.

(2) MC is a local criterion of similarity which could attenuate the large error terms effectively. The proposed SAE model adopting MC as the loss function reduces the poor influence of outliers effectively, compared with adopting MSE as the loss function.

In summary, the proposed SAE model can be suitably applied for analyzing the dynamic behaviors of the 1000 MW USC boiler-turbine system.

Author Contributions: Conceptualization: H.Z. and X.L.; methodology: H.Z. and X.K.; project administration: X.L. and X.K.; software: H.Z.; supervision: X.L. and K.Y.L.; writing—original draft: H.Z.; writing—review and editing: X.L., X.K., and K.Y.L.

Funding: This research was funded by the National Natural Science Foundation of China (grant numbers U1709211, 61673171, 61533013, 61603134), and the Fundamental Research Funds for the Central Universities (grant numbers 2017MS033, 2017ZZD004, 2018QN049).

Conflicts of Interest: The authors declare no conflict of interest.

Abbreviations

The following abbreviations are used in this manuscript:

USC	Ultra-supercritical
SAE	Stacked auto-encoder
CO_2	Carbon dioxide
SO_2	Sulfur dioxide
NO_X	Nitrogen oxides
NN	Neural network
DNN	Deep neural network
AE	Auto-encoder
MSE	Mean square error
MC	Maximum correntropy
HP	High-pressure
IP	Intermediate-pressure
LP	Low-pressure
RMSE	Root mean squared error
MLP	Multi-layer perceptron

References

1. National Bureau of Statistics of China. Available online: http://www.stats.gov.cn/tjsj/zxfb/201803/t2018039_1588744.html (accessed on 25 August 2019).
2. Zhao, Y.; Wang, S.; Duan, L.; Lei, Y.; Cao, P.; Hao, J. Primary air pollutant emissions of coal-fired power plants in China: Current status and future prediction. *Atmos. Environ.* **2008**, *42*, 8442–8452. [CrossRef]
3. Zhao, Z.; Chen, Y.; Chang, R. How to stimulate renewable energy power generation effectively? China's incentive approaches and lessons. *Renew. Energy* **2016**, *92*, 147–156. [CrossRef]
4. Wang, C.; Zhao, Y.; Liu, M.; Qiao, Y.; Chong, D.; Yan, J. Peak shaving operational optimization of supercritical coal-fired power plants by revising control strategy for water-fuel ratio. *Appl. Energy* **2018**, *216*, 212–223. [CrossRef]
5. János, M.B. High efficiency electric power generation: The environmental role. *Prog. Energy Combust.* **2007**, *33*, 107–134.
6. Liu, J.; Yan, S.; Zeng, D.; Hu, Y.; Lv, Y. A dynamic model used for controller design of a coal fired once-through boiler-turbine unit. *Energy* **2015**, *93*, 2069–2078. [CrossRef]
7. Alobaid, F.; Mertens, N.; Starkloff, R.; Lanz, T.; Heinze, C.; Epple, B. Progress in dynamic simulation of thermal power plants. *Prog. Energy Combust.* **2016**, *59*, 79–162. [CrossRef]
8. Sun, L.; Hua, Q.; Li, D.; Pan, L.; Xue, Y.; Lee, K.Y. Direct energy balance based active disturbance rejection control for coal-fired power plant. *ISA Trans.* **2017**, *70*, 486–493. [CrossRef]
9. Sun, L.; Li, D.; Lee, K.Y.; Xue, Y. Control-oriented modeling and analysis of direct energy balance in coal-fired boiler-turbine unit. *Control Eng. Pract.* **2016**, *55*, 38–55. [CrossRef]
10. Kocaarslan, I.; Cam, E. Experimental modeling and simulation with adaptive control of power plant. *Energy Convers. Manag.* **2007**, *48*, 787–796. [CrossRef]
11. Åström, K.J.; Bell, R.D. Simple Drum-Boiler Models. In Proceedings of the IFAC Power Systems Modeling and Control Applications, Brussels, Belgium, 5–8 September 1988; Volume 21, pp. 123–127.

12. Liu, X.; Guan, P.; Chan, C. Nonlinear multivariable power plant coordinate control by constrained predictive scheme. *IEEE Trans. Control Syst. Technol.* **2010**, *18*, 1116–1125. [CrossRef]

13. Wu, X.; Shen, J.; Li, Y.; Lee, K.Y. Data-driven modeling and predictive control for boiler-turbine unit. *IEEE Trans. Energy Convers.* **2013**, *28*, 470–481. [CrossRef]

14. Moradi, H.; Alasty, A.; Vossoughi, G. Nonlinear dynamics and control of bifurcation to regulate the performance of a boiler-turbine unit. *Energy Convers. Manag.* **2013**, *68*, 105–113. [CrossRef]

15. Bracco, S.; Troilo, M.; Trucco, A. A simple dynamic model and stability analysis of a steam boiler drum. *J. Power Energy* **2009**, *223*, 809–820. [CrossRef]

16. Alobaid, F.; Postler, R.; Ströhle, J.; Epple, B.; Gee, K.H. Modeling and investigation start-up procedures of a combined cycle power plant. *Appl. Energy* **2008**, *85*, 1173–1189. [CrossRef]

17. Alobaid, R.; Ströhle, J.; Epple, B.; Gee, K.H. Dynamic simulation of a supercritical once-through heat recovery steam generator during load changes and start-up procedures. *Appl. Energy* **2009**, *86*, 1274–1282. [CrossRef]

18. Benato, A.; Bracco, S.; Stoppato, A.; Mirandola, A. Dynamic simulation of combined cycle power plant cycling in the electricity market. *Energy Convers. Manag.* **2016**, *107*, 76–85. [CrossRef]

19. Tian, Z.; Yuan, J.; Zhang, X.; Kong, L.; Wang, J. Modeling and sliding mode predictive control of the ultra-supercritical boiler-turbine system with uncertainties and input constraints. *ISA Trans.* **2018**, *76*, 43–56. [CrossRef]

20. Fan, H.; Zhang, Y.; Su, Z.; Wang, B. A dynamic mathematical model of an ultra-supercritical coal fired once-through boiler-turbine unit. *Appl. Energy* **2017**, *189*, 654–666. [CrossRef]

21. Irwin, G.; Brown, M.; Hogg, B.; Swidenbank, E. Neural network modeling of a 200 MW boiler system. *IEE Proc. Control Theory Appl.* **1995**, *142*, 529–536. [CrossRef]

22. Suresh, M.V.J.J.; Reddy, K.S.; Kolar, A.K. ANN-GA based optimization of a high ash coal-fired supercritical power plant. *Appl. Energy* **2011**, *88*, 4867–4873. [CrossRef]

23. Lu, S.; Hogg, B.W. Dynamic nonlinear modeling of power plant by physical principles and neural networks. *Int. J. Electr. Power Energy Syst.* **2000**, *22*, 67–78. [CrossRef]

24. Rusinowski, H.; Stanek, W. Neural modelling of steam boilers. *Energy Convers. Manag.* **2007**, *48*, 2802–2809. [CrossRef]

25. Smrekar, J.; Pandit, D.; Fast, M.; Assadi, M.; De, S. Prediction of power output of a coal-fired power plant by artificial neural network. *Neural Comput. Appl.* **2010**, *19*, 725–740. [CrossRef]

26. Liu, X.; Kong, X.; Hou, G.; Wang, J. Modeling of a 1000 MW power plant ultra super-critical boiler system using fuzzy-neural network methods. *Energy Convers. Manag.* **2013**, *65*, 518–527. [CrossRef]

27. Zhang, F.; Wu, X.; Shen, J. Extended state observer based fuzzy model predictive control for ultra-supercritical boiler-turbine unit. *Appl. Therm. Eng.* **2017**, *118*, 90–100. [CrossRef]

28. Hou, G.; Du, H.; Yang, Y.; Huang, C.; Zhang, J. Coordinated control system modeling of ultra-supercritical unit based on a new T-S fuzzy structure. *ISA Trans.* **2018**, *74*, 120–133. [CrossRef]

29. Hinton, G.E.; Salakhutdinov, R.R. Reducing the dimensionality of data with neural networks. *Science* **2006**, *313*, 504–507. [CrossRef]

30. Bengio, Y.; Lamblin, P.; Popovici, D.; Larochelle, H. Greedy layer-wise training of deep networks. In Proceedings of the 20th Annual Conference on Neural Information Processing Systems, Vancouver, BC, Canada, 3–6 December 2006; Volume 19, pp. 153–160.

31. Schmidhuber, J. Deep learning in neural networks: An overview. *Neural Netw.* **2015**, *61*, 85–117. [CrossRef]

32. Sun, W.; Shao, S.; Zhao, R.; Yan, R.; Zhang, X.; Chen, X. A sparse auto-encoder-based deep neural network approach for induction motor faults classification. *Measurement* **2016**, *89*, 171–178. [CrossRef]

33. Wang, L.; Zhang, Z.; Chen, J. Short-term electricity price forecasting with stacked denoising autoencoders. *IEEE Trans. Power Syst.* **2016**, *32*, 2673–2681. [CrossRef]

34. Khodayar, M.; Kaynak, O.; Khodayar, M.E. Rough deep neural architecture for short-term wind speed forecasting. *IEEE Trans. Ind. Inform.* **2017**, *13*, 2770–2779. [CrossRef]

35. Liu, W.; Pokharel, P.P.; Principe, J.C. Correntropy: Properties and applications in non-Gaussian signal processing. *IEEE Trans. Signal Process.* **2007**, *55*, 5286–5298. [CrossRef]

Article

Modelling and Analysis of Plate Heat Exchangers for Flexible District Heating Systems

Serafym Chyhryn

Department of Electrical Engineering, Technical University of Denmark, 2800 Lyngby, Denmark; serchy@elektro.dtu.dk; Tel.: +45-5355-0185

Received: 11 September 2019; Accepted: 27 October 2019; Published: 30 October 2019

Abstract: Seamless integration of district heating (DH) and power systems implies their flexible operation, which extends their typical operational boundaries and, thus, affects performance of key components, such as plate heat exchangers (PHXs). Despite that the heat transfer in a PHX is regulated by mass flows, flexible operation and demand variations cause shifts in temperature levels, which affects the system operation and must be efficiently accounted for. In this paper, an overall heat transfer coefficient (OHTC) model with direct relation to temperature is proposed. The model is based on a linear approximation of thermophysical components of the forced convection coefficient (FCC). On one hand, it allows to account for temperature variations as compared to mass flow-based models, thus, improving accuracy. On the other hand, it does not involve iterative lookup of thermophysical properties and requires fewer inputs, hence, reducing computational effort. The proposed linear model is experimentally verified on a laboratory PHX against estimated correlations for FCC. A practical estimation procedure is proposed based on component data. Additionally, binding the correlation to one of varying parameters shows reduction in the heat transfer error. Finally, operational optimization test cases for a basic DH system demonstrate better performance of the proposed models as compared to those previously used.

Keywords: heat exchanger; forced convection; film coefficient; heat transfer; water properties; integrated energy system; operational optimization

---

1. Introduction

Ongoing energy systems integration requires additional flexibility from district heating (DH) to enhance accommodation of intermittent energy sources. However, DH systems have operational constraints imposed by components. Mathematical models of these components are used in optimization procedures to achieve a minimal operation cost at a specific quality of heat supply, reflected in sufficient consumer temperature levels. Provision of flexibility to power system will entail greater fluctuations in temperatures and mass flows, thus, affecting operational boundaries and the efficiency of the system. Heat exchangers (predominantly of the plate type) are the key components in DH systems operation, allowing controllable heat exchange between transmission and distribution pipelines. Since return temperatures in a transmission system affect heat source efficiency and distribution, supply temperatures reflect the quality of the heat supply, and effective modelling of plate heat exchangers (PHXs) becomes crucial.

Models for PHXs can be classified on how estimation of the overall heat transfer coefficient (OHTC) is addressed. Models for DH operational optimization are considered in several works, for instance, in [1] an OHTC is modelled as dependent on mass flows and compared to its constant version, while the focus is on dynamic optimization of supply temperatures and optimal load distribution. In [2], a mass flows model is described to obtain a solution for the primary return temperature with the further focus on minimization of the heat loss and pumping cost, whereas the thermophysical component is also constant. In a more recent study [3], the coefficient is assumed constant, whilst the goal is to minimize the heat loss and pumping cost in a steady state. The problem of optimal combined power and heat dispatch is addressed in [4], involving dynamic variations of both temperatures and mass flows in a district heating network to provide flexibility to a power system utilizing a pipeline for energy storage. However, physical limitations of heat transfer in heat exchangers are not considered and the end nodes are merely represented by their heat demands. A dynamic simulation study was conducted in [5] to evaluate PHXs performance in a ring DH network, where the heat transfer was modelled dynamically and OHTC calculated using thermophysical properties. Besides system level studies, in [6] the coefficient model includes recalculated thermophysical components for the purpose of stability analysis in a DH substation. The same model was used in [7] for robust control design, as well as in [8] for a proportional-integral control design. In [9], performance of a PHX in a small geothermal heating system was analysed based on a thermophysical model, while a mass flow-only model was concluded to be more appropriate for the system conditions. Finally, in [10] OHTC was modelled in order to predict the temperature of cooling water. Temperature dependence of OHTC was accounted for and the thermophysical part was recalculated iteratively. Additionally, estimation of all correlation values was performed by constrained nonlinear optimization.

Since modern DH systems are needed to operate with variable temperatures and mass flows to provide flexibility, temperature variations must be accounted for in OHTC models in order to achieve better performance of the PHX control. The existing OHTC models require lookup of thermophysical properties to account for those variations, thus, increasing computational effort, especially when operational optimization is performed. A direct relation with temperature in an OHTC model can improve its computational efficiency, especially when addressing flexible DH system problems, such as described in Section 2.

In this paper, it is shown that OHTC can be directly related to temperature, as forced convection coefficients (FCCs) of water have a temperature dependence close to linear. This allows approximating the FCCs directly from temperature, without involving thermophysical properties. The approximation is described in Section 3 and then experimentally verified in Section 4 against previously used models on a laboratory PHX. Additionally, a simplified estimation procedure of a better fitting correlation for a particular PHX is proposed. Performance of the proposed models is demonstrated in Section 5 on optimization test cases for simple DH systems, where the aim is to find such hot circuit mass flow, supply and return temperatures that would yield the minimum electric power consumed. Besides the DH sector, the obtained results can be useful in other applications, for instance, wastewater heat recovery.

2. Flexible District Heating Systems and the Main Operational Problem

Operation and design of a flexible DH system, as well as the main operational problem description are given in this section. A flexible DH system is defined in the first subsection as capable of providing auxiliary services to the power system by varying temperatures and mass flows. This capability is essential for DH system functioning within an integrated energy system as a whole and limited by physical constraints from system components, which are reflected in their models. Mathematical formulation of the objective function and constraints for the nonlinear optimization problem, which contain component models, are given in the second subsection.

2.1. Description of a Flexible District Heating System in an Integrated Energy System Context

District heating combines heat sources, heat transmission systems and heat distribution systems to which end consumers are connected. DH coupling with an electric power system in an integrated energy system context is shown in Figure 1.

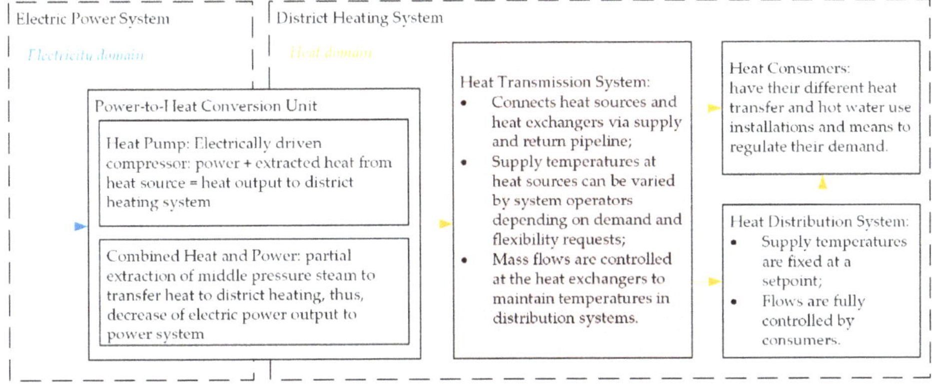

Figure 1. Integrated energy system including district heating (DH) and power systems.

The transmission system is hydraulically decoupled from each distribution system by a heat exchanger, typically of plate type and counter-current flow arrangement. This decoupling is necessary due to the need to control temperature at sources depending on the expected demand, hydraulic stability, too high temperature of supply water, etc. [11]. The system with such a decoupling allows greater flexibility in the transmission part, meaning that the electric power consumption can be altered to provide auxiliary services to electric power system, while satisfying DH system operational constraints and the heat demand.

In the distribution system, end consumers regulate mass flows according to their demands. At the time, their supply temperature is kept constant via flow control on the transmission side of the heat substation. The setpoint of the temperature depends on the distribution system design and preferences of the end consumers, as indicated in [12]. The return temperature level reflects the behaviour of end consumers and the distribution network. In the case of diverse consumption patterns, end equipment and complex distribution networks, this temperature is challenging to predict, unless the distribution part is fully modelled. Numerous works contribute to solving this problem, for instance, in [13] DH temperature dynamics are addressed and modelling of an existing DH system was performed.

Besides demand changes, deviations in the secondary supply temperature are caused by changes in the primary inlet temperature. These are caused by output temperature changes at the heat source or fluctuations of heat loss in the transmission pipeline due to the mass flow control. All the described processes consequently affect the primary return temperature. Badly optimized supply temperatures will lead to frequent flow control actions in order to maintain the necessary temperature level in the distribution network. Such processes can yield in unpredictable behaviour of the system and cause additional wear on components.

Both supply and return temperatures, as well as the mass flow affect in a non-linear fashion the efficiency of different power-to-heat sources, such as large heat pump (HP) systems or combined heat and power plants (CHP). For electrically driven HPs the ratio of heat output ($\dot{Q}$) to electric power input (P_{hp}) is commonly considered, called the total coefficient of performance (COP_t). Given in its simplest form, it is sufficient to express the nonlinearity of the process [14]:

$$COP_t = \frac{\dot{Q}}{P_{hp}} = \frac{c_p \dot{m}(T_s - T_r)}{P_{hp}} = \frac{\frac{T_s + T_r}{2}\eta_{hp}}{\left(\frac{T_s + T_r}{2} - T_0\right)}, \tag{1}$$

noting that the temperatures in Equation (1) must be used in the absolute scale (Kelvin).

The power consumption of circulation pumps is significantly lesser than that of heat sources. Their output is controlled against the pressure changes caused by flow control valves at substations. Nevertheless, their power consumption contributes to operational costs and is approximately related to the mass flow rate cubed. Pumping power required for low temperature systems is usually higher, as the supply temperature level at the end consumer has to be maintained.

One of the key problems in DH operation is to find such optimal temperatures and mass flows for the transmission part of the system. This is an operational optimization problem. To formulate the problem mathematically, the following components are the most relevant to model: heat sources, circulation pumps, supply and return pipelines, as well as plate heat exchangers.

2.2. Operational Optimization Problem

Operational optimization in district heating is often regarded as a time-dependent process due to delays (which comprise the flow history) in the network. Moreover, the return temperature is not only affected by delay and demand, but also defined by the past supply temperature. This makes the process of optimization computationally heavy when the mass flow, as well as both supply and return temperatures, are optimized in short time steps (dictated by flexibility provision with power-to-heat solutions) over a certain time horizon.

For the sake of simplicity, in test cases the demand and the price for the electric power are assumed as constants, so the problem transforms into minimization of consumption power in a steady state. This power is divided into power for hot water production (P_{hp}) and pumping power (P_{cirp}), which together form the objective function. The optimized variables are supply and return temperatures at the source and the transmission mass flow. All three of those variables are essential to include in the optimization as they are essential to quantify available power for system flexibility.

The objective function for the optimization problem is formed as

$$\min(P) = P_{hp} + P_{cirp}. \tag{2}$$

The power required to produce heat can be obtained from the expression for coefficient of performance (1) as

$$P_{hp} = \frac{\dot{Q}}{COP_t} = \frac{c_p \dot{m}_h (T_s - T_r)(T_s + T_r - 2T_0)}{(T_s + T_r)\eta_{hp}}. \tag{3}$$

Power of the circulation pump is the sum of pressure drops over components of the system, including the PHX, supply and return pipes, as well as the condenser of the HP. For the sake of simplicity, the return pipe is assumed to have the same pressure drop as the supply and the condenser is assumed to have the same pressure drop as in the PHX, consisting of port and channel pressure drops. A general expression given as follows:

$$P_{cirp} = 2\frac{\dot{m}_h}{\eta_{cirp}\rho_w}(\Delta p_{pipe} + \Delta p_{ch} + \Delta p_{port}). \tag{4}$$

The pressure drop in pipes is obtained via the Darcy-Weisbach equation, where the friction factor for pipes is determined by an approximation given in [15]. Equations for the pressure drop in a circuit of PHX are found in [16]. Thus, Equation (4) can be re-written as follows:

$$P_{cirp} = 2\frac{\dot{m}_h}{\eta_{cirp}\rho_w}\left(f\frac{8L\dot{m}_h^2}{\pi^2 D^5 \rho_w} + f_{ch}\frac{2L_{ch}}{D_{hydr}\rho_w}\left(\frac{\dot{m}_h}{A_{ch}}\right)^2 + 1.4\frac{8\dot{m}_h^2}{\pi^2 D_{port}^4 \rho_w}\right). \tag{5}$$

The objective is subjected to constraints, which include maximum and minimum limits for the mass flow and temperatures, as well as heat transfer capability of the heat exchanger and the energy balance between primary and secondary sides:

$$\min(\dot{m}_h) \leq \dot{m}_h \leq \max(\dot{m}_h)$$
$$\min(T_s) \leq T_s \leq \max(T_s)$$
$$\min(T_r) \leq T_r \leq \max(T_r)$$
$$UAT_{LMTD} - \dot{Q}_d = UA\frac{(T_{hi}-T_{co})-(T_{ho}-T_{ci})}{\ln\left(\frac{T_{hi}-T_{co}}{T_{ho}-T_{ci}}\right)} - \dot{Q}_d = 0,$$
$$(T_{hi} - T_{ho})\dot{m}_h c_{ph} - \dot{Q}_d = 0. \tag{6}$$

The constraints indicating minimum and maximum limits of optimized mass flow and temperatures are of linear inequality type. The non-linear equality constraints represent the transferred heat and the heat balance between the sides. The inlet and outlet temperatures of the hot circuit are related to the optimized supply and return temperatures at the heat source via the equation for temperature drop in the pipe, which also includes the mass flow as a part of time delay calculation [1,2] as follows:

$$T_{hi} = T_{amb} + (T_s - T_{amb}) \cdot e^{-K_{pipe} \cdot \frac{L}{\dot{m}_h c_{ps}}},$$
$$T_{ho} = T_{amb} + (T_r - T_{amb}) \cdot e^{K_{pipe} \cdot \frac{L}{\dot{m}_h c_{pr}}}. \tag{7}$$

Thermophysical properties, especially viscosity, vary depending on the temperature of water. These variations cause changes in OHTC U, which must be efficiently accounted for. An inaccurate OHTC model can result in non-optimal system operation and frequent control actions, which can further cause temperature and flow oscillations in the system. Thus, detailed modelling and analysis of this coefficient is given in the following section.

3. Modelling and Analysis of Heat Transfer in Plate Heat Exchangers

Modelling and analysis of OHTC is essential to identify models, relevant for the application. The first subsection defines two models, with the difference in whether both cold and hot circuits are treated separately or not, resulting in four approaches to OHTC calculation, two iterative and two with constant thermophysical properties. The second subsection gives analysis on FCC of a circuit, which shows that the thermophysical component can be related directly with temperature. Finally, the third subsection describes the two new models based on results of FCC analysis.

3.1. Modeling of Overall Heat Transfer Coefficient

The calculation of OHTC comprises FCCs from each circuit (H_c and H_h) and thermal resistances of plate material and fouling. A complete equation can be written as [16]:

$$U = \frac{1}{\frac{1}{H_h} + \frac{1}{H_c} + R_{tot}}. \tag{8}$$

Forced convection coefficients H_c and H_h are related to mass flows and mean temperatures (via thermophysical properties) in cold and hot circuits, respectively. Using a generalized form of the Dittus-Boelter equation for the correlation of Nusselt number, the expression for the convection coefficient [16] is

$$H = \frac{Nu \cdot k}{D_{hydr}} = \frac{CRe^n Pr^m k}{D_{hydr}} = \frac{C\left(\frac{\dot{m}D_{hydr}}{A_{ch}\mu}\right)^n \left(\frac{c_p\mu}{k}\right)^m k}{D_{hydr}}. \tag{9}$$

Different values for m, n and C are used depending on plate design and prevailing flow conditions. Typically, transition to turbulence in PHX occurs at a low Reynold number. Even at low mass flow rate, the Reynolds number is quite high.

Similar to [6], after rearranging Equation (9) to get temperature dependent and mass flow dependent parts separated, the expression becomes

$$H = \frac{C}{A_{ch}^{n} D_{hydr}^{1-n}} \left(\mu^{-n+m} c_p{}^m k^{1-m} \right) \dot{m}^n = K \cdot B \cdot \dot{m}^n,$$ (10)

where

$$B = \mu^{-n+m} c_p{}^m k^{1-m}.$$ (11)

Following the stated, OHTC Equation (8) can be re-written as

$$U = \frac{K \cdot B_h \left(\dot{m}_h \right)^n B_c \left(\dot{m}_c \right)^n}{B_h \left(\dot{m}_h \right)^n + B_c \left(\dot{m}_c \right)^n + K \cdot B_h \left(\dot{m}_h \right)^n B_c \left(\dot{m}_c \right)^n R_{tot}}.$$ (12)

Consequently, if thermophysical properties are assumed to be found for the mean temperature between circuits ("coupling" their mean temperatures into one), Equation (12) turns into

$$U = \frac{K \cdot B \left(\dot{m}_h \right)^n \left(\dot{m}_c \right)^n}{\left(\dot{m}_c \right)^{-n} + \left(\dot{m}_h \right)^{-n} + K \cdot B \cdot \left(\dot{m}_c \right)^{-n} \left(\dot{m}_h \right)^{-n} R_{tot}}.$$ (13)

Both Equations (12) and (13) can be used iteratively in the optimization problem. Therefore, (12) and (13) are yielding four approaches for calculation:

- Non-iterative coupled based on (13): approach 1.1;
- Iterative coupled based on (13): approach 1.2;
- Non-iterative decoupled based on (12): approach 2.1;
- Iterative decoupled based on (12): approach 2.2.

Approaches 1.1 and 2.1, applied in a district heating context, do not account for temperature changes in the system and define B for the most common temperatures. approaches 1.2 and 2.2 will adjust U iteratively for every temperature change. Lookup tables are used to link temperatures with thermophysical properties, which are further complicating the calculation procedure. It is also worth noting that the resistance component can often be neglected as in [1].

Direct relation of FCC to a temperature would be more convenient to use. To obtain it, the temperature dependent part (denoted as B) is investigated to use temperature directly instead of thermophysical properties.

3.2. Temperature Dependency of Forced Convection Coefficient

Given the values of the thermophysical properties for water [17] and applicable values of n and m [18] we can evaluate the function B from temperature. The behavior of B for $m = 0.4$ and $n = (0.6, 0.7, 0.8, 0.9)$ is demonstrated in Figure 2. Note that thermophysical properties remain nearly constant for the same temperature for any possible pressure in the DH system (the values are taken for $p = 1$ atm).

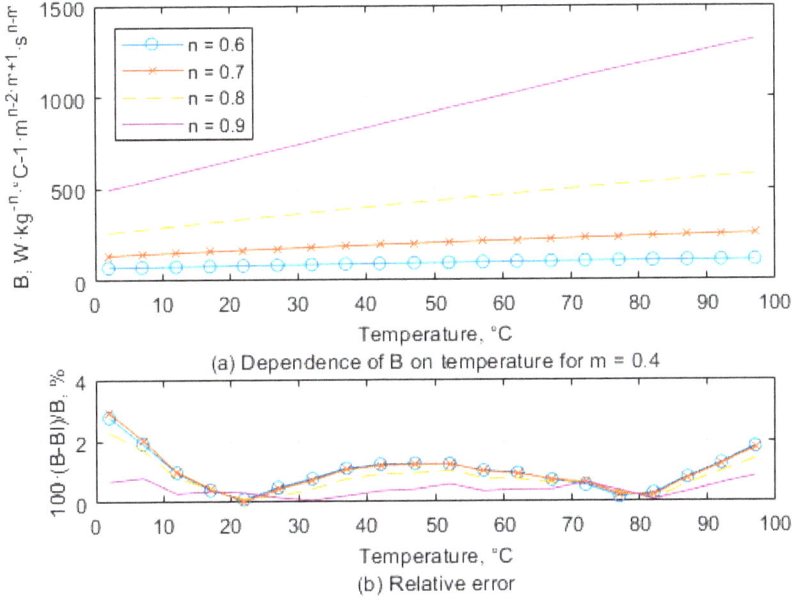

(a) Dependence of B on temperature for m = 0.4

(b) Relative error

Figure 2. Temperature dependent component of the forced convection coefficient (FCC) (**a**) and the relative error of its linear fit (**b**).

As it can be seen from the Figure 2, the behaviour of function B is very close to linear. Linear regression of B can be applied:

$$B_l = \alpha + \beta T. \tag{14}$$

A similar relation for water is described, for instance, in [19] using the original Dittus–Boelter equation for the case of flow in a circular pipe, where $n = 0.8$ and $m = 0.4$. Linearization coefficients α and β for different n and m combinations are summarized in the Table 1.

Table 1. Linearization coefficients for different m and n.

n	$m = 0.3$		$m = 0.4$		$m = 0.5$	
	α	β	α	β	α	β
0.6	57.27	0.538	73.84	0.435	94.39	0.290
0.7	107.08	1.436	139.45	1.263	179.65	1.099
0.8	197.82	3.666	260.98	3.390	339.53	2.963
0.9	359.62	9.082	482.70	8.683	636.00	7.999

To characterize the goodness of the fit, coefficient of determination R^2 can be found for each approximation (summarized in Table 2) as

$$R^2 = 1 - \frac{\sum_{i=1}^{20}\left(B_l - B\right)^2}{\sum_{i=1}^{20}\left(B - \overline{B}\right)^2}. \tag{15}$$

The R^2 values confirm that the approximation is very close to the original as they are approaching 1, especially for higher values of n.

Table 2. The relative error and R^2 for different n and m.

n	$m = 0.3$		$m = 0.4$		$m = 0.5$	
	mean e, %	R^2	mean e, %	R^2	mean e, %	R^2
0.6	1.025	0.9962	0.9918	0.9928	0.715	0.9889
0.7	0.810	0.9985	0.9924	0.9962	0.938	0.9931
0.8	0.419	0.9997	0.7945	0.9985	0.958	0.9963
0.9	0.496	0.9998	0.4143	0.9997	0.778	0.9985

Thus, it is reasonable to consider that the FCC depends linearly on temperature, while relation with the mass flow is the power of n, which gives us the following expression:

$$H = K\dot{m}^{n}B_l = K\dot{m}^{n}(\alpha + \beta T). \tag{16}$$

This expression can be used for the FCC estimation using temperature directly, without considering thermophysical properties.

The FCC of a circuit to be found as mean of inlet and outlet coefficients, which in their turn are calculated using inlet and outlet temperatures or directly from the mean temperature of the circuit.

3.3. Resulting Models for Overall Heat Transfer Coefficient

The obtained Equation (16) for the direct relation of FCC with temperature, mapped through a fixed PHX design (represented by K, α and β coefficients) allows transforming temperature decoupled (12) and coupled models (13), respectively:

$$U = \frac{K(\dot{m}_h)^{n}(\dot{m}_c)^{n}(\alpha + \beta T_c)(\alpha + \beta T_h)}{((\dot{m}_c)^{n}(\alpha + \beta T_c) + (\dot{m}_h)^{n}(\alpha + \beta T_h) + K(\dot{m}_h)^{n}(\dot{m}_c)^{n}(\alpha + \beta T_c)(\alpha + \beta T_h)R_{tot})}, \tag{17}$$

$$U = \frac{K(\dot{m}_h)^{n}(\dot{m}_c)^{n}(\alpha + \beta T_m)}{((\dot{m}_c)^{n} + (\dot{m}_h)^{n} + K(\dot{m}_h)^{n}(\dot{m}_c)^{n}(\alpha + \beta T_m)R_{tot})}, \tag{18}$$

where

$$T_h = \left(\frac{T_{hi} + T_{ho}}{2}\right), \quad T_c = \left(\frac{T_{ci} + T_{co}}{2}\right), \quad T_m = \left(\frac{T_{ci} + T_{co} + T_{hi} + T_{ho}}{4}\right)$$

Therefore, two approaches in addition to abovementioned four (1.1, 1.2, 2.1, 2.2) can be outlined:

- Coupled linear (18): approach 1.3;
- Decoupled linear (17): approach 2.3.

Since the correlations used for the coefficients are approximate in their nature, they perform differently from one system to another. All the described and obtained heat transfer models require experimental verification before applying them in practice. Such verification is provided in the following section.

4. Experimental Verification of Heat Transfer Models

In this section focus is brought to the laboratory PHX as the main component of the setup and the subject of the study. The first subsection describes the setup, measurement sets and datasheet parameters, further continued to the estimation procedure for the correlation and its results in the second subsection. The verification of the models is provided in the third subsection and the simplest means to reduce the correlation error are given in the last subsection.

4.1. Experimental Setup, Parameters and Measured Data

The experimental setup (Figure 3) consists of a laboratory PHX (Figure 4) of the counter-flow type, operating in a small heat substation. Hot water in the primary circuit is supplied from the "oil stove" heat source and the heat load (on the secondary side) is represented by two small houses, as well as two small and one large air coolers (dump loads).

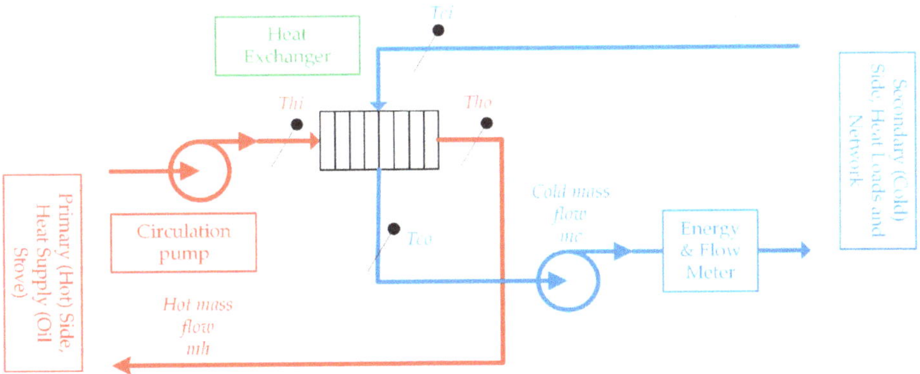

Figure 3. Layout of the experimental setup.

Figure 4. Small laboratory plate heat exchanger (PHX).

PHX parameters, given in the datasheet (manufacturer SPXFlow APV) along with that calculated [16] and found in the manufacturer's handbook [18] are summarized in Table 3.

Since manufacturers typically do not provide exact values for n, m and C, they have to be estimated within the given range, based on datasheet values and measurements. These include the heat supplied (Figure 5) and all four temperatures and mass flow in the cold circuit (Figure 6), collected by a Kamstrup Multical 602 meter in one-second intervals. Temperatures were measured by thermocouples 745690-J001 (Iron/Constantan).

Table 3. PHX parameters.

Parameters from Manufacturer's Datasheet		Calculated and Handbook Parameters	
Effective area, m^2	1.127	Enlargement factor (calculated)	1.0678
Number of plates	65	Plate gap, mm (calculated)	1.422
Plate thickness, mm	0.4	Fouling resistance, $m^2 \cdot K/W$	8×10^{-6}
Plate material	SS AISI 316	C (range)	0.15–0.45
Plate conductivity, W/m·K	16.3	n (range)	0.65–0.85
Port radius, mm	19.05 (3/4 in)	m (range)	0.3–0.45
Plate (chevron) angle, degree	30		
Plate pack length, mm	117		
Horizontal port distance, mm	54		
Vertical port distance, mm	220		
Design thermal power, kW	25		

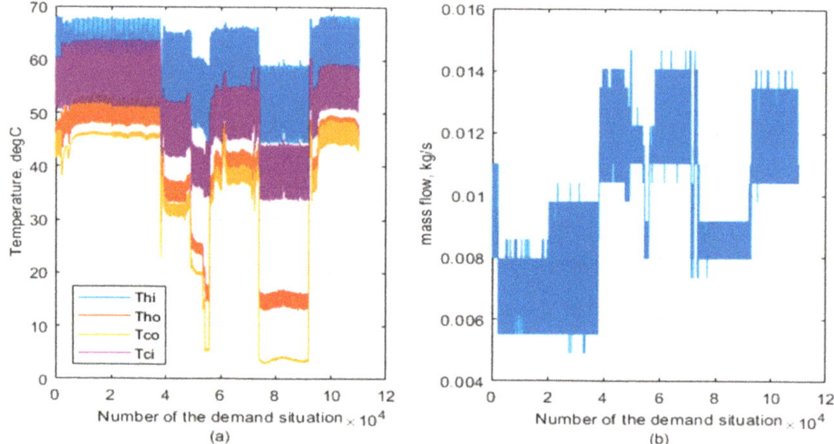

Figure 5. Measured temperatures (**a**) and mass flow (**b**) for all the demand situations (concatenated).

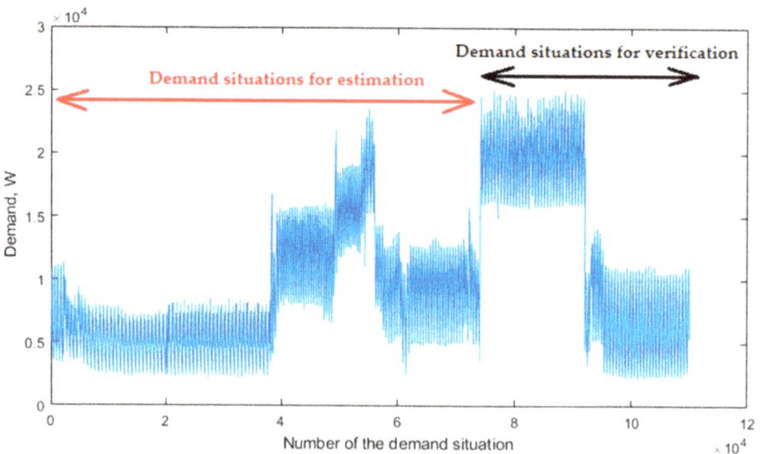

Figure 6. Demand values for all the demand situations (concatenated).

The whole data consists of five datasets, each obtained over approximately five-hour intervals on a separate day. The datasets were concatenated together. Every time step of the data (1 s) corresponds to a different heat demand situation, and every situation lasts one second and is independent from the other.

The oscillatory behaviour is caused by the heat source operation due to features of the system and crude control. One period corresponds to travel time of water through the primary circuit of the PHX, pipeline and heat source, and can last up to 10 min. The travel time of the secondary circuit (transient) can last from 10 to 30 min, or even more as the circuits of the houses are the longer ones and the circuitry of coolers is the shorter one. The very low secondary inlet (primary outlet) temperature is due to all loads operating during a cold day together with the most powerful cooler.

For estimation purposes, measurements from Datasets 1–3 are used, and for verification purposes—from Datasets 4 and 5.

4.2. Estimation of the Correlation Values

The estimation procedure involves setting up a nonlinear constrained optimization problem (Figure 7). It was concluded previously that n has a greater effect on the calculation process than the other two parameters [9]. It also can be seen from Equation (10) that m does not affect the mass flow contribution to heat transfer calculation, while C is a multiplier for both mass flow and thermophysical components. Based on that, we can simplify the procedure by specifying them as constant for all measured sets, i.e., $C = 0.15$ and $m = 0.375$. Thus, the problem is limited to finding optimal n. A more general and complex approach is used in [10], where all three values are optimized.

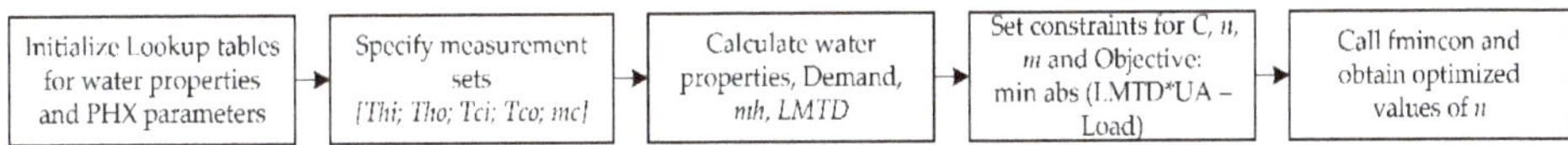

Figure 7. Estimation procedure to obtain optimal n for each demand situation.

For that purpose, the MATLAB (R2017b, MathWorks, Natick, MA, USA) function *fmincon* [20] was used and the objective is to minimize the difference between the transferred heat calculated by Model (12) and the measured demand.

Resulting optimal values of n for all demand situations are displayed on Figure 8.

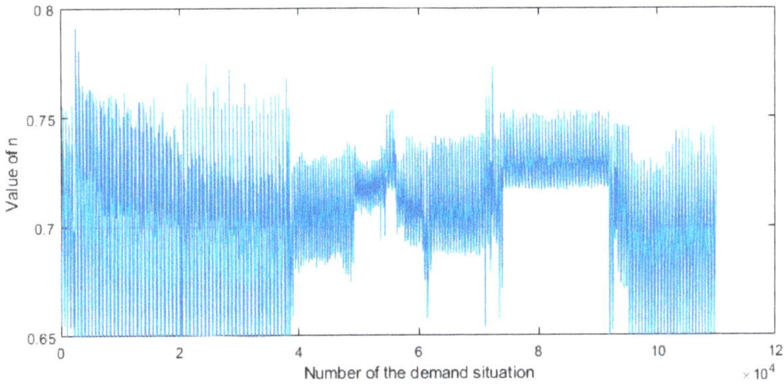

Figure 8. Optimized values of n.

Values of n vary from 0.65 to 0.78 and the weighted mean for estimation datasets is $n = 0.71$, which corresponds to $\beta = 1.4465$ and $\alpha = 138.9041$ after the regression. Those will be used in the following subsection for model verification.

4.3. Verification of Models

To verify the models, the relative errors calculated by each approach of transferred heat power were compared to the measured one on the secondary side, which is the actual demand. In terms of the relative error,

$$q = \frac{Q_{transf<Model>} - Q_{demand}}{Q_{demand}} \cdot 100\% \tag{19}$$

Performance of each approach in terms of a relative error is shown in Figure 9, using the demand situations from the verification datasets.

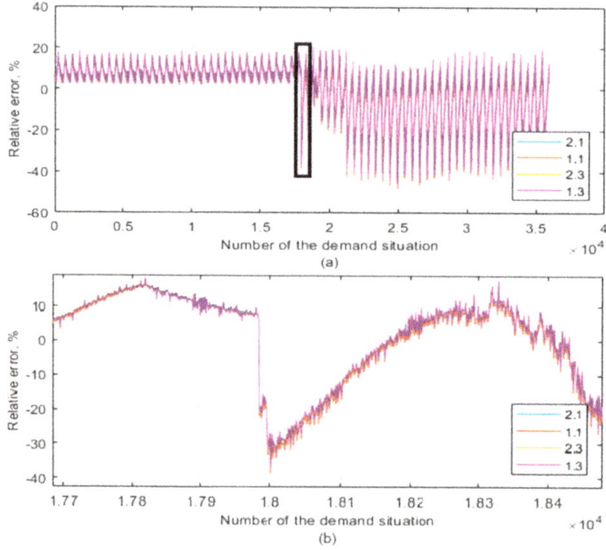

Figure 9. Relative error of calculated transferred heat for each model: (**a**) whole range and (**b**) zoomed region.

The calculated transferred heat differs from the actual values for the following reasons:

- The correlation is an approximation of the complex process of heat transfer;
- Measurements themselves have smalls error;
- Variable time delay between inlets and outlets in both circuits.

The relative errors between the models almost do not differ:

- Thermophysical approach 2.1, a mean temperature for each circuit separately—10.91%;
- Thermophysical approach 1.1, one mean temperature for both circuits—10.63%;
- Linear regression approach 2.3, a mean temperature for each circuit separately—10.65%;
- Linear regression approach 1.3, one mean temperature for both circuits—10.54%.

A negligible difference between all models confirms that linear regression models can be used in practice.

Despite the error being quite high, in general, it can be reduced by collecting more data to cover the whole demand range as well as setting up a "cleaner" experiment overall, e.g., well-tuned controls and properly sized equipment. This is, unfortunately, impossible for the current system due to limited additional demand options, time of the year and continuous operation during certain season.

Further error reduction under existing conditions, without significant complications of the models, can be achieved by changing the value n depending on one of the varying parameters.

4.4. Variable Correlation as a Way to Improve Accuracy

One way to improve accuracy is to further relate n with one or more of varying parameters:

- One of mass flows when temperature variations are not significant;
- One of circuit temperatures when mass flows variations are not significant;
- Mass flow and temperature on one side when parameters on another side vary correspondingly;
- Heat load itself, in case most of temperatures and flows are affected by fluctuations.

Since the system behaviour is quite unstable and all parameters are affected by large variations, the value n is related to the heat demand. To relate n with the transferred heat, let us first re-arrange the obtained values of n from Datasets 1–3 correspondingly to values of the heat load in an ascending order, as shown in Figure 10.

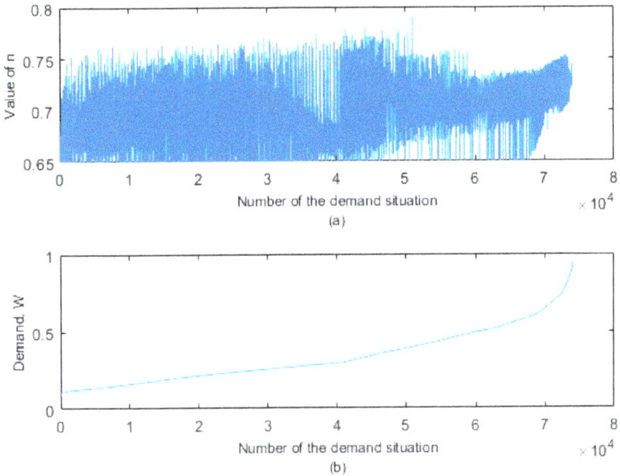

Figure 10. Re-arranged values of n (**a**) according to respective demand in ascending order (**b**).

Similar to the previous section, linear regression (since simplified solutions are sought) of n from Q will result in the expression

$$n_l = 0.0396 \cdot \frac{Q}{Q_{max}} + 0.6899 = n_{l2} \cdot \frac{Q}{Q_{max}} + n_{l1}. \tag{20}$$

Consequently, values α and β can be tied to demand for this particular PHX to enable the use of regression models. Using five points from min (20%) to max (100%) heat load in the expression (20), the respective α and β are summarized in the Table 4.

Table 4. Calculated coefficients varying with the heat demand.

n	Q	α	β
0.6978	20%	128.91	1.285
0.7057	40%	135.49	1.391
0.7137	60%	142.40	1.504
0.7216	80%	149.66	1.627
0.7295	100%	157.27	1.759

This correlation is shown to be nearly linear as we can see from Figure 11.

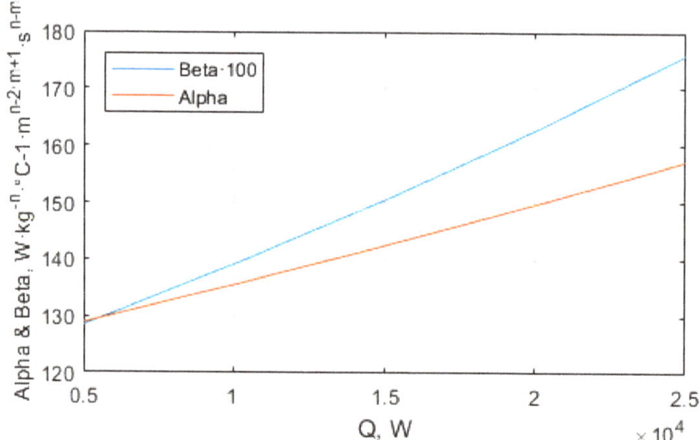

Figure 11. Alpha and beta are changing almost linearly with the demand.

A further linear regression will yield

$$\beta_l = 0.5924 \cdot \frac{Q}{Q_{max}} + 1.1578 = \beta_{l2} \cdot \frac{Q}{Q_{max}} + \beta_{l1},$$
$$\alpha_l = 35.4464 \cdot \frac{Q}{Q_{max}} + 121.4771 = \alpha_{l2} \cdot \frac{Q}{Q_{max}} + \alpha_{l1}. \tag{21}$$

Since Equation (20) is a very "specific" approximation, as well as derived Equation (21), the fit will obviously be poor; however, the relative error for models with those expressions is evaluated and displayed Figure 12 along with errors from previous models.

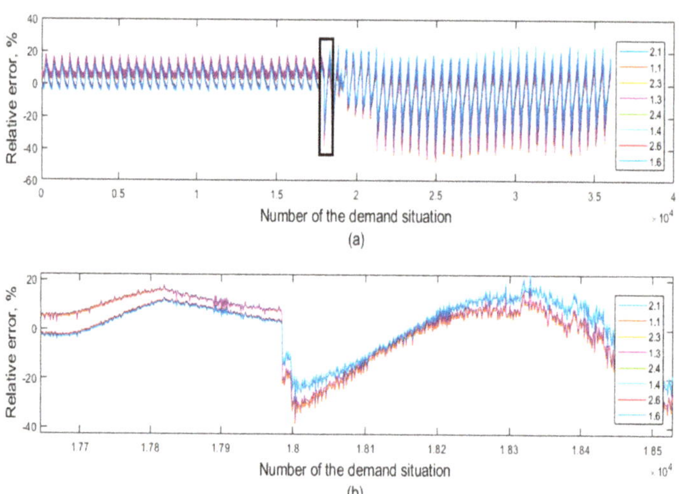

Figure 12. Relative error of calculated transferred heat for each model: (**a**) whole range and (**b**) zoomed region.

The mean relative errors for the variable exponent model with linear regression:

- Thermophysical approach 2.4, a mean temperature for each circuit separately—7.40%;
- Thermophysical approach 1.4, one mean temperature for both circuits—7.45%;
- Linear regression approach 2.6, a mean temperature for each circuit separately—7.67%;

-    Linear regression approach 1.6, one mean temperature for both circuits—7.61%.

For the variable exponent model with thermophysical properties the error is approximately the same. The iterated versions of approaches 1.4 and 2.4 are denoted as 1.5 and 2.5, respectively.

Mean values of relative errors show that varying the correlation exponent n with a certain parameter (in our case it is heat load) can yield better accuracy. The typical error interval for correlations is ±10%.

The ways of improving heat transfer models remain an open research area and the proposed model is situational. To find the most accurate model, extensive tests are required, covering the entire range for each varying parameter.

5. Results and Discussion on an Operational Optimization Test Case

To demonstrate performance of the obtained models on DH operation, an operational optimization (2) and (6) was performed in a basic heat transmission system under several demand situations. In the first subsection, parameters, configurations and heat demand situations were described. The second subsection presented optimization results, containing values from the reference approach and absolute errors of other approaches as compared to the reference one. The obtained results were followed by a discussion. Finally, the computational performance of each approach is evaluated in the last subsection by optimizing the system with an increasing number of branches.

5.1. System Parameters of the Test Case

The diagram of the test system is shown in Figure 13. In the first configuration, the PHX is connected to the source via a pipeline and in the second configuration, two identical PHXs are connected via identical separate pipelines and circulation pumps.

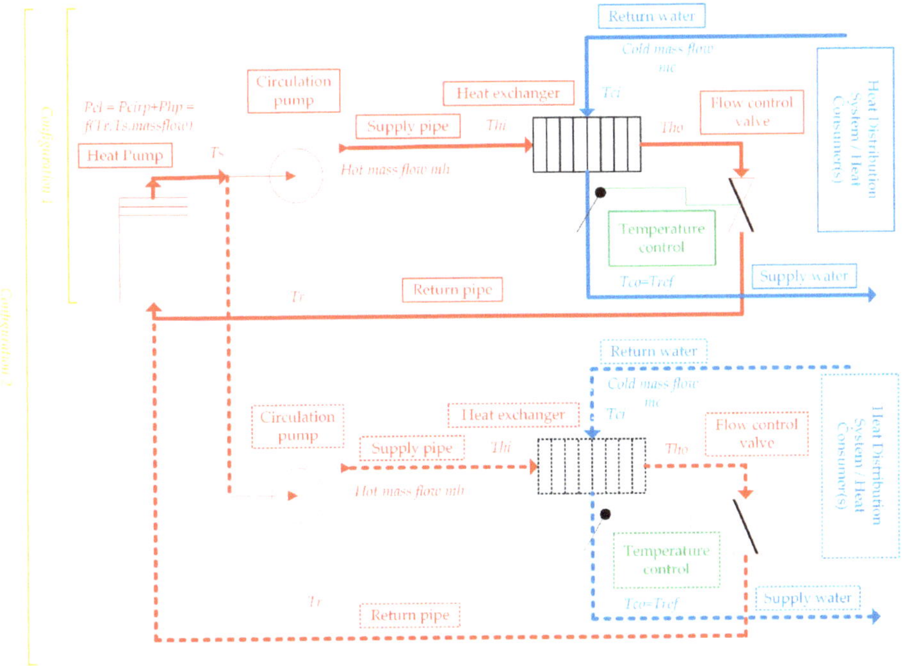

Figure 13. HP-pipeline-PHX heat transmission test system.

Parameters of the test system are listed in the Table 5. The supply and return pipes are connecting the inlet and outlet of a (each) PHX. A large HP system (power-to-heat source) is extracting heat from the heat source at 6.85 °C (for instance, sea or lake water). The maximum mass flow in all circuits is equally limited (this corresponds to full valve opening). For any PHX in both configurations return cold inlet temperatures and cold mass flows are assumed to get three values as indicated in Table 5 to comprise a heat demand (thermal load).

Table 5. Parameters for components.

Plate Heat Exchangers, Mass Flows and Heat Demand Parameters							
L_{ch}, m	W_{pl}, m	b, m	N	T_{co}, °C	Enlargement Factor	Fouling, m^2·K/W	Conductivity, W/m·K(Plate)
1.5	0.7	0.004	141	50	1.12	0.000006	16.3
Thickness, m (plates)	$\dot{m}_{cLOW}$, kg/s	$\dot{m}_{cCOM}$, kg/s	$\dot{m}_{cmax}$, kg/s	$\dot{m}_{hmax}$, kg/s	T_{ciLOW}, °C	T_{ciHIGH}, °C	T_{ciCOM}, °C
0.0005	20	37	74	74	15	35	25

Pipelines				
L, m	D, m	K_{pipe}, W/°C·m	T_{amb}, °C	k_{rough}, mm
1000	0.25	0.51	8	0.5

Heat Pump and Circulation Pumps		
T_0, (°C)	η_{hp}	η_{cirp}
6.85	0.7	0.75

Thus, five cases of heat demand are considered for a PHX in Configuration 1:

-   Common heat demand, when temperature is medium, as well as mass flow;
-   (2 cases) unbalanced heat demand, when secondary return temperatures and mass flows are both low and high, respectively;
-   (2 cases) low or high heat demand, which are low mass flow and high temperature or high mass flow and low temperature, respectively.

As for Configuration 2, in order to observe how PHX models perform when several components are present, the following five heat demand cases are considered:

-   Common heat demand at both PHXs;
-   (2 cases) unbalanced heat demand with low or high temperatures/flows at one PHX and common at another;
-   unbalanced heat demand at each PHX, low temperature at one and low mass flow at another;
-   low heat demand at one PHX and high heat demand at another.

The optimization is performed by each approach, given in previous sections. Since it has been demonstrated that the variable correlation model is the most accurate, it is used as a reference. All the correlation values are taken the same as in previous section and applied for a larger PHX.

Firstly, approach 2.5 is run for the common demand and obtained temperatures are used to find mean temperatures $(T_{ci} + T_{co} + T_{hi} + T_{ho})/4$ and $(T_{hi} + T_{ho})/2$ (defining thermophysical properties) for approaches that do not account for temperature variation. This will yield better performance for them.

5.2. Optimization Results

Optimization procedure is performed using the *fmincon* function in MATLAB. The procedure follows the problem, described in Section 2. We will consider approach 2.5 as our reference since it was shown that thermophysical models with variable correlation are performing better than others. Optimization results for Configurations 1 and 2 are shown in Table 6.

Other models' approaches are compared in relation to the reference. Two key indicators are the absolute deviation from the secondary (cold) supply temperature setpoint $T_{co} = 50$, assumed ideally coinciding with approach 2.5, and the absolute deviation of total electric power consumed as compared to the one for approach 2.5. Obviously, the deviation of the cold outlet (secondary) temperature will cause a respective deviation of the hot outlet and, moreover, control actions to bring the temperature back to 50 °C (in case of a poorly performing approach) will cause a hot return temperature to further rise and, consequently, affect total electric power consumed.

Procedure of temperature deviation calculation for other approaches related to approach 2.5:

1. Calculating the OHTC using the complete Equation (12) from provided optimization results, cold side temperatures and mass flow;
2. Finding the logarithmic mean temperature difference from the given demand, surface area and the OHTC;
3. Finally, finding such outlet temperatures, which yield equality of the expression for logarithmic mean temperature difference to the found one.

For convenience, two plots are used per one indicator per configuration: for variable correlation approaches and for a fixed correlation. Thus, for Configuration 1 with one PHX and one source there will be four plots total, displayed in Figure 14.

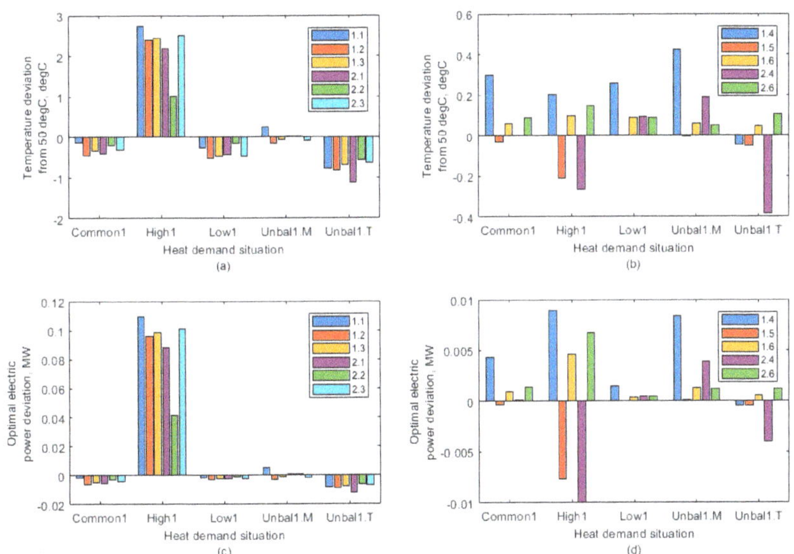

Figure 14. Configuration 1: Temperature (**a,b**) and total power (**c,d**) deviations.

As we can see from the Figure 14, the non-variable correlation approaches perform poorly at the second load situation (highest demand), overheating the water by up to 2.5 °C (at least 1 °C for iterative approach 2.2) and drawing up to 100 kW more of electric power. Performance at temperature unbalanced (high cold return temperature/low flow) heat demand is better, but still the temperature deviation is between 0.5 and 1 °C, this time under the setpoint. For common and low heat demands, the temperature deviation is within 0.5 °C, and for mass flow and unbalanced heat demand its deviations are minimal.

Table 6. Optimization results with a reference approach: Five demand situations per configuration.

Configuration	Heat Demand Type	Electric Power (Objective), MW	Demand Power, MW	$\dot{m}_c$, kg/s	(op)$\dot{m}_h$, kg/s	(op)T_{hi}, °C	(op)T_{ho}, °C	T_{ci}, °C
1	Common1	0.7255	3.8647	37	41.16	58.86	36.4	25
	High1	2.0668	10.821	74	74	66.11	31.13	15
	Unbal1.T	0.5070	2.9246	20	24.28	58.84	30.02	15
	Unbal1.m	0.9430	4.6376	74	60.6	60.01	41.7	35
	Low1	0.2397	1.2534	20	19.51	55.48	40.12	35
2	Common21	1.451	3.8647	37	41.16	58.86	36.4	25
			3.8647	37	41.16	58.86	36.4	25
	Common22	0.9679	3.8647	37	44.26	57.96	37.07	25
			1.2534	20	16.02	57.71	38.99	35
	Common23	2.8108	3.8647	37	27.81	65.95	32.71	25
			10.821	74	74	66.11	31.13	15
	Unbal2.	1.4723	2.9246	20	23.16	59.58	29.38	15
			4.6376	74	61.85	59.75	41.82	35
	LowHigh2	2.3247	10.821	74	74	66.11	31.13	15
			1.2534	20	10.48	65.53	36.94	35

All the non-variable correlation approaches perform inconsistently, but approach 2.2 is more robust due to being a non-variable version of the reference approach 2.5.

Variable correlation approaches perform significantly better than their non-variable versions. The maximum temperature deviation is 0.4 °C by approach 1.4 at Unbal.1M heat demand. For 1.4 and 2.4 the temperature dependence is not included (thermophysical properties are constant, similar to approaches 1.1 and 2.1) and they perform generally worse.

Amongst other approaches, iterative 1.5 performs better, but has a 0.2 °C deviation at high load, whilst approaches 1.6 and 2.6 are more consistent. The highest deviation in electric power consumption among all five is 10 kW by approach 2.4 at high heat demand (corresponds to around a −0.27 °C temperature deviation).

For Configuration 2, there are six plots total, displayed in Figure 15. This study case is demonstrating how approaches perform in a system. Four of them are temperature deviations at two PHXs and two are the deviations of total electric power consumption.

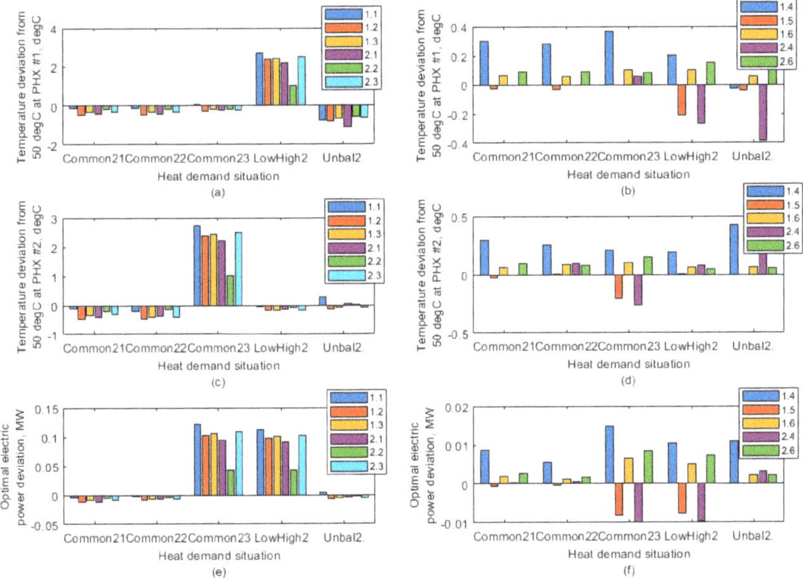

Figure 15. Configuration 2: Temperature (**a–d**) and total power (**e,f**) deviations.

For non-variable correlation approaches, the temperature and power consumption deviations look similar when both PHX are at common heat demand, also for Configuration 1. Similar to Configuration 1. the trend exhibits PHX (#2) at an unbalanced heat demand with a high temperature difference, which was even amplified by another PHX (#2), unbalanced the other way around and having a low temperature difference but high mass flow on its secondary side. Highest deviations are occurring at high heat demands, at least at one of the PHXs. The magnitude of the deviation is approximately the same as in Configuration 1, with approach 2.2 performing better than other approaches.

For variable correlation approaches, the pattern is similar to Configuration 1, with 1.6 and 2.6 performing the best, 1.4 performing the worst, and 1.5 and 2.4 are being inconsistent.

Generally, the magnitude of errors in Configuration 2 is slightly higher than in Configuration 1, for unbalanced heat demands in particular. Relative errors would be nearly of the same values. Nevertheless, in a real DH system all the pipeline and pumping system, as well as secondary side values and PHXs themselves can differ a lot to one another. Note, those are study aspects (number of which can be dramatically high) for every particular system and will certainly affect approaches performance.

5.3. Computational Performance

Another important criterion (if not the most) is computation time for each approach used. This is becoming crucial when the system is optimized on a short time scale over a greater time horizon (the travel times remain very large). Therefore, the faster we are able to optimize our system, the more flexible it becomes for integration with other energy systems.

We will calculate computation time for each approach versus the system with a number of identical branches with PHX at the end, having the same heat demand for each.

The times can be evaluated in MATLAB using tic-toc commands around the optimization code for each approach. The obtained computation times trend is displayed in Figure 16: (a) for non-variable correlation approaches and (b) for variable correlation approaches.

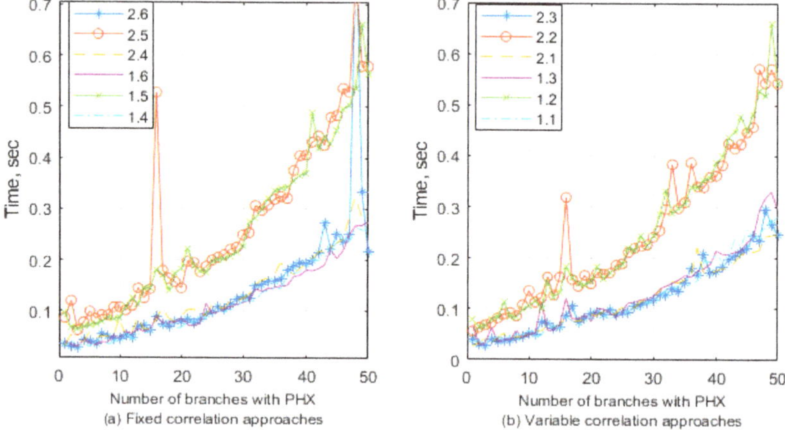

Figure 16. Comparison of computation times: (**a**) constant correlation and (**b**) variable correlation.

As we can see, optimization time grows exponentially with the number of branches in the system. Both variable and fixed correlation approaches take nearly the same time to converge, which favours relating the correlation to a varying parameter, in the case of a heat demand.

Thermophysical iterative approaches (1.2, 2.2, 1.5, 2.5) take twice longer to converge, which means they need at least one more iteration. On the other hand, approaches with linear approximation of temperature dependence take nearly the same time as thermophysical non-iterative approaches, while accuracy is higher and close to the iterative versions as mentioned early in this work.

Additionally, each provided branch has a specific design and setpoints, and the difference in approaches can further raise in favour of the ones with a linear approximation.

6. Conclusions

Flexible operation of DH systems entails significant variations in parameters of DH transmission systems. Defining optimal parameters for the transmission system is the key to maintain sufficient heat quality and minimal operation costs. Even in a relatively small DH system, precise modelling of each individual component can become crucial as the optimization problem includes such models as constraints. Precise modelling of PHX is the most critical for the operation of a transmission system. The better the model fits the reality, the less control effort is required at the heat transfer station afterwards and, therefore, secondary temperature deviations from setpoints will be minimal, especially when demand is unpredictable and flexibility is provided (altering of setpoints). This also means that the primary return temperature and mass flow will be more predictable. Poorly performing models can cause frequent control actions, which result in oscillations in the system, as was seen from the experimental part of this work.

Both mass flows and temperatures affect heat transfer capability of plate heat exchangers. Thus, at various demands plate heat exchangers will produce different return temperatures, which in its turn affects the efficiency of a power-to-heat source (e.g., the heat pump). Especially, this has a greater effect when mass flows and temperatures vary across a wide range, and PHXs, as well as their respective distribution systems, differ significantly in their design (for instance, rating) and demand patterns.

Analysis of heat transfer through DH PHXs has shown that the temperature dependent component of FCC for water has nearly linear behaviour. This persists for the whole range of possible temperatures and covers most of the design correlations, as well as for a large range of pressure since it does not affect thermophysical properties of DH water. This behaviour allows approximating it as a linear function of temperature $\alpha + \beta T$ where α and β are fixed and defined by the design, which was validated experimentally on a laboratory PHX.

The described linear approximation eliminates the use of lookup tables for thermophysical properties for the calculation of OHTC and, therefore, removes the iterative process from the optimization problem. The approximation use is not limited to district heating only and can be applied whenever the temperature dependent component of FCC for a fluid exhibits linear behaviour.

Correlation estimation procedures, outlined in the experimental part of the work, enables finding the most suitable correlation for a particular PHX from temperature and mass flow measurements. This can be used in practice when inspection of a PHX model is performed. It shows that the correlation accuracy can be further improved by relating its coefficient to one (or several) of the varying parameters.

The two operational optimization study cases show that variable correlation models with linearized temperature dependence perform generally better than the original fixed-temperature thermophysical models, having the same computation time and higher accuracy. The proposed models require at least a two-times lower computation time than variable temperature thermophysical models, while having the same accuracy.

The choice between the models described is made depending on their performance for a specific system and engineering judgement. These, due to some models, are not being consistently good at all heat demands. In real systems, however, with dozens of heat exchangers and multiple power-to-heat sources, the optimization (non-linear control) problem would result in a large number of simulations if all constraints must be satisfied within one optimization interval (time step). Besides, the delays in a DH transmission system can reach a dozen of hours, therefore yielding an enormous number of optimization time steps within the given time horizon. Thus, computational performance is critical.

Detailed modelling and control of power to heat sources, i.e., the coupling units between the sectors, is of high interest, which will be the focus of future work.

Funding: This work is funded by the "Enhancing wind power integration through optimal use of cross-sectoral flexibility in an integrated multi-energy system (EPIMES)" project granted by the Danish Innovation Finding (No. 5185-00005A).

Acknowledgments: The author would like to thank Tue V. Jensen for all the brainstorming sessions, Henrik W. Bindner, Shi You and Yi Zong for their guidance and comments and Daniel Arndtzen for setting up the measurements.

Conflicts of Interest: The author declares no conflict of interest.

Nomenclature

Abbreviations:

DH	District Heating;
HP	Heat Pump;
PHX	Plate Heat Exchanger;
OHTC	Overall Heat Transfer Coefficient;
FCC	Forced Convection Coefficient;

Additional terms:

Circuit	a hydraulic path within a PHX;
Terminal	where circuit meets external pipeline;
Hot/Cold circuit	Primary/Secondary circuit;

Additional subscripts:

$LMTD$	logarithmic mean;
c	cold circuit;
h	hot circuit;
i	inlet of a circuit;
o	outlet of a circuit;
l	linearized quantity;
m	arithmetic mean;
f_{ch}	friction factor in a PHX channel;

Heat and circulation pump:

COP_t	total coefficient of performance [-];
P_{cirp}	electric power consumption of a circulation pump [W];
P_{hp}	HP electric power consumption [W];
$\dot{Q}$	thermal power output [W];
T_0	mean temperature in the evaporator [$^\circ$C];
η_{hp}	total efficiency of a HP [-];
η_{cirp}	total efficiency of a circulation pump [-];

Pipeline:

D	diameter [m];
L	length [m];
K_{pipe}	thermal conductivity [W·m/$^\circ$C];
T_{amb}	ambient temperature of the soil [$^\circ$C];
k_{rough}	roughness [m];
Δp_{pipe}	pressure loss [Pa];

Plate heat exchanger:

A	heat transfer area of the plate pack [m^2];
A_{ch}	cross-section of a channel [m];
s	supply (temperature at the HP);
r	return (temperature at the HP);

Complex dimension numbers:

B	thermoph. component of FCC [W·m^{n-2m+1}·s^{n-m}/kg^n.$^\circ$C];
α	linearization coefficient (free) [W·m^{n-2m+1}·s^{n-m}/kg^n.$^\circ$C];
β	linearization coefficient [W·m^{n-2m+1}·s^{n-m}/kg^n.$^\circ$C];
K	function of PHX design [1/m^{n-1}];

Dimensionless numbers:

Nu	Nusselt number;
Pr	Prandt number;
Re	Reynolds number;
C	Nusselt number correlation constant;
f	friction factor;
m	Prandt number exponent;
n	Reynolds number exponent;
D_{hydr}	hydraulic diameter of a channel [m];
D_{port}	diameter of the port [m];
H	FCC [W/$^\circ$C·m^2];
L_{ch}	length of a channel [m];

$\dot{Q}_d$	demand thermal power [W];
R_{tot}	thermal resistance of plates [°C/W·m];
T	water temperature [°C];
U	OHTC [W/°C·m^2];
W_{pl}	Plate width [m];
Δp_{ch}	pressure loss in a channel [Pa];
Δp_{port}	pressure loss in the port [Pa];
b	channel width [m];
v	velocity of the flow in the channel [m/s];

Water properties and global variables:

c_p	specific heat [W/kg·°C];
k	thermal conductivity [W·m/°C];
$\dot{m}$	water mass flow [kg/s];
μ	dynamic viscosity [kg/m·s];
ρ	density [kg/m^3];

References

1. Benonysson, A. Dynamic Modelling and Operational Optimization of District Heating Systems. Ph.D. Thesis, Laboratory of Heating and Air Conditioning, Technical University of Denmark, Lyngby, Denmark, 1991.
2. Pálsson, H.; Larsen, H.V.; Bøhm, B.; Ravn, H.F.; Zhou, J.J. *Equivalent Models of District Heating Systems for On-Line Minimization of Operational Costs of the Complete District Heating System*; Technical University of Denmark and Risø National Laboratory: Roskilde, Denmark, August 1999.
3. Jie, P.; Zhu, N.; Li, D. Operation optimization of existing district heating systems. *Appl. Therm. Eng.* **2015**, *78*, 278–288. [CrossRef]
4. Mitridati, L.; Taylor, J.A. Power systems flexibility from district heating networks. In Proceedings of the 2018 Power Systems Computation Conference (PSCC), Dublin, Ireland, 11–15 June 2018.
5. Kuosa, M.; Aalto, M.; Assad, M.E.H.; Mäkilä, T.; Lampinen, M.; Lahdelma, R. Study of a district heating system with the ring network technology and plate heat exchangers in a consumer substation. *Energy Build.* **2014**, *80*, 276–289. [CrossRef]
6. Wang, Y.; You, S.; Zhang, H.; Zheng, X.; Wei, S.; Miao, Q.; Zheng, W. Operation stability analysis of district heating substation from the control perspective. *Energy Build.* **2017**, *154*, 373–390. [CrossRef]
7. Wang, Y.; You, S.; Zheng, W.; Zhang, H.; Zheng, X.; Miao, Q. State space model and robust control of plate heat exchanger for dynamic performance improvement. *Appl. Therm. Eng.* **2018**, *128*, 1588–1604. [CrossRef]
8. Bastida, H.; Ugalde-Loo, C.E.; Abeysekera, M.; Qadrdan, M. Dynamic modeling and control of a plate heat exchanger. In Proceedings of the 2017 IEEE Conference on Energy Internet and Energy System Integration (EI2), Beijing, China, 26–28 November 2017.
9. Karlsson, T. Numerical evaluation of plate heat exchanger performance in geothermal district heating systems. *Proc. Inst. Mech. Eng. Part A J. Power Energy* **1996**, *210*, 139–147. [CrossRef]
10. Guo, Y.; Wang, F.; Jia, M.; Niu, D. A novel modelling method for plate heat exchanger to predict the outlet cooling water temperature. *Can. J. Chem. Eng.* **2019**, *97*, 1809–1820. [CrossRef]
11. Olsen, P.K. Guidelines for Low-Temperature District Heating. In *EUDP 2010-II: Full-Scale Demonstration of Low-Temperature District Heating in Existing Buildings*; Cowi Holding A/S: Lyngby, Denmark, 2014.
12. Benonysson, A.; Boysen, H. Optimum control of heat exchangers. Danfoss technical paper. In Proceedings of the 5th International Symposium on Automation of District Heating Systems, Helsinki, Finland, 20–23 August 1995; p. 10.
13. Larson, G. On Dynamics in District Heating Systems. Ph.D. Thesis, Chalmers University of Technology, Gothenburg, Sweden, 1999.
14. Granryd, E.; Ekroth, I.; Lundqvist, P.; Melinder, Å.; Palm, B.; Rohlin, P. *Refrigeration Engineering*; Royal Institute of Technology, KTH, Department of Energy Technology, Division of Applied Thermodynamics and Refrigeration: Stockholm, Sweden, 2009.
15. Papaevangelou, G.; Evangelides, C.; Tzimopoulos, C. A new explicit equation for the friction coefficient in the Darcy-Weisbach equation, Proceedings of the Tenth Conference on Protection and Restoration of the Environment: PRE10, 6–9 July 2010. *Greece Corfu.* **2010**, *166*, 1–7.

16. Kakaç, S.; Liu, H. *Heat Exchangers: Selection, Rating and Thermal Design*, 2nd ed.; CRC PRESS: Boca Raton, FL, USA, 2002; p. 501.
17. Schmidt, E. *Properties of Water and Steam in SI-Units. 0-800 C, 0-1000 Bar*; Springer: München, Germany, 1969; p. 205.
18. SPXFlow. *APV Heat Transfer Handbook. A History of Excellence*; SPX: Charlotte, NC, USA, 2008; p. 66.
19. McAdams, W.H. *Heat Transmission*, 2nd ed.; McGRAW-HILL: Brooklyn, NY, USA, 1942; p. 459.
20. The MathWorks Inc. Nonlinear Programming Solver Fmincon. Available online: https://se.mathworks.com/help/optim/ug/fmincon.html (accessed on 23 August 2019).

Article

Supplementary Control of Air–Fuel Ratio Using Dynamic Matrix Control for Thermal Power Plant Emission

Taehyun Lee [1], Eungsu Han [2], Un-Chul Moon [2,*] and Kwang Y. Lee [3]

[1] School of Mechanical System Engineering, ChungAng University, HukSuk-dong DongJak-Gu, Seoul 06974, Korea; yu0744@naver.com

[2] School of Electrical and Electronics Engineering, ChungAng University, HukSuk-dong DongJak-Gu, Seoul 06974, Korea; dmdtnwkd@naver.com

[3] Department of Electrical and Computer Engineering, Baylor University, Waco, TX 76798-7356, USA; kwang_y_lee@baylor.edu

* Correspondence: ucmoon@cau.ac.kr; Tel.: +82-10-8775-0173

Received: 27 July 2019; Accepted: 6 September 2019; Published: 2 January 2020

Abstract: This paper proposes a supplementary control for tighter control of the air–fuel ratio (AFR), which directly affects the environmental emissions of thermal power plants. Dynamic matrix control (DMC) is applied to the supplementary control of the existing combustion control loops and the conventional double cross limiting algorithm for combustion safety is formulated as constraints in the proposed DMC. The proposed supplementary control is simulated for a 600-MW drum-type power plant and 1000 MW ultra-supercritical once-through boiler power plant. The results show the tight control of the AFR in both types of thermal power plants to reduce environmental emissions.

Keywords: air–fuel ratio; combustion control; dynamic matrix control; power plant control

1. Introduction

Currently, environmental emissions from thermal power plants have drawn much concern. Various environmental emissions such as carbon monoxide (CO) and nitrogen oxide (NOx), are released from thermal power plants. To meet the current stringent environmental standards, combustion conditions should be maintained tightly to reduce these emissions [1,2].

Environmental emissions in thermal power plants, such as CO and NOx, are directly influenced by the air-fuel ratio (AFR). Figure 1 shows the quantity of emissions and thermal efficiency as a function of the AFR in a typical furnace combustion operation [3,4]. High CO is observed in the low AFR range because of the incomplete combustion due to the lack of combustion air or excessive fuel. Meanwhile, the stack heat loss in the high AFR range increases NOx.

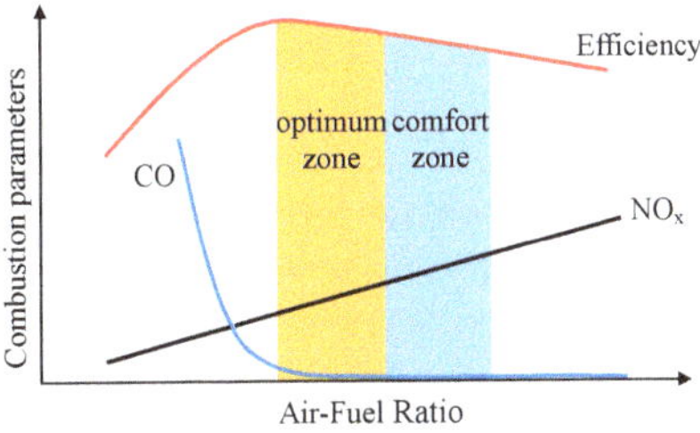

Figure 1. Emissions and efficiency as a function of air–fuel ratio.

An ideal AFR is a function of plant characteristics, condition and electric load. In practice, this ideal AFR is predetermined and updated by offline experiments and used in the combustion control. Therefore, maintaining the ideal AFR in the thermal power plant combustion is crucial in reducing the environmental emissions of power plants.

In practice, many power plant manufacturers have similar philosophies for thermal power plant control. For combustion control, the unit load demand and main steam pressure signal generate the boiler master demand (BMD) signal. The BMD signal generates the air flow demand signal and fuel flow demand signal based on the ideal AFR [5–7].

The concept of cross limiting technique is used in the conventional combustion control, which is designed to limit the fuel demand in order to prevent the lack of combustion air and the extinguishing of the boiler firing. That is, the fuel flow demand signal is limited by the current combustion air and conversely, the air flow demand signal is limited by the current fuel in the furnace. This "AFR with cross limiting override control" has been a standard concept for large capacity boilers [5,8,9]. Although this conventional control system is well developed in practice, the performance of AFR control tends to be degraded in the transient state [8].

Recently, Bhowmick and Bera [9] pointed out the weakness of the cross-limiting technique in abnormal condition, and Liu, He, and Wang [10] and Zanoli et al. [11] proposed the double cross-limiting (DCL) strategy that generalizes the cross limiting technique. In the DCL algorithm, the fuel and air demand signals are limited within a reasonable band of corresponding combustion air and fuel, respectively. Although DCL can assist the AFR control, because it is a simple static mapping, it has a limitation to improve the transient or dynamic response of the AFR.

Most modern control technologies require a good-quality model for the object system, which is not an easy task for a practical plant. One of the approaches to avoid this difficulty is model free control (MFC) [12]. The main advantage of the MFC technique is that the process model is approximated through a fast estimator using an approximation of the process model, which is locally valid and, furthermore, on a relatively short time window. The MFC techniques were applied to a wide range of processes, which include immune systems [13], robot systems [14,15], twin rotor aerodynamic systems (TRASs) [16–18], aircraft system [19] and servo systems [20].

Another approach to avoid first principles or complex identification is dynamic matrix control (DMC) with a step-response model which can be easily obtained by plant step test. DMC is a proven algorithm in the model predictive control (MPC) for dynamic systems [21]. It computes optimized control inputs using linear programming or quadratic programming while considering the constraints at every sampling time [22,23]. Moon and Lee applied DMC to a simple thermal power plant model in [24], and the DMC was also successfully applied to the power plant coordinated control [25,26].

Herein, a DMC was designed to generate supplementary signals to the existing combustion control to maintain the ideal AFR. The optimal supplementary signals were generated in the optimization window constrained by the DCL strategy. This supplementary control structure over the existing combustion control is very practical and easy to implement because it can be easily bypassed in the case of an emergency.

2. DMC Combustion Control

2.1. Conventional Boiler Combustion

Figure 2 shows a typical conventional combustion control with DCL strategy of a coal-fired power plant [5,10,11]. In the figure, each block of $F(x)$ represents a look-up table or static mapping between the input and output. The BMD signal is the output of $F_1(x)$ corresponding to the power load demand, electric power output and main steam pressure. Subsequently, $F_2(x)$ and $F_3(x)$ generate the air flow demand signal and fuel flow demand signal, respectively, based on the predetermined ideal AFR. From the air flow demand signal, the air controller drives the forced draft (FD) fan to control the amount of combustion air into the furnace. Further, with the fuel flow demand signal, the fuel controller drives the primary air fan and pulverizer, accordingly.

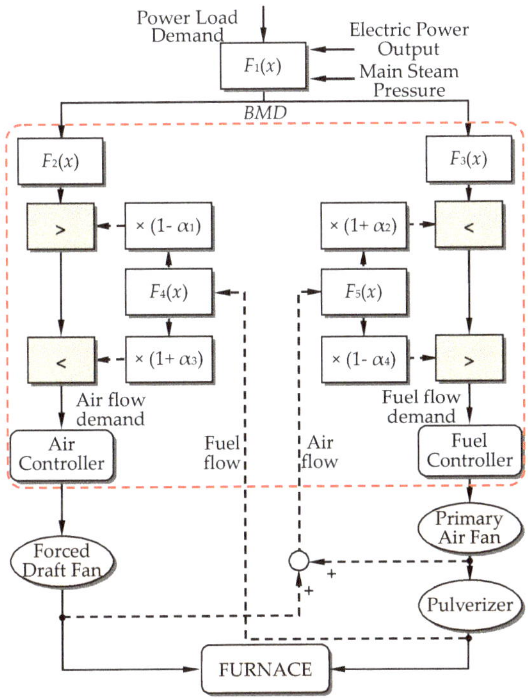

Figure 2. Typical power plant combustion with double cross-limiting strategy.

The feedback loops with dotted black lines in Figure 2 represent the DCL strategy that prevents the incomplete burning and the excessive combustion air at various loads. Parameters α_1, α_2, α_3 and α_4 are fixed constants of several percent to allow a small margin. This DCL structure is the updated form of the conventional cross-limiting technique that uses α_1 and α_2 loops only, and α_3 and α_4 loops are ignored [5].

The output of $F_2(x)$, the air flow demand, is compared with the output of $F_4(x)$, which is the conversion of fuel flow to air flow. The larger value between the output of $F_2(x)$ and $F_4(x) \times (1-\alpha_1)$ was selected to prevent the lack of combustion air in the furnace. In addition, this selector output was compared with $F_4(x) \times (1 + \alpha_3)$ and the smaller value was finally selected as the air flow demand to prevent the overfeeding of combustion air in the furnace. In the fuel side, the output of $F_3(x)$ was similarly compared with the output of $F_5(x)$, which is the conversion of air flow to fuel flow. Subsequently, the final fuel flow demand is determined between two margins, α_2 and α_4. The values of α are typically selected between 2% and 5% [5,11]. Therefore, the DCL strategy prevents high or low AFR by selecting appropriate air and fuel flow demands.

2.2. Supplementary DMC for AFR

We implemented the tighter control of the AFR while minimizing its effects on the performance of the existing combustion control system. The DMC supplementary control is applied to the air and fuel flow demands of the conventional multi-loop control, thereby adjusting the amounts of combustion air and fuel to maintain the ideal AFR.

Figure 3 shows the structure of the supplementary control for the AFR using DMC, which is a replacement of the dotted red rectangle in Figure 2. In Figure 3, AFR_k is the air-fuel ratio at the k-th time step, which is the output or controlled variable (CV), and $\text{AFR}_{ref,k}$ is the reference air-fuel ratio at the k-th step; $\widetilde{u}_{a,k}^{\text{DMC}}$ and $\widetilde{u}_{f,k}^{\text{DMC}}$ are the plant inputs or manipulated variables (MV) of the proposed DMC, which are the supplementary air flow demand and supplementary fuel flow demand at the k-th step, respectively.

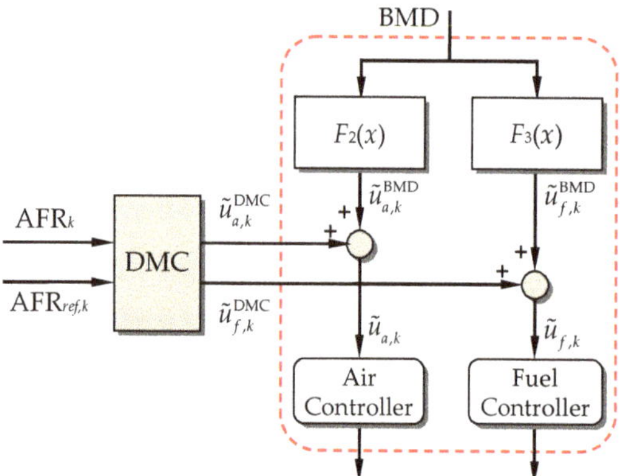

Figure 3. Proposed control system configuration.

These signals are added to $\widetilde{u}_{a,k}^{\text{BMD}}$ and $\widetilde{u}_{f,k}^{\text{BMD}}$, which are, respectively, the air flow and fuel flow demands from the BMD of the conventional multi-loop control. Therefore, the air flow demand $\widetilde{u}_{a,k}$ and fuel flow demand $\widetilde{u}_{f,k}$ are the sum of the BMD signals and the supplementary DMC signals as follows:

$$\widetilde{u}_{a,k} = \widetilde{u}_{a,k}^{\text{BMD}} + \widetilde{u}_{a,k}^{\text{DMC}} \tag{1}$$

$$\widetilde{u}_{f,k} = \widetilde{u}_{f,k}^{\text{BMD}} + \widetilde{u}_{f,k}^{\text{DMC}} \tag{2}$$

The output, AFR_k, is then defined as the ratio:

$$\text{AFR}_k = u_{a,k}/u_{f,k} \tag{3}$$

where $u_{f,k}$ is the fuel flow at the k-th step, which is the output of the pulverizer, and $u_{a,k}$ is the air flow at the k-th step, which is the sum of the two air flows, from FD fan and primary air fan in Figure 2.

From the viewpoint of practical implementation, this supplementary control structure over the existing multi-loop control logic is very realistic and easy to implement. In an emergency, this type of supplementary control can be easily removed and returned to the conventional multi-loop control system, with which the plant operators are familiar.

Herein, the standard form of the DMC algorithm is used [27]. If we use the standard notation of the plant input and plant output,

$$y_k = \text{AFR}_k \tag{4}$$

$$\overline{u}_k = [\widetilde{u}_{a,k}^{\text{DMC}}, \widetilde{u}_{f,k}^{\text{DMC}}]^T \tag{5}$$

Subsequently, the prediction equation is:

$$\overline{Y}_{k+1|k} = \overline{Y}_{k+1|k-1} + \overline{S}\Delta\overline{U}_k + \overline{Y}_{k+1|k}^d \tag{6}$$

where

$$\overline{Y}_{k+1|k} = [\text{AFR}_{k+1|k}\, \text{AFR}_{k+2|k} \cdots \text{AFR}_{k+p|k}]^T \tag{7}$$

$$\overline{Y}_{k+1|k-1} = [\text{AFR}_{k+1|k-1}\ \text{AFR}_{k+2|k-1}\ \cdots\ \text{AFR}_{k+p|k-1}]^T \tag{8}$$

$$\begin{aligned}
\Delta \overline{U}_k &= [\Delta \overline{u}_k\ \Delta \overline{u}_{k+1}\ \cdots\ \Delta \overline{u}_{k+m-1}]^T \\
&= [(\Delta \widetilde{u}^{\text{DMC}}_{a,k}, \Delta \widetilde{u}^{\text{DMC}}_{f,k})\ \cdots\ (\Delta \widetilde{u}^{\text{DMC}}_{a,k+m-1}, \Delta \widetilde{u}^{\text{DMC}}_{f,k+m-1})]^T
\end{aligned} \tag{9}$$

Here, p is the prediction horizon and m is the control horizon; $\overline{Y}_{k+1|k}$ is a $p \times 1$ vector, the future output of AFR trajectory at $t = k$; $\overline{Y}_{k+1|k-1}$ is a $p \times 1$ vector, an open-loop prediction of the future output when input u remains at the previous step value u_{k-1}; $\overline{Y}^d_{k+1|k}$ is a $p \times 1$ vector, an estimate of the unmeasured disturbance; $\Delta \overline{U}_k$ is a $2\,m \times 1$ input adjustment vector; and $\overline{S}$ is a $p \times 2\,m$ dynamic matrix including step responses as follows:

$$\overline{S} = \begin{bmatrix}
\overline{s}_1 & \overline{0} & \cdots & \overline{0} \\
\overline{s}_2 & \overline{s}_1 & \ddots & \vdots \\
\vdots & \vdots & \ddots & \overline{s}_1 \\
\vdots & \vdots & \ddots & \vdots \\
\overline{s}_p & \overline{s}_{p-1} & \cdots & \overline{s}_{p-m+1}
\end{bmatrix} \tag{10}$$

$$\overline{s}_i = \left(s^a_i\ \ s^f_i \right) \tag{11}$$

where $s_i{}^a$ and $s_i{}^f$ are the step response coefficients of the AFR from the incremented air flow demand and fuel flow demand, respectively, at the i-th sampling step.

To calculate the input adjustment vector, an online quadratic optimization with constraints was performed at every sampling step:

$$\min_{\Delta \overline{U}_k} \|\overline{E}_{k+1|k}\|_\Lambda + \|\Delta \overline{U}_k\|_\Gamma \tag{12}$$

where:

$$\overline{E}_{k+1|k} = \overline{Y}_{k+1|k} - \overline{R}_{k+1|k} = [\ e_{k+1|k}\ \ e_{k+2|k}\ \ \cdots\ \ e_{k+p|k}\ ]^T \tag{13}$$

$$\overline{R}_{k+1|k} = [\ \text{AFR}_{ref,k+1|k}\ \ \text{AFR}_{ref,k+2|k}\ \ \cdots\ \ \text{AFR}_{ref,k+p|k}\ ]^T \tag{14}$$

Here, $\overline{E}_{k+1|k}$ is a $p \times 1$ error vector, and $\overline{R}_{k+1|k}$ is a $p \times 1$ desired trajectory output vector, and Λ and Γ are the weight matrices for the corresponding vectors in the quadratic optimization.

2.3. Constraints of Proposed DMC

An important benefit of using the DMC is the handling of constraints in optimization. Because the DMC herein is supposed to be a supplementary control, additional large changes in the air and fuel demand signal could significantly influence the power output and main steam pressure. Therefore, to minimize its effect on the existing control, the output of the DMC is limited to:

$$-\beta_a \le \widetilde{u}^{\text{DMC}}_{a,k} \le \beta_a \quad k = 1 \cdots m \tag{15}$$

$$-\beta_f \le \widetilde{u}^{\text{DMC}}_{f,k} \le \beta_f \quad k = 1 \cdots m \tag{16}$$

where, β_a and β_f are constants to limit the supplementary air and fuel adjustments, respectively.

It is noteworthy that this supplementary DMC could violate the conventional cross-limit or DCL strategy in Figure 2. Therefore, herein, the DCL strategy is formulated as the constraint in the DMC optimization, i.e., the DCL strategy in the proposed control system is represented as the optimization window at the k-th step as follows:

$$(1 - \alpha_1)F_4(u_{f,k}) \le \widetilde{u}^{\text{BMD}}_{a,k} + \widetilde{u}^{\text{DMC}}_{a,k} \le (1 + \alpha_3)F_1(u_{f,k}) \tag{17}$$

$$(1 - \alpha_4)F_5(u_{a,k}) \leq \widetilde{u}_{f,k}^{\mathrm{BMD}} + \widetilde{u}_{f,k}^{\mathrm{DMC}} \leq (1 + \alpha_2)F_5(u_{a,k}) \tag{18}$$

Combining (15)-(18), the two constraints at the k-th step are combined into the following form:

$$\bar{u}_{\min,k} \leq \bar{u}_k \leq \bar{u}_{\max,k} \tag{19}$$

where:

$$\bar{u}_{\min,k} = \begin{pmatrix} \max\left\{-\beta_a, \ (1 - \alpha_1)F_4(u_{f,k}) - \widetilde{u}_{a,k}^{\mathrm{BMD}}\right\} \\ \max\left\{-\beta_f, \ (1 - \alpha_4)F_5(u_{a,k}) - \widetilde{u}_{f,k}^{\mathrm{BMD}}\right\} \end{pmatrix} \tag{20}$$

$$\bar{u}_{\max,k} = \begin{pmatrix} \min\{\beta_a, \ (1 + \alpha_3)F_4(u_{f,k}) - \widetilde{u}_{a,k}^{\mathrm{BMD}}\} \\ \min\{\beta_f, \ (1 + \alpha_2)F_5(u_{a,k}) - \widetilde{u}_{f,k}^{\mathrm{BMD}}\} \end{pmatrix} \tag{21}$$

Therefore, the constraints for the $2\,m \times 1$ input vector is

$$\begin{bmatrix} \bar{u}_{\min,k} \\ \begin{pmatrix} -\beta_a \\ -\beta_f \end{pmatrix} \\ \vdots \\ \begin{pmatrix} -\beta_a \\ -\beta_f \end{pmatrix} \end{bmatrix} \leq \bar{U}_k \leq \begin{bmatrix} \bar{u}_{\max,k} \\ \begin{pmatrix} \beta_a \\ \beta_f \end{pmatrix} \\ \vdots \\ \begin{pmatrix} \beta_a \\ \beta_f \end{pmatrix} \end{bmatrix} \tag{22}$$

For the quadratic programming problem of (12), the constraints need to be changed into the standard linear inequality form:

$$\bar{C}_k \Delta \bar{U}_k \geq \bar{D}_k \tag{23}$$

where $\bar{C}_k$ and $\bar{D}_k$ are constant matrixes. Therefore, (22) should be represented in the form of (23). From the definition of difference,

$$\bar{u}_k = \bar{u}_{k-1} + \Delta \bar{u}_k \tag{24}$$

$$\bar{u}_{k+l} = \bar{u}_{k-1} + \sum_{i=0}^{l} \bar{u}_{k+i}, \quad l = 0, 1, \cdots, m-1 \tag{25}$$

(22) is represented as:

$$\begin{bmatrix} \bar{u}_{\min,k} - \bar{u}_{k-1} \\ -\begin{pmatrix} \beta_a \\ \beta_f \end{pmatrix} - \bar{u}_{k-1} \\ \vdots \\ -\begin{pmatrix} \beta_a \\ \beta_f \end{pmatrix} - \bar{u}_{k-1} \end{bmatrix} \leq \begin{bmatrix} \Delta \bar{u}_k \\ \sum_{i=0}^{l} \Delta \bar{u}_{k+i} \\ \vdots \\ \sum_{i=0}^{l} \Delta \bar{u}_{k+i} \end{bmatrix} \leq \begin{bmatrix} \bar{u}_{\max,k} - \bar{u}_{k-1} \\ \begin{pmatrix} \beta_a \\ \beta_f \end{pmatrix} - \bar{u}_{k-1} \\ \vdots \\ \begin{pmatrix} \beta_a \\ \beta_f \end{pmatrix} - \bar{u}_{k-1} \end{bmatrix} \tag{26}$$

The middle term of (26) is represented as:

$$\begin{bmatrix} \Delta \bar{u}_k \\ \sum_{i=0}^{1} \Delta \bar{u}_{k+i} \\ \vdots \\ \sum_{i=0}^{l} \Delta \bar{u}_{k+i} \end{bmatrix} = \begin{bmatrix} \bar{I} & \bar{0} & \cdots & \bar{0} \\ \bar{I} & \bar{I} & \bar{0} & \vdots \\ \vdots & \vdots & \ddots & \bar{0} \\ \bar{I} & \bar{I} & \bar{I} & \bar{I} \end{bmatrix} \begin{bmatrix} \Delta \bar{u}_k \\ \Delta \bar{u}_{k+1} \\ \vdots \\ \Delta \bar{u}_{k+m-1} \end{bmatrix} = \bar{I}_L \begin{bmatrix} \Delta \bar{u}_k \\ \Delta \bar{u}_{k+1} \\ \vdots \\ \Delta \bar{u}_{k+m-1} \end{bmatrix} \tag{27}$$

where $\bar{I}$ is a 2×2 identity matrix, and $\bar{I}_L$ is a $2\,m \times 2\,m$ matrix. Finally, (26) is represented as follows:

$$\begin{bmatrix} \bar{I}_L \\ -\bar{I}_L \end{bmatrix} \Delta \bar{U}_k \geq \begin{bmatrix} \bar{u}_{\min,k} - \begin{pmatrix} \tilde{u}^{\mathrm{DMC}}_{a,k-1} \\ \tilde{u}^{\mathrm{DMC}}_{f,k-1} \end{pmatrix} \\ \begin{pmatrix} -\beta_a - \tilde{u}^{\mathrm{DMC}}_{a,k-1} \\ -\beta_f - \tilde{u}^{\mathrm{DMC}}_{f,k-1} \end{pmatrix} \\ \vdots \\ \begin{pmatrix} -\beta_a - \tilde{u}^{\mathrm{DMC}}_{a,k-1} \\ -\beta_f - \tilde{u}^{\mathrm{DMC}}_{f,k-1} \end{pmatrix} \\ \begin{pmatrix} \tilde{u}^{\mathrm{DMC}}_{a,k-1} \\ \tilde{u}^{\mathrm{DMC}}_{f,k-1} \end{pmatrix} - \bar{u}_{\max,k} \\ \begin{pmatrix} \tilde{u}^{\mathrm{DMC}}_{a,k-1} - \beta_a \\ \tilde{u}^{\mathrm{DMC}}_{f,k-1} - \beta_f \end{pmatrix} \\ \vdots \\ \begin{pmatrix} \tilde{u}^{\mathrm{DMC}}_{a,k-1} - \beta_a \\ \tilde{u}^{\mathrm{DMC}}_{f,k-1} - \beta_f \end{pmatrix} \end{bmatrix} \tag{28}$$

Therefore, both the DCL strategy and the supplementary control limits are represented in the standard form of (23) in the proposed DMC. The constraint of (28) in the proposed supplementary DMC is used in the standard online optimization of (12) at every sampling step.

3. Applications to Power Plants

3.1. 600-MW Drum-Type Thermal Power Plant

Figure 4 shows the boiler system of a 600-MW oil-fired drum-type thermal power plant model. Each component of the model is developed with mass, momentum, and energy balance equations. This nonlinear power plant model was applied and validated in many studies [25,28,29].

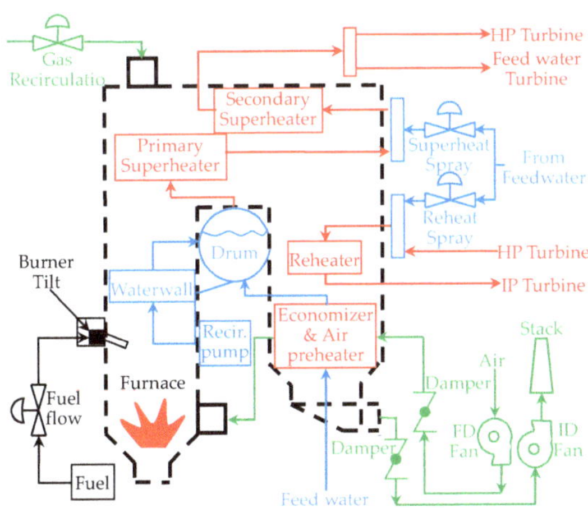

Figure 4. Boiler system of a 600-MW drum-type thermal power plant.

The combustion control system of this plant primarily follows the standard structure of the coal-fired plant in Figure 2. Because this model is an oil-fired plant, the fuel flow demand signal manipulates the fuel flow valve in Figure 4 instead of the pulverizer and primary air fan in Figure 2.

Therefore, $u_{f,k}$ is the output of the fuel flow valve in this plant. The conventional cross-limiting algorithm is equipped in this model, and both α_1 and α_2 were set to 5%.

Figure 5 shows the response of the AFR as a result of the step increments of $\tilde{u}_a^{DMC}$ and $\tilde{u}_f^{DMC}$ at $t = 0$ from the steady state of 450 MW. The 1% of the normal operation range is used independently for the two step inputs. Because this is a closed-loop test for supplementary control, these transient responses include the dynamics of not only the power plant model but also the existing multi-loop control logics. In Figure 5, from the initial value 15.35, the AFR is finally increased/decreased to 15.57/15.12 due to the increase in combustion air/fuel, respectively.

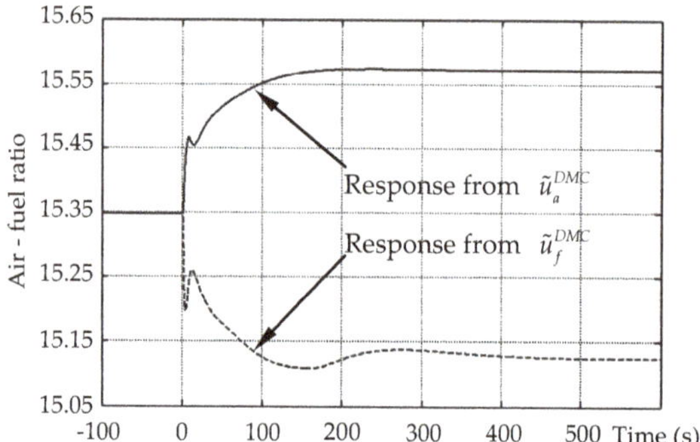

Figure 5. Step responses in a 600 MW power plant.

The tuning of the DMC parameters is important. Theoretically, a small sampling time and a large prediction and control horizons are desirable from the view point of control performance. However, this increases the computational burden in practice. The sampling time is determined as 1 s. Therefore, the responses of Figure 5 were sampled at every 1 s and stored in the step-response matrix $\overline{S}$. The error and input increments in (12) are:

$$\|e_{k+1|k}\| = [e_{k+1|k}]^T [\Lambda] [e_{k+1|k}] \tag{29}$$

$$\|\Delta\bar{u}_k\| = \begin{bmatrix} \Delta\tilde{u}_{a,k}^{DMC} \\ \Delta\tilde{u}_{f,k}^{DMC} \end{bmatrix}^T \begin{bmatrix} \Gamma_a & 0 \\ 0 & \Gamma_f \end{bmatrix} \begin{bmatrix} \Delta\tilde{u}_{a,k}^{DMC} \\ \Delta\tilde{u}_{f,k}^{DMC} \end{bmatrix} \tag{30}$$

where Λ is the weight of the AFR error; Γ_a is the weight of $\Delta\tilde{u}_{a,k}^{DMC}$; and Γ_f is the weight of $\Delta\tilde{u}_{f,k}^{DMC}$. The DMC parameters are listed in Table 1. The p is selected as 300 [s] for AFR response to settle down in Figure 5. The m is selected as 100 [s] considering computational burden. A small β limits the performance of proposed supplementary control, while a large β can affect the other control loops. By trial and error, to limit excessive control action, β_a and β_f were selected as 2% and 1.25% of their normal operation ranges, respectively. Because of the small β_f, this supplementary DMC primarily manipulates the combustion air flow demand rather than the fuel demand.

Table 1. DMC parameters of 600 MW drum-type thermal power plant.

Λ	Γ_a	Γ_f	p	m	β_a	β_f
1	1	10	300	100	2%	1.25%

3.2. 1000-MW Once-Through Type Thermal Power Plant

Figure 6 is the dynamic boiler simulation model (DBSM) of the 1000 MW ultra-supercritical (USC) coal-fired once-through type model. This nonlinear model was also developed based on mass, momentum, and energy balances. It is a field-proven simulator for the power plant control logic design in industry [26,30,31].

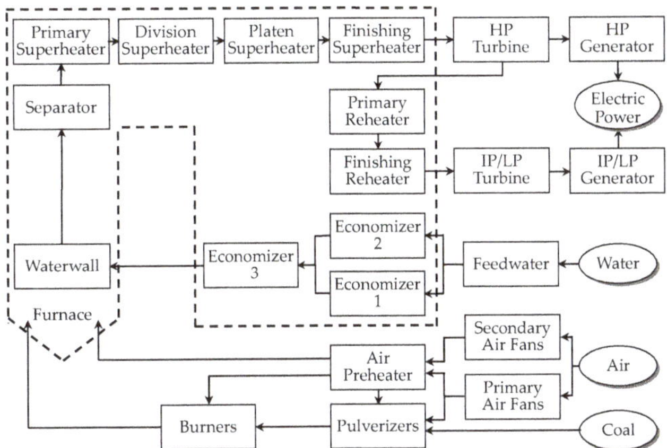

Figure 6. Schematic of a 1000 MW once-through-type thermal power plant.

In Figure 6, the secondary air fan represents the FD fan in Figure 2 to control the amount of combustion air. The combustion control system of this plant primarily follows the standard structure in Figure 2. The conventional cross-limiting algorithm is equipped in this model, and the values of α_1 and α_2 were set to 5%.

Figure 7 shows the step response of the AFR as a result of the step increase of $\widetilde{u}_a^{DMC}$. One percent of the normal operation range of $\widetilde{u}_a^{DMC}$ was applied at the steady state of 825 MW. The AFR increased from 11.12 to 11.26 due to the increase in combustion air. A faster transient response was observed compared to the response in Figure 5. This faster response can be interpreted as the faster dynamics of the once-through-type boiler compared to that of the drum-type boiler.

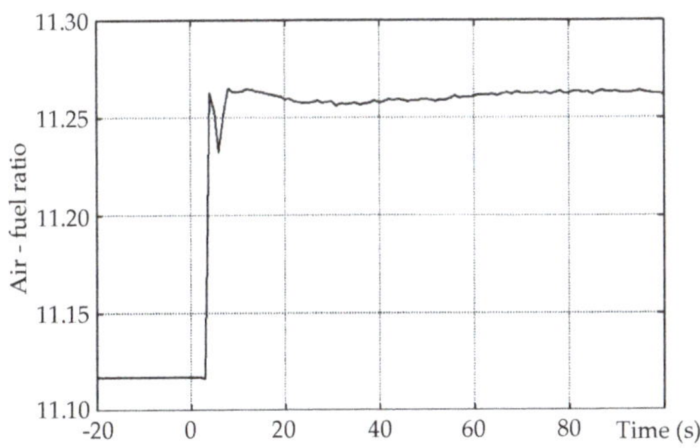

Figure 7. Response of air–fuel ratio in 1000 MW plant due to step increase of $\widetilde{u}_a^{DMC}$.

Figure 8 shows the step response of the AFR for a 1% step increase of $\widetilde{u}_f^{DMC}$ in the steady state of 825 MW. The AFR is decreased from 11.12 to 10.97 in steady state. The slower response than that of Figure 7 can be attributed to the slow dynamics of the pulverizer. Therefore, in this study, $\widetilde{u}_a^{DMC}$ is primarily used for MV to achieve a fast control response.

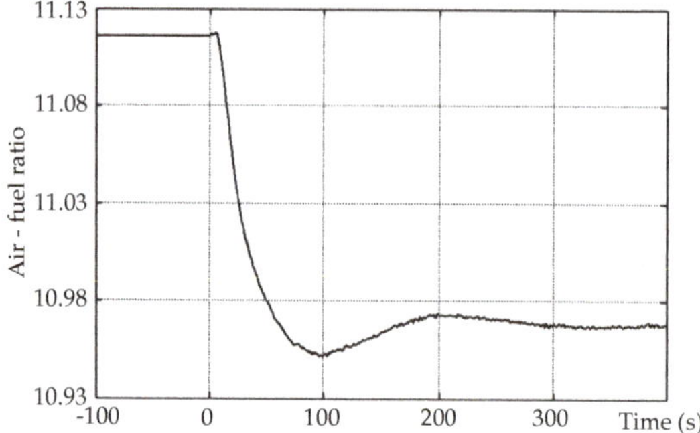

Figure 8. Response of air–fuel ratio in 1000 MW plant due to step increase of $\widetilde{u}_f^{DMC}$.

The same structure of the DMC was applied to this plant. The sampling time was selected as 1 s and the DMC weight parameters are shown in Table 2. The p was selected as 300 s for AFR to settle down in Figure 8. The m was also selected as 100 s considering computational burden. For this plant, β_a and β_f were selected as 1.36% and 0.02% by trial and error, respectively. Because of the small β_f, this supplementary DMC primarily manipulates the combustion air flow demand rather than the fuel demand.

Table 2. DMC parameters of 1000 MW once through-type thermal power plant.

Λ	Γ_a	Γ_f	p	m	β_a	β_f
10	1	0.1	300	100	1.36%	0.02%

4. Simulation Results

4.1. Simulation of 600-MW Drum-Type Thermal Power Plant

The DMC supplementary control is developed in the MATLAB environment. In this simulation, the control performance (12) with constraints (28) is optimized by the quadratic programming, MATLAB function "quadprog ( )", at every sampling time.

The simulation scenario has two step changes for the electric power load demand in Figure 9. The load demand change is limited to 0.5 MW/s, which is 5% per minute of the total load. In this simulation, for simplicity, the ideal AFR of the drum-type plant model is assumed to be constant, i.e., 15.35 at every electric power load.

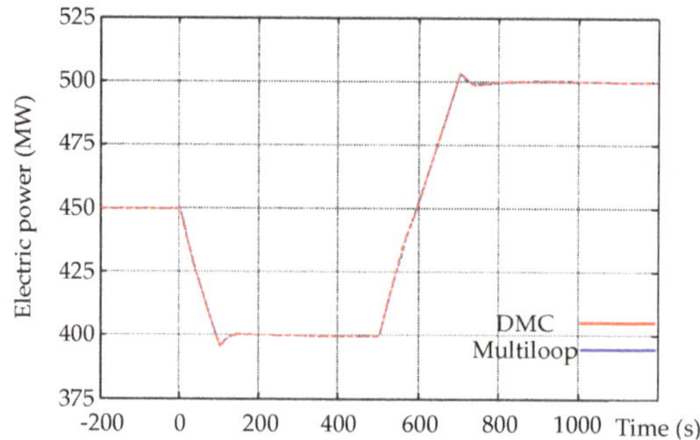

Figure 9. Comparison of electric power output of 600-MW plant.

To confirm the effect of the proposed supplementary control on the main control loop, two responses of the electric power output are compared in Figure 9, with and without DMC. Because the supplementary control signal is limited by β_a and β_f, which are only 2% and 1.25% of their normal operation ranges, respectively, two responses are almost the same, and the proposed supplementary control does not affect the existing multi-loop control system operation.

Figure 10 shows the comparison of the AFR between the conventional multi-loop control and the proposed DMC supplementary control during the transient. Although the responses of the electric power output are similar, the AFR of the proposed DMC shows a tighter control during the transient of the load changes. Table 3 shows that the squared error sum of the conventional control can be significantly reduced, to 4.93%, by the proposed supplementary control. Therefore, the environmental emissions by the conventional control during the transient of the load change can be effectively reduced by the proposed supplementary DMC.

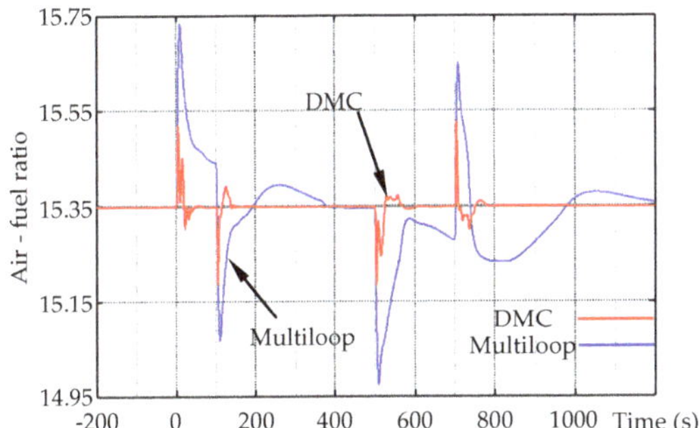

Figure 10. Comparison of AFR in 600-MW plant.

Table 3. Squared error sum comparison of AFR in 600 MW plant.

Multi-Loop	23.069
DMC	1.137
Percentage of DMC/Multi-loop	4.93%

Figures 11 and 12 show the movements of the DMC control signals $\widetilde{u}_{a,k}^{DMC}$ and $\widetilde{u}_{f,k}^{DMC}$, respectively. In the figures, supplementary control limits of ±2% and ±1.25% are represented as ±12.50 kg/s and ±0.51 kg/s, respectively. During the first 50 [s] in Figure 10, a large AFR was expected by the DMC prediction of (6). Then, optimization of (12) with constraint of (28) was calculated to keep the AFR to be 15.35, based on the step-response model of (10) which is developed from Figure 5. As a result of optimization of (12), the supplementary signals $\widetilde{u}_{a,k}^{DMC}$ and $\widetilde{u}_{f,k}^{DMC}$ were calculated. The $\widetilde{u}_{a,k}^{DMC}$ is negative in Figure 11, while $\widetilde{u}_{f,k}^{DMC}$ is positive in Figure 12 to reduce the AFR. This control process is repeated at every sampling step in DMC. Accordingly, the AFR of the proposed control is reduced in the first 50 s as shown in Figure 10, and excessive NOx and stack heat loss can be reduced with the proposed supplementary control.

To confirm the DCL logic of the proposed DMC, the air side optimization window (17) in the first 200 s is represented in Figure 13. In the figure, the lower bound "DCL min" is $(1-\alpha_1) F_4(u_{f,k})$, the upper bound "DCL max" is $(1 + \alpha_3) F_4(u_{f,k})$, and $\widetilde{u}_{a,k}$ is $\widetilde{u}_{a,k}^{BMD} + \widetilde{u}_{a,k}^{DMC}$. At approximately 10 s and 110 s, the amplitude of $\widetilde{u}_{a,k}^{DMC}$ was effectively restricted for $\widetilde{u}_{a,k}$ to stay within the optimization window constrained by the DCL logic in (28).

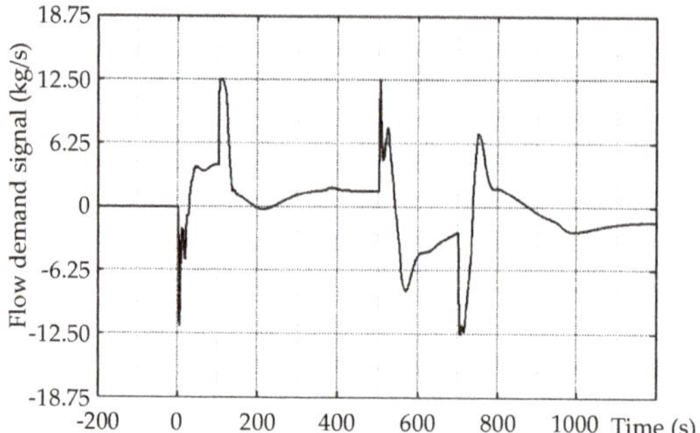

Figure 11. DMC supplementary air flow demand of 600-MW plant.

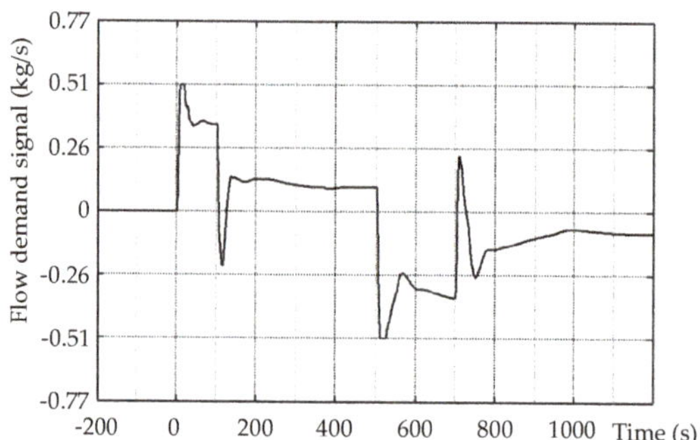

Figure 12. DMC supplementary fuel flow demand of 600-MW plant.

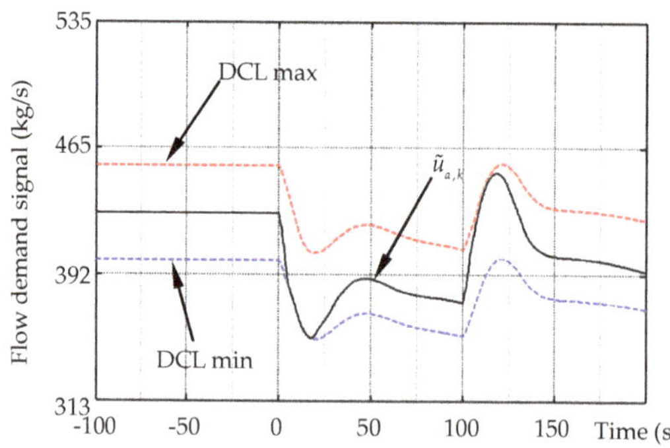

Figure 13. Optimization window for air flow demand signal of DMC for 600-MW plant.

4.2. Simulation of 1000-MW Once-Through Type Thermal Power Plant

The DMC supplementary control was also developed for the 1000-MW once-through type unit in MATLAB and linked with the DBSM simulator. The simulation scenario contains two step changes of the power load demand, where the load demand is reduced from 825 MW to 750 MW at 0 s, and increased to 950 MW at 1500 s. The load demand change at each sampling time is also restricted by the internal logic of the DBSM. Unlike the 600-MW drum-type plant, the ideal AFR of the DBSM is not a constant but is specified by the internal logic of the DBSM as a function of the load. In this simulation, the ideal AFR of the DBSM is used as the $\text{AFR}_{ref,k}$ for the DMC.

Though they are not represented in this paper, the power output and main steam pressure of the two controls are almost the same, therefore, the proposed supplementary control does not affect the existing multi-loop control system operation. Figure 14 shows the ideal AFR of the DBSM, the AFR of the conventional multi-loop control, and the AFR of the proposed DMC. Figure 15 shows the variation of $\widetilde{u}_{a,k}^{DMC}$. In the figure, the supplementary control of $\widetilde{u}_{a,k}^{DMC}$ is limited with ±14.30 kg/s, which is ±1.36% of the normal operation range. The variation of $\widetilde{u}_{f,k}^{DMC}$ is not represented because it is limited with a small β_f. Although it is not shown, the DMC controls satisfies the optimization window constrained by the DCL logic in (28).

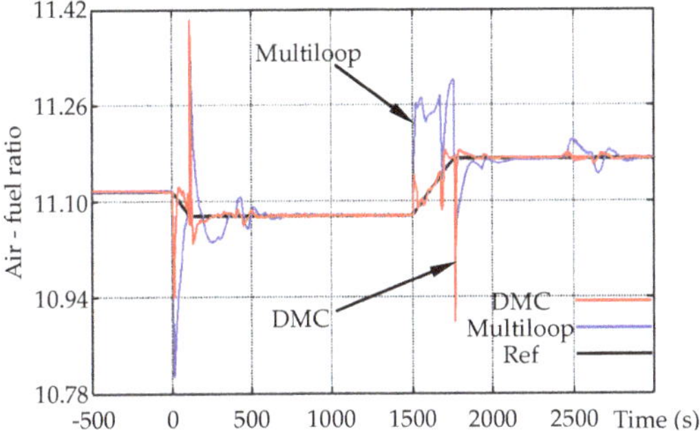

Figure 14. Comparison of air in 1000-MW plant.

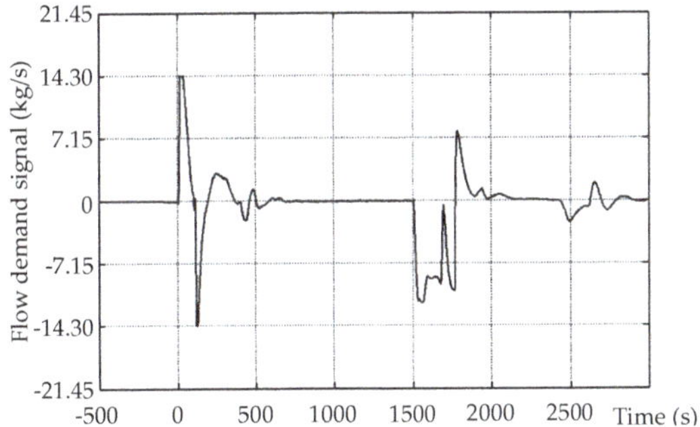

Figure 15. DMC supplementary air flow demand of 1000 MW plant.

In Figure 14, the responses of the second step change between 1500 s and 1750 s show the clear comparison between the two controls. A large AFR of the conventional control can be reduced by the proposed DMC. Accordingly, NOx emissions can be reduced at the same time. The numerical comparison is shown in Table 4. For the once-through type plant, the squared error sum was reduced to 14.36% with the proposed supplementary DMC.

Table 4. Squared error sum comparison of AFR in 1000 MW plant.

Multi-loop	8.613
DMC	1.237
Percentage of DMC/ Multi-loop	14.36%

5. Conclusions

In this paper, we proposed a supplementary DMC for tighter control of the AFR to reduce the environmental emissions of thermal power plants. Two manipulated variables of the DMC are the air flow demand and fuel flow demand. The amplitudes of the supplementary control signals are limited by the DMC constraints to minimize the effect on the main control loops of power plants, while optimized within the optimization window constrained by the conventional DCL logic.

Simulations considered two different types of power plants, which are 600-MW drum-type oil fired plant and 1000 MW once-through type coal-fired plant. Without affecting existing power plant operation, the proposed supplementary DMC shows very tight control of the AFR in transient period. Therefore, the reduction in environmental emissions in various thermal power plants can be expected.

Because the supplementary structure maintains the existing multi-loop control, the proposed control can be directly applied to the currently operating power plant. In addition, since the plant operator can easily return to the original multi-loop control logic, it can cope easily with emergency situations.

For practical implementation, the number of adjustable parameters of DMC is quite high and an additional computer server might be necessary over the exiting DCS (distributed control system). Therefore, future research could investigate a simpler control structure as a supplementary control which can be implemented into DCS of practical power plant.

Author Contributions: U.-C.M. and T.L. designed the basic idea and the principles of the overall work. T.L. and E.H. realized simulations. U.-C.M. prepared the initial draft of the paper and K.Y.L. proposed some technical comments and edited the final draft of the paper.

Funding: This research was supported by the Chung-Ang University Graduate Research Scholarship in 2018.

Conflicts of Interest: The authors declare no conflict of interest.

References

1. Lewtas, J. Air pollution combustion emissions: Characterization of causative agents and mechanisms associated with cancer, reproductive, and cardiovascular effects. *Rev. Mutat. Res.* **2007**, *636*, 95–133. [CrossRef] [PubMed]
2. Streets, D.G.; Waldhoff, S.T. Present and future emissions of air pollutants in China: SO2, NOx and CO. *Atmos. Environ.* **2000**, *34*, 363–374. [CrossRef]
3. Wiatros-Motyka, M. *Optimising Fuel Flow in Pulverised Coal and Biomass-Fired Boilers*; IEA Clean Coal Centre: London, UK, 2016.
4. Li, S.; Xu, T.; Hui, S.; Wei, X. NOx emission and thermal efficiency of a 300MWe utility boiler retrofitted by air staging. *Appl. Energy* **2009**, *86*, 1797–1803. [CrossRef]
5. Dukelow, S.G. *The Control of Boilers*; Instrument Society of America, Research Triangle Park: North Carolina, USA, 1986.
6. Wang, Y.; Yu, X. New coordinated control design for thermal-power-generation units. *IEEE Trans. Ind. Electron.* **2010**, *57*, 3848–3856. [CrossRef]
7. Wei, L.; Fang, F. H∞-LQR-based coordinated control for large coal-fired boiler–turbine generation units. *IEEE Trans. Ind. Electron.* **2017**, *64*, 5212–5221. [CrossRef]
8. Takiyama, T.; Shiomi, E.; Morita, S. Air–fuel ratio control system using pulse width and amplitude modulation at transient state. *JSAE Rev.* **2001**, *22*, 537–544. [CrossRef]
9. Bhowmick, M.; Bera, S.C. An approach to optimum combustion control using parallel type and cross-limiting type technique. *J. Process Control* **2012**, *22*, 330–337. [CrossRef]
10. Liu, Z.H.; He, G.X.; Wang, L.Z. Optimization of furnace combustion control system based on double cross-limiting strategy. In Proceedings of the 2010 International Conference on Intelligent Computation Technology and Automation, Changsha, China, 11–12 May 2010; pp. 858–861.
11. Zanoli, S.M.; Barchiesi, D.; Astolfi, G.; Barboni, L. Advanced control solutions to increase efficiency of a furnace combustion process. In Proceedings of the European Control Conference (ECC), Zurich, Switzerland, 17–19 July 2013; pp. 4316–4321.

12. Hou, Z.; Xiong, S. On Model-Free Adaptive Control and its Stability Analysis. *IEEE Trans. Auto. Control* **2019**. [CrossRef]

13. Bara, O.; Fliess, M.; Join, C.; Day, J.; Djouadi, S.M. Toward a model–free feedback control synthesis for treating acute inflammation. *J. Theor. Biol.* **2018**, *478*, 26–37. [CrossRef]

14. Safaei, A.; Mahyuddin, M.N. Optimal model–free control for a generic MIMO nonlinear system with application to autonomous mobile robots. *Int. J. Adapt. Control Signal Process.* **2018**, *32*, 792–815. [CrossRef]

15. Lee, S.D.; Bae, Y.G.; Jung, S. A Decentralized Model Identification Scheme by Random-walk RLS Process for Robot Manipulators: Experimental Studies. *J. Control Autom. Syst.* **2019**, *17*, 1856–1865. [CrossRef]

16. Roman, R.C.; Precup, R.E.; Radac, M.B. Model–free fuzzy control of twin rotor aerodynamic systems. In *2017 25th Mediterranean Conference on Control and Automation (MED)*; IEEE: Valletta, Malta, 2017; pp. 559–564.

17. Precup, R.E.; Radac, M.B.; Roman, R.C.; Petriu, E.M. Model–free sliding mode control of nonlinear systems: Algorithms and experiments. *Inf. Sci.* **2017**, *381*, 176–192. [CrossRef]

18. Roman, R.C.; Precup, R.E.; David, R.C. Second Order Intelligent Proportional-Integral Fuzzy Control of Twin Rotor Aerodynamic System. *Procedia Comput. Sci.* **2019**, *139*, 372–380. [CrossRef]

19. Kim, D.; Oh, H.S.; Moon, I.C. Black-box Modeling for Aircraft Maneuver Control with Bayesian Optimization. *J. Control Autom. Syst.* **2019**, *17*, 1558–1568. [CrossRef]

20. Roman, R.C.; Radac, M.B.; Precup, R.E.; Petru, E.M. Virtual reference feedback tuning of model–free control algorithms for servo systems. *Machines* **2017**, *5*, 25. [CrossRef]

21. Qin, S.; Badgwell, T. A survey of industrial model predictive control technology. *Control Eng. Pract.* **2003**, *11*, 733–764. [CrossRef]

22. Gracia, C.E.; Prett, D.M.; Morari, M. Model predictive control: Theory and practice—A survey. *Automatica* **1989**, *25*, 335–348. [CrossRef]

23. Lee, J.H. Model predictive control in the process industries: Review, current status and future outlook. In Proceedings of the 2nd Asian Control Conference, Seoul, Korea, 22–25 July 1997; pp. 435–438.

24. Moon, U.C.; Lee, K.Y. Step-response model development for dynamic matrix control of a drum-type boiler-turbine system. *IEEE Trans. Energy Convers.* **2009**, *24*, 423–430. [CrossRef]

25. Moon, U.C.; Lee, Y.; Lee, K.Y. Practical dynamic matrix control for thermal power plant coordinated control. *Control Eng. Pract.* **2018**, *71*, 154–163. [CrossRef]

26. Lee, Y.; Yoo, E.; Lee, T.; Moon, U.C. Supplementary control of conventional coordinated control for 1000MW ultra-supercritical thermal power plant using dynamic matrix control. *J. Electr. Eng. Technol.* **2018**, *13*, 97–104.

27. Moon, U.C.; Lee, K.Y. An adaptive dynamic matrix control with fuzzy-interpolated step-response model for a drum-type boiler-turbine system. *IEEE Trans. Energy Convers.* **2011**, *26*, 393–401. [CrossRef]

28. Usoro, P.B. Modeling and Simulation of a Drum Boiler-Turbine Power Plant under Emergency State Control. Master's Thesis, Massachusetts Institute. Technology, Cambridge, MA, USA, May 1977.

29. Heo, J.S.; Lee, K.Y. A multiagent-system-based intelligent reference governor for multiobjective optimal power plant operation. *IEEE Trans. Energy Convers.* **2008**, *23*, 1082–1092. [CrossRef]

30. Lee, K.Y.; Van Sickel, J.H.; Hoffman, J.A.; Jung, W.H.; Kim, S.H. Controller design for a large-scale ultrasupercritical once-through boiler power plant. *IEEE Trans. Energy Convers.* **2010**, *25*, 1063–1070. [CrossRef]

31. Go, G.; Moon, U.C. A water-wall model of supercritical once-through boilers using lumped parameter method. *J. Electr. Eng. Technol.* **2014**, *9*, 1900–1908. [CrossRef]

Article

Uncertainties in the Testing of the Coefficient of Thermal Expansion of Overhead Conductors

Miren T. Bedialauneta [1,*], Igor Albizu [2,*], Elvira Fernandez [1,*] and A. Javier Mazon [1,*]

[1] Department of Electrical Engineering, School of Engineering Bilbao, University of the Basque Country UPV/EHU, Plaza Ingeniero Torres Quevedo, 1, 48013-Bilbao, Spain
[2] Department of Electrical Engineering, School of Engineering Gipuzkoa, University of the Basque Country UPV/EHU, Otaola Etorbidea, 29, 20600-Eibar, Spain
* Correspondence: miren.bedialauneta@ehu.eus (M.T.B.); igor.albizu@ehu.eus (I.A.); elvira.fernandezh@ehu.eus (E.F.); javier.mazon@ehu.eus (A.J.M.)

Received: 13 December 2019; Accepted: 13 January 2020; Published: 14 January 2020

Abstract: Overhead lines can be replaced by high temperature low sag (HTLS) conductors in order to increase their capacity. The coefficients of thermal expansion (CTE) of the HTLS conductors are lower than the CTE of conventional conductors. The utilities and conductor manufacturers usually carry out the verification of the CTE of the overhead conductors in an actual size span. The verification is based on the observation of the change of the conductor length as a result of the conductor temperature change. This process is influenced by the coefficient of thermal expansion to be verified. However, there are other factors that also affect it. This paper analyzes the effect of some of the uncertainty sources in the testing of the coefficient of thermal expansion of the overhead conductors. Firstly, the thermal expansion process is described and the uncertainty sources related to the conductor and the line section are identified. Then, the uncertainty sources and their effect on the CTE testing are quantified.

Keywords: high temperature low sag conductor; coefficient of thermal expansion; overhead conductor; low sag performance

---

1. Introduction

Overhead lines can be replaced by high temperature low sag (HTLS) conductors in order to increase their capacity without the need to reinforce the towers [1].

The conventional conductors work at lower temperatures than HTLS conductors, and their coefficients of thermal expansion (CTE) are higher than the CTE of the HTLS conductors [2,3].

CIGRÉ has written a guide providing suggestions for methods and testing for qualifying HTLS conductors [4]. One of the main characteristics to be checked is the low sag performance. The low sag performance of HTLS conductors is due to a lower thermal expansion of the conductor. The CTE value of the strands can be determined by tests that are carried out according to certain standards [5–7]. However, there is no standard test for checking the value of the whole conductor thermal expansion coefficient. Usually, no verification is carried out and it is calculated by the method given in [8]. The conductor thermal expansion coefficient α ($°C^{-1}$) is obtained from the core and aluminum thermal expansion coefficients α_{core} and α_a ($°C^{-1}$), the elastic modulus E_{core} and E_a (kg/m^2), and the areas A_{core} and A_a (m^2) (1).

$$\alpha = (E_a \times A_a \times \alpha_a + E_{core} \times A_{core} \times \alpha_{core}) / (E_a A_a + E_{core} A_{core}) \tag{1}$$

The verification of the thermal expansion of the conductors requires changing the conductor temperature and measuring the related change of the conductor length. The conductor temperature increases with the current intensity due to the heat generated by Joule losses. The utilities and conductor manufacturers usually carry out this verification with the conductor installed in an actual size span. This span could be indoors, but due to the required size, it is usually outdoors. There are two types of outdoor testing: the operating line tests and the outdoor laboratory tests. In an operating line test, the conductor is at the total line voltage and the current through the line depends on the fluctuations of the power consumed by the customers and cannot be controlled, whereas in an outdoor laboratory test the conductor is at low voltage and it is possible to control the injected current. The best option for the verification of the low sag performance is an outdoor laboratory, because it is possible to control the injected current. This test is short because in a few hours the whole temperature range can be obtained. An example of an outdoor laboratory is the Powerline Conductor Accelerated Testing facility (PCAT) at Oak Ridge National Laboratory in USA [9]. For this verification the operating line test is also useful. When a utility installs the HTLS conductor in a line, some sensors should be installed so that it can evaluate the conductor performance. Moreover, the information provided by the same dynamic line rating (DLR) systems that are installed for the evaluation of the ampacity can be used for the evaluation of the conductor performance [10]. The dynamic line rating systems measure weather magnitudes such as wind speed, air temperature and solar radiation, as well as the conductor temperature and the conductor tension or sag values. The main disadvantage of the operating line test is that the current intensity, and consequently the conductor temperature, depends on the power flow of the line. Some examples of operating lines tests are shown in [11–14].

The estimation of the coefficient of thermal expansion requires quantifying the conductor length change with temperature. It is difficult to measure directly the conductor length in the catenary. When the conductor length increases, the conductor tension decreases and the sag increases. Therefore, both the conductor tension and the sag can be used to quantify the conductor length change with temperature. It is easier to measure the conductor tension, especially in an operating line. For this reason, in the analysis presented in the paper, the tension is the chosen magnitude for representing the conductor length change.

The low sag performance evaluation is achieved by comparing the conductor tension and temperature measured values with the theoretical tension–temperature performance. Besides the CTE influence of the conductor, other factors also affect the tension–temperature performance. For example, an error in the conductor weight or in the span length can result in a deviation of the tension–temperature performance that could be wrongly attributed to the CTE of the conductor if these errors are not considered. Because of this, it is necessary identify and quantify the influence of the factors that affect the tension–temperature performance.

The conductor tension and temperature measured by a monitoring system in a line section depends on several factors (Figure 1). Obviously, the characteristics of the line section are the most determinant factors. The conductor characteristics such as the weight, the modulus of elasticity and the coefficient of thermal expansion affect the sag–tension performance. The span length and the number of suspension towers between the two tension dead-ends also affect the sag–tension performance. Apart from the physical configuration of the line, other factors affect the measured performance. The weather conditions during the measurement period affect the measured tension values because of the overload due to the wind, the rain and the ice. Finally, the uncertainties of the measurement systems influence the measured performance.

Besides, some parameters such as the wind speed and the conductor temperature may vary along the line section and, for this reason, the limitation of a local measurement should be taken into account. Furthermore, midspan joints or conductor clamps may introduce a discontinuity in the conductor temperature due to a distortion in the thermal balance.

This paper presents the results of the first analysis towards the quantification of the effects of the uncertainties in the testing of the CTE of overhead conductors. It analyzes the effect of the uncertainties due to the conductor and the line section in the testing of the CTE of the overhead conductors. Firstly, the thermal expansion process is described and the uncertainty sources related to the conductor and the line section are identified. Then, the uncertainty sources and their effect on the CTE testing are quantified. The objective of the study is to quantify the effect of these uncertainties so that they are identified as important uncertainty sources or sources that can be neglected.

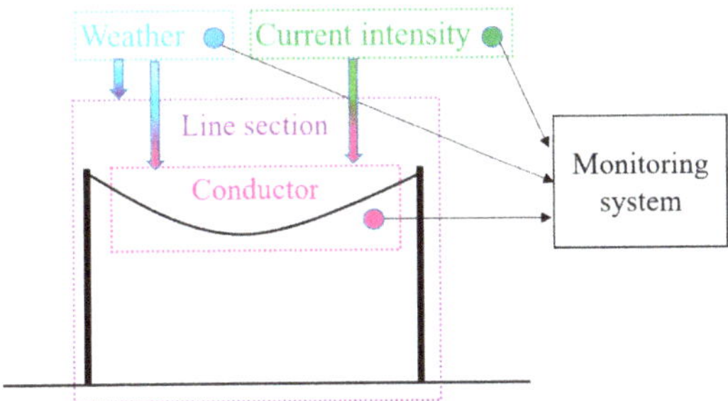

Figure 1. Line section monitoring.

2. The Thermal Expansion Process

Thermal expansion is the result of a temperature change of the conductor. This temperature change is due to current intensity or weather magnitude changes. The thermal expansion process is shown in Figure 2. The first result is a change of the conductor length (first step). The consequence of the conductor length change is a change in the catenary shape that results in a tension change (second step). The tension change affects the conductor length because of the elastic performance of the conductor (third step). The elastic performance counteracts the thermal expansion length change. For example, when the temperature increases, the conductor length increases (first step), the tension decreases (second step) and the conductor length decreases (third step), reducing the initial conductor length increase. As a result of the conductor length change (first step), a change in the catenary shape occurs that results in a tension change (second step), that results in a conductor length change (third step), that results in a catenary shape change (second step), etc.; in other words, the second and third steps influence each other until some final tension and length values are obtained, where neither more conductor length change nor catenary shape change occurs.

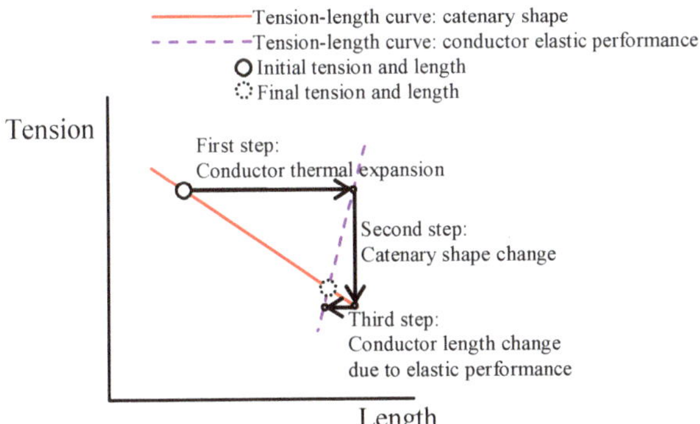

Figure 2. Conductor tension and length change due to the conductor temperature change.

The described process depends on the conductor characteristics. The parameters that affect each step are shown in Table 1. The conductor characteristics that influence the thermal expansion process are the weight, the CTE and the modulus of elasticity. In the case of non-homogeneous conductors, the CTE and the modulus of elasticity have different values for the core and the outer aluminum.

Table 1. Conductor parameters that affect the thermal expansion process.

Step	Conductor Parameter Related to the Step
1. Conductor thermal expansion	Coefficient of thermal expansion
2. Catenary shape change	Weight
3. Conductor elastic performance	Modulus of elasticity

The higher the coefficient of thermal expansion, the higher the increase in length of the first step. As a result, the higher the coefficient of thermal expansion, the greater the influence of this parameter in the expansion process. For the modulus of elasticity, the higher the value is, the lower the length recovery of the third step. As a result, the higher the modulus of elasticity is, the lower the influence of this parameter in the expansion process.

The catenary length L (m) is given by Equation (2). It depends on the horizontal tension H (kg), the conductor weight per unit length ω (kg/m) and the span length a (m).

$$L = 2 \times c \times \sin h\left(\frac{a/2}{c}\right) \tag{2}$$

$$c = \frac{H}{\omega} \tag{3}$$

The relation between the tension and the length of the catenary is not linear (Figure 3a). For a certain length change, the tension change is higher for high tension values. As a consequence, the same temperature change results in a lower tension decrease at lower tension values (Figure 3b). In Figures 2 and 3b, the tension–length curve of the catenary is linearized around the initial tension–length value in order to simplify the explanation of the process.

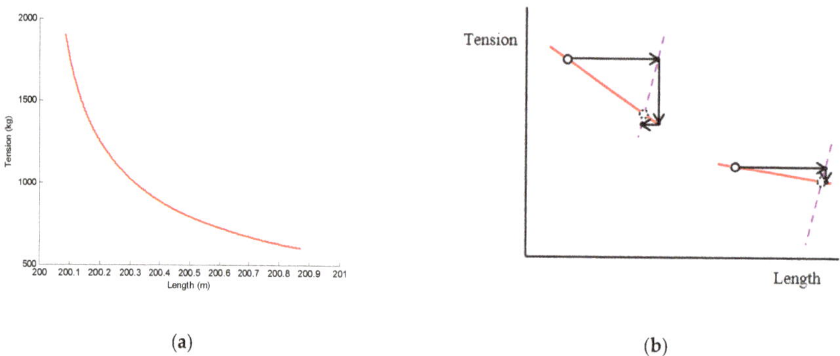

Figure 3. (**a**) Length–tension curve for a catenary (a = 200 m, ACSR Hawk). (**b**) Tension change due to the same conductor temperature change.

3. Identifying Uncertainty Sources in the Tension–Temperature Performance

The measured tension and temperature values are compared with the values given by a theoretical model. There is a deviation between the measured and calculated values if the parameter values assumed in the model, such as the conductor weight or modulus of elasticity, differ from the actual values. There is also a deviation if the model does not take into account the factors that affect the tension–temperature performance.

3.1. Parameters

The parameters that are used in the sag–tension calculation methods are the conductor parameters (weight, modulus of elasticity, CTE) and the span length [15].

The influence of the error of each conductor parameter can be analyzed by taking into account the process described in Figure 2. A higher CTE value (positive error) results in a higher change of the conductor length (first step). As a consequence, the tension change due to the catenary shape change is higher too, and the final tension change is also higher. A positive error in the weight results in a lower change of the conductor tension due to a lower slope of the catenary length–tension curve. As a consequence, the final tension change is lower. A positive error in the modulus of elasticity results in a lower change in the conductor length (third step). As a consequence, the tension change due to the catenary shape change is lower and the final tension change is higher. Table 2 shows the relation between the positive and negative errors of the parameters and the lower and higher tension change result.

Table 2. Tension deviation as a function of the parameter error.

	Positive Error (Higher Value)	Negative Error (Lower Value)
Weight	Lower tension change	Higher tension change
Modulus of elasticity	Higher tension change	Lower tension change
CTE	Higher tension change	Lower tension change
Span length	Lower tension change	Higher tension change

According to Equation (2), the relation between the tension and the length depends on the span length *a*. If there is an error in the assumed span length value, the observed tension change due to the thermal expansion is different from the expected value. The results show that the effect of the span length error is similar to the effect of the conductor weight (Table 2). This is because the same percentage error in both cases causes a similar deviation of the slope of the catenary length–tension curve.

Supposing a tension–temperature reference, the tension value depends on the parameter error, at a different conductor temperature (Figure 4). The solid curve corresponds to the performance without errors. The dotted curve corresponds to a performance where the tension change with temperature change is lower. This performance can be for example due to a positive error in the weight or the span length. The dashed curve corresponds to a performance where the tension change with temperature change is higher.

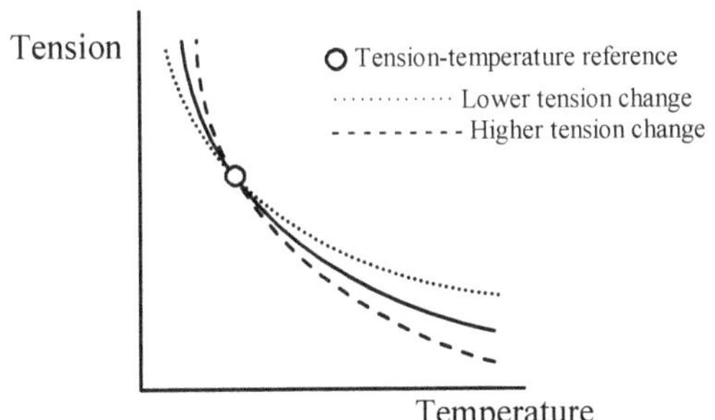

Figure 4. Tension–temperature curve with conductor parameter errors.

3.2. Model

Usually, some assumptions are made in order to simplify the sag–tension models. One assumption assumes that the catenary end points are fixed: no movement of the strain towers is assumed. Another assumption is the modelling of the whole catenary length with the conductor parameters: the insulating string, midspan joints and conductor clamps are not taken into account.

3.2.1. Strain Tower Deflection

In the second step of the thermal expansion process, the catenary shape change results in a conductor tension change. This tension change is also noticed in the strain tower where the conductor is attached. Several conductors are attached in a strain tower, three conductors per circuit and span. The tension change can be different between contiguous spans due to different span lengths or installation tensions. As a result, small movements on the cross-arms can occur. These movements affect the catenary shape and the conductor tension.

3.2.2. Insulating Strings, Midspan Joints and Conductor Clamps

In the first step of the thermal expansion process, the conductor length changes as a result of a temperature change. The length change is proportional to the conductor length. In a simplified model, the conductor length is assumed to be the same as the catenary length. However, the insulating strings are part of the catenary. As a result, the conductor length is lower than the catenary length. For this reason, the length change that results from a temperature change is lower than that assumed in the simplified model.

Besides, midspan joints or conductor clamps introduce a distortion in the temperature due to a different thermal performance compared with the conductor [16].

4. Quantifying the Source Errors

Besides identifying the uncertainty sources that affect the sag–tension performance, it is important to quantify the value of the uncertainty so that their effect can be quantified.

4.1. Conductor Weight

According [15], the conductor weight is one of the basic error sources. Typically, the conductor weight exceeds the nominal value by 0.2% to 0.6%. Because of tarnishing effects, the conductor's weight increases slightly during its lifetime.

The conductor manufacturing norms limit the conductor weight deviation. In accordance [17–19], the conductor linear weight cannot vary more than 2% from the nominal value. In the course of the manufacturing process, the conductor weight is tested. According to the information provided by different manufacturers, the weight typically does not vary from the nominal value more than 0.5%. Besides, the deviation inside series production is around 0.2%.

4.2. Conductor Modulus of Elasticity

During the conductor manufacturing process, it is not mandatory to test the modulus of elasticity. It is calculated according the method given in [8]. The conductor elastic modulus E (kg/m^2) is obtained from core E_{core} (kg/m^2) and aluminum E_a (kg/m^2) elastic modulus and their respective areas, A_{core} and A_a (m^2) (4).

$$E = (E_a \times A_a + E_{core} \times A_{core}) / (A_a + A_{core})$$ (4)

The conductor modulus of elasticity depends on the modulus of elasticity of the wires. However, as the wires are helically wound, the strain in the conductor axis is higher than the strain in the wire axis [8]. Therefore, the modulus of elasticity of the conductor is lower than the modulus of elasticity of the wires. The manufacturing norms define a maximum and a minimum value for the pitch diameter ratio of each conductor layer. According to the information obtained from a manufacturer, the modulus of elasticity can vary 4% between two conductors with different pitch diameter ratios. The conductor with the lower pitch diameter ratio has a lower modulus of elasticity value and vice versa.

The manufacturer and the purchaser, in order to verify the actual value of the modulus of elasticity, can accord to carry out a stress–strain test [17,20]. In this test, the conductor is loaded increasing and decreasing the load in several cycles. The modulus of elasticity is obtained from the stress–strain slope of the load decreasing period. However, the obtained value of the modulus of elasticity is different depending on the load decreasing period chosen (30%, 50%, 70% or 80% of the Rated Breaking Strength, (RBS)). The slope is higher if the load value of the cycle is low. In other words, the modulus of elasticity obtained from the unloading of the 30% RBS cycle is higher than that obtained from the 80% RBS cycle. Furthermore, it depends on the number of points taken in the decreasing period because the slope is not constant. According to the information obtained from a manufacturer, the modulus of elasticity can vary 5% depending on the period of the test considered for the calculation.

4.3. Span Length

The error in the span length is another typical error source. Because of changes associated with the installation process, the actual span length can differ from the length of the original project design. Therefore, it is recommended that a topographer measures the actual length and reflects it in a new document. For this reason, it could be an error if the span length is taken from the line project information given by the utility and the topographer correction is not included.

In this project, in order to evaluate the uncertainties of the testing of the conductor CTE, three different lines in operation have been monitored. The number of monitored spans is four because in one of the lines two contiguous spans have been monitored. Table 3 shows the span length values given by the utility and the values measured by the topographer. In most of the cases the correspondence is good but in one case an error of 16.8% has been obtained. This shows the assumed risk if the span length is not measured. There are uncertainties in tower placing and it is necessary to have GIS map of all towers.

Table 3. Measured span lengths.

Span Length Given by the Utility (m)	Measured Span Length (m)	Span Length Error (%)
202	168	−16.8
100	100.85	0.8
89	88.95	−0.1
284	282.49	−0.5

4.4. Strain Tower Deflection

Because of imbalances between adjacent spans the strain towers experience forces. These forces origin deflections that depend on the stiffness of the attachment point. The strain tower deflection calculation is complex. It depends on the design and characteristics of the tower and the characteristics and operation performance of the conductors attached to it. It requires a detailed mechanical model of the tower and the use of advanced calculation tools such as the PLS-CADD.

4.5. Insulating String, Midspan Joints and Conductor Clamps

The length of the insulating strings depends on the line voltage. The higher the voltage is, the longer the insulation string. Similarly, the higher the voltage is, the longer the span. The relation between the insulating string and the span length depends of each case, but the insulating string length is usually around 2% of the span length.

The length of the midspan joints and conductor clamps is lower than the length of the insulating strings. Compared with the length of the conductor, the length of the midspan joints and conductor clamps is negligible.

5. Quantifying the Effect of the Source Errors: the Equivalent Coefficient of Thermal Expansion (CTE) Error

In order to quantify and compare the effect of the different error sources, the equivalent CTE error is defined. The equivalent CTE error of an error source is the CTE error value that results in the same tension deviation. For example, the effect of a 2% positive error of the conductor weight is equivalent to the effect of a 1% negative error of the CTE if both errors result in the same tension deviation.

The equivalent CTE error is not a constant value. It depends on the span length value, the tension value and the characteristics of the conductor. In the case of the span length, the effect of a certain error—for example a 5% positive error in the modulus of elasticity—is different if the span length is 150 m or 300 m. The equivalent CTE error also depends on the tension value because when the tension is low, the tension change is lower, as has been shown above (Figure 3) and this affects the effect of the error source. Finally, the effect of a certain error is different depending on the characteristics of the conductor.

A span of 200 m with an ACSR Hawk conductor has been chosen as the base case for the analysis. Figure 5 shows the CTE equivalent error of the weight, the modulus of elasticity and the span length errors, and the insulating string.

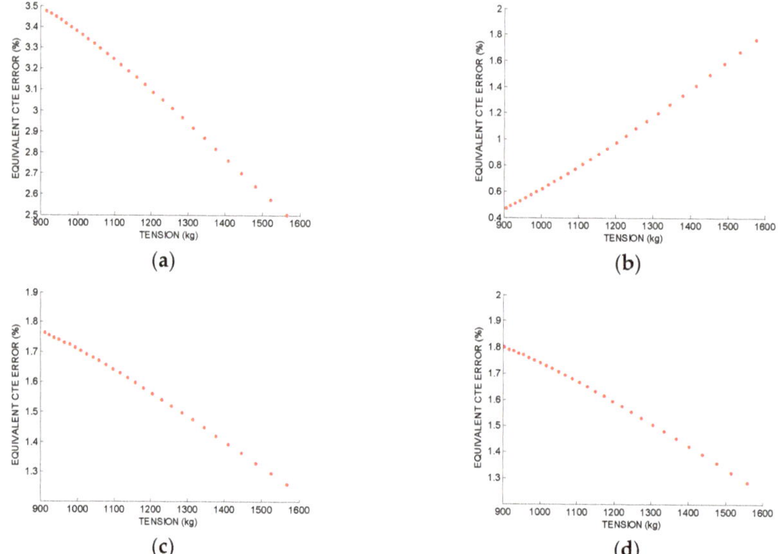

Figure 5. Equivalent coefficients of thermal expansion (CTE) error. (**a**) 2% weight error; (**b**) 5% modulus of elasticity error; (**c**) 1% span length error; (**d**) 2% insulating string length.

In Figure 5a, the equivalent CTE error for a 2% error in the weight is shown. The tension range corresponds to the tension values between −5 °C and 80 °C when the conductor is installed at the 15% of the Rated Breaking Strength at 15 °C. Therefore, the average CTE equivalent error is 3.1%. If the weight error is 0.2%, the average CTE equivalent error is 0.3%.

In this analysis, a value of 5% error has been assumed in order to quantify the effect of the modulus of elasticity error. In Figure 5b, the equivalent CTE error for a 5% error in the modulus of elasticity is shown. The average CTE equivalent error is 0.9%.

In Figure 5c, the equivalent CTE error for a 1% error in the span length is shown. The average CTE equivalent error is 1.6%.

In order to quantify the effect of the insulating string, a modified model where the conductor length and the catenary length are different is used. Then, the simplified model is used in order to calculate the equivalent CTE error that gives the same result. In Figure 5d, the equivalent CTE error for a 2% insulating string length is shown. The average equivalent CTE error is 1.6%.

Table 4 shows the equivalent CTE error for several error sources: weight error, modulus of elasticity error, span length error, and insulating string length error. The average value and the change of the equivalent CTE error in the temperature range between −5 °C and 80 °C is shown. The equivalent CTE errors obtained for the same weight and span length errors are the same. There is also a relationship between the effect of these errors and the effect of the insulating string length. In this case, the effect of a 2% insulating string length is equivalent to the effect of a 1% weight or span error.

Table 4. Equivalent CTE error.

Error Sources	(%)	Average Equivalent CTE Error (%)	Equivalent CTE Error Change in Temperature Range (%)
Weight error	0.2	0.3	0.1
	1	1.6	0.5
	2	3.1	1
Modulus of elasticity error	2.5	0.5	0.6
	5	0.9	1.3
	7.5	1.4	1.9
Span length error	1	1.6	0.5
	2	3.1	0.9
	3	4.6	1.4
Insulating string length	1	0.8	0.3
	2	1.6	0.5
	4	2.9	1.6

5.1. Equivalent CTE Error of the HTLS Conductors

The HTLS conductors have either low knee-point temperatures or core materials with low CTE values. The knee-point temperature is the temperature where the aluminum gets slack. Above the knee-point temperature, the modulus of elasticity and CTE values are those of the core. Hence, the CTE and the modulus of elasticity of a conductor changes depending on the temperature value. For this reason, the equivalent CTE error has different values above and below the knee-point.

In the case of the weight error, above the knee-point, the equivalent CTE error decreases. Figure 6a shows the equivalent CTE error change for a GTACSR conductor in a 200 m span. The low tension values correspond to the conductor above the knee-point temperature. Similar results are obtained for the span length error and the insulating string (Figure 6c,d).

Figure 6b shows the equivalent CTE error change, assuming an error of 5% of the modulus of elasticity both above and below the knee-point. Above the knee-point, the equivalent CTE error increases.

Table 5 shows the equivalent CTE errors for different HTLS conductors, when they are installed at 15% RBS at 15 °C in a 200 m span. Although there are some differences, the obtained results are similar for all the conductors.

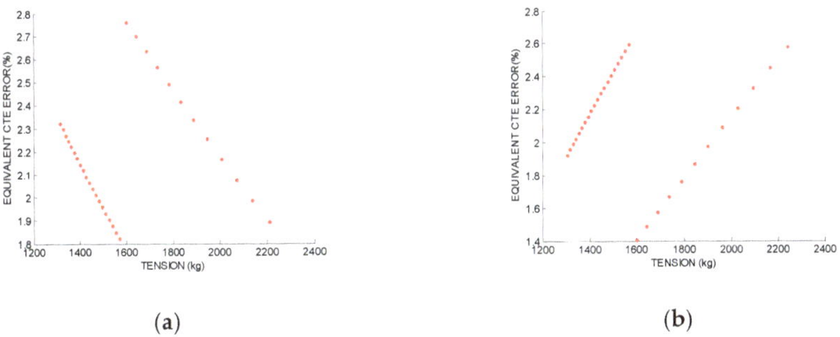

(a)   (b)

Figure 6. *Cont.*

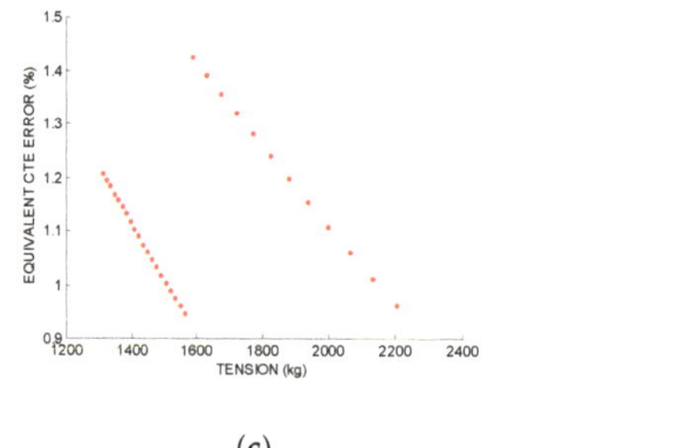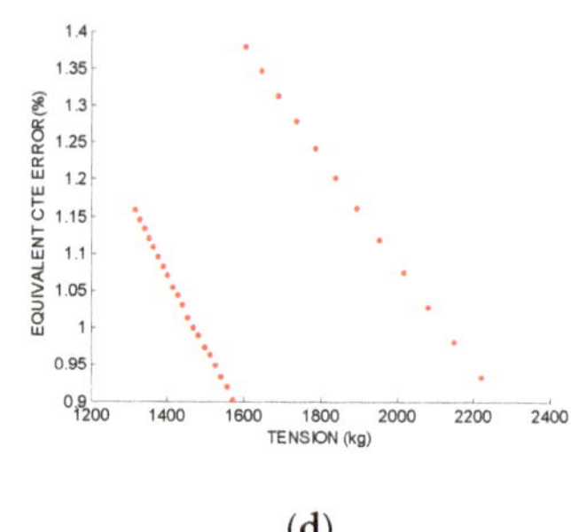

(c)   (d)

Figure 6. Equivalent CTE error of a GTACSR (Gap Type Thermal-Resistant Aluminum Alloy Conductor Steel Reinforced) conductor above and below the knee-point. (**a**) 2% weight error; (**b**) 5% modulus of elasticity error; (**c**) 1% span length error; (**d**) 2% insulating string length.

Table 5. Equivalent CTE errors for different HTLS conductors.

	2% Weight Error		5% Modulus of Elasticity Error		1% Span Length Error		2% Insulating String Length	
	Below the Knee-Point	Above the Knee-Point	Below the Knee-Point	Above the Knee-Point	Below the Knee-Point	Above the Knee-Point	Below the Knee-Point	Above the Knee-Point
ACSR	3	2.8	1.1	1.3	1.5	1.4	1.5	1.5
GTACSR	2.4	2.1	1.9	2	1.2	1	1.2	1.1
ACSS	3.3	3.1	0.7	1	1.7	1.6	1.7	1.6
ZTACIR	3	2.4	1.1	1.7	1.5	1.2	1.5	1.3
ACCR	2.7	2.7	1.5	1.4	1.4	1.4	1.5	1.4
ACCC	2.5	1.9	1.7	2.2	1.3	1	1.4	1.1

ACSR—Aluminium Conductor Steel Reinforced; GTACSR—Gap Type Thermal-Resistant Aluminum Alloy Conductor Steel Reinforced; ACSS—Aluminum Conductor Steel Supported; ZTACIR—Super Thermal-Resistant Aluminum Alloy Conductor Invar Reinforced; ACCR—Aluminum Conductor Composite Reinforced; ACCC—Aluminum Conductor Composite Core.

5.2. Influence of Span Length Error in the Equivalent CTE Error

The effect of a certain error is different depending of the value of the span length.

In the case of the conductor weight error, the equivalent CTE error increases with the span length (Figure 7a). However, the influence of the span length is low.

In the case of the conductor modulus of elasticity error, the equivalent CTE error decreases with the span length (Figure 7b).

In the case of the span length error, the effect is double. On the one hand, a certain span length deviation, for example 1 m, has a lower relative value for a long span. For example, it is a 1% error for a 100 m span but a 0.5% error for a 200 m span. On the other hand, the same percentage error results in a different equivalent CTE error depending on the span length (Figure 7c). However, the influence of the span length is low.

In the case of the insulating string length, the effect is also double. On the one hand, for the same insulating string, the relative value has a lower value for a long span. On the other hand, the same percentage error results in a different equivalent CTE error depending on the span length (Figure 7d). However, the influence of the span length is low.

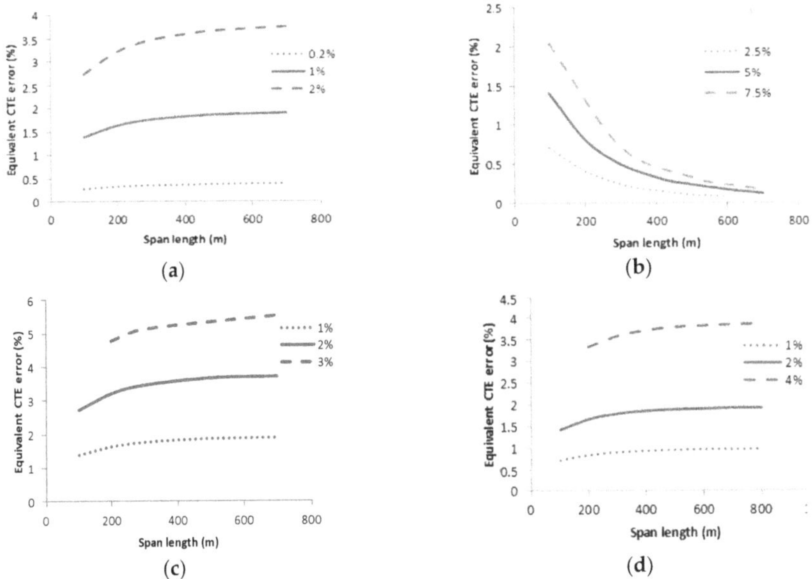

Figure 7. Equivalent CTE error as a function of the span length. (**a**) Weight error; (**b**) Modulus of elasticity error; (**c**) Span length error; (**d**) Insulating string length.

5.3. Analysis of the Results

The results obtained for the conductor weight, modulus of elasticity and span length errors, and the insulating string length are analyzed in order to obtain some practical conclusions.

In the case of the weight, the equivalent CTE error of a 2% weight error is around 3%. The obtained result is similar for all the HTLS overhead conductors. The influence of the weight error increases at high temperature (low tension) values but decreases above the knee-point temperature. The maximum allowable temperature of high temperature conductor is between 150 and 250 °C [2]. Anyway, the change in the equivalent CTE error for these reasons is low, below 1%. The influence of the span length in the effect of the weight error is negligible. As a result, the only way to minimize the effect of the weight error is to minimize the source error. This can be achieved if the weight is measured in the manufacturing process. In this case, the source error is assumed to be around 0.2% and the equivalent CTE error is reduced in proportion to the source error reduction.

In the case of the modulus of elasticity, the equivalent CTE error of a 5% error is around 1%. The obtained result is similar for all the HTLS overhead conductors. The influence of the modulus of elasticity decreases at high temperature (low tension) values but increases above the knee-point temperature. Regardless, the change in the equivalent CTE error for these reasons is low, below 1%. The influence of the span length in the effect of the modulus of elasticity error is high. The effect of the modulus of elasticity error is reduced if a high span length is chosen for the test. Besides, a stress–strain test of the conductor can be carried out in order to reduce the uncertainty of the error source. In this case, the error is assumed to be around 5%.

In the case of the span length error, the influence is qualitatively and quantitatively the same as the weight error. As a result, the only way to minimize the effect is to minimize the source error. This can be achieved if the span length error is measured by a topographer. In this case, the source error and its effect are negligible.

In the case of the insulating string effect, its influence is removed if it is taken into account in the sag–tension model. Therefore, a model that takes it into account should be used. This model assumes that the conductor is shorter than the catenary.

As a result, if the conductor weight is tested (e.g., 0.2% error), the modulus of elasticity is tested (e.g., 5% error) and the span length is measured (0% error), if the span length is at least 200 m, and the insulating string is included in the model, the equivalent CTE error is around 1%. Therefore, the influence of these error sources is low. However, if the conductor weight is not measured (e.g., 2% error), the modulus of elasticity is not tested (e.g., 7.5% error), the span length is not measured (e.g., 1% error), if the span length is 100 m, and the insulating string is not included in the model (e.g., 2% insulating string length), then the equivalent CTE error is around 7%.

6. Conclusions

The thermal expansion process of an overhead conductor has been described in detail. As a result, the uncertainty sources of the sag-tension performance related to the thermal expansion process have been identified. The conductor weight, the conductor modulus of elasticity, the span length, the insulating string and the tower deflection affect the sag-tension performance.

Once they have been identified, the uncertainty sources have then been quantified. The conductor uncertainties are easy to quantify. The weight uncertainty is a manufacturing requirement. The maximum accepted deviation is 2%. The modulus of elasticity can be tested too. However, there is an uncertainty that is assumed to be below 5%. The span length uncertainty is removed if a topographer measures its value. The tower deflection quantification requires a detailed model of the tower and it is the most difficult to quantify. The insulating string length is usually a 2% of the span length.

Once they have been quantified, their effect in the sag-tension performance has been analyzed. The results related to the conductor weight, modulus of elasticity, span length and insulating string have been presented. The quantification of the effect of the tower deflection requires a specific modelling of the towers.

In order to quantify and compare the effect of the different error sources, the equivalent CTE error of an error source is defined as the CTE error value that results in the same tension deviation. The equivalent CTE error depends on the tension value, the span length value, and the characteristics of the conductor. The influence of the tension value is low—below 1%. The obtained results are similar for all the analyzed HTLS conductors, and although there is an influence of the knee-point, its influence is low—below 1%. The span length value considerably affects the effect of the modulus of elasticity error, and long spans decrease the effect of the modulus of elasticity error.

The analysis presented in the paper is the first step in evaluating the uncertainties in the testing of the CTE of overhead conductors. The influence of the strain tower deflection, the weather and the monitoring system has to be considered in order to obtain a whole picture of the testing uncertainties. The authors have installed some pilot systems in order to address these issues. The obtained results will be presented in a future paper.

Author Contributions: Conceptualization, M.T.B., I.A. and E.F.; methodology, M.T.B., I.A. and E.F.; software, M.T.B.; validation, M.T.B., I.A., E.F. and A.J.M.; formal analysis, M.T.B., E.F. and I.A.; investigation, M.T.B., E.F. and I.A.; data curation, M.T.B.; writing—original draft preparation, M.T.B., E.F., I.A. and A.J.M.; writing—review and editing, M.T.B., I.A. and E.F.; supervision, M.T.B., E.F., I.A. and A.J.M.; funding acquisition, M.T.B., E.F., I.A. and A.J.M. All authors have read and agreed to the published version of the manuscript.

Funding: This research was funded by the MINISTERIO DE ECONOMÍA, INDUSTRIA Y COMPETITIVIDAD, Spain, DPI2016-77215-R (AEI/FEDER, UE).

Conflicts of Interest: The authors declare no conflict of interest.

References

1. *CIGRE Brochure (Ref. No. 353): Guidelines for Increased Utilization of Existing Overhead Transmission Lines;* CIGRE: Paris, France, 2008.
2. *CIGRE Brochure (Ref. No. 763): Conductors for the Uprating of Existing Overhead Lines;* CIGRE: Paris, France, 2019.

3. Albizu, I.; Mazon, A.J.; Valverde, V.; Buigues, G. Aspects to take into account in the application of mechanical calculation to high-temperature low-sag conductors. *IET Gener. Transm. Distrib.* **2010**, *4*, 631–640. [CrossRef]

4. *CIGRE Brochure (Ref. No. 426): Guide for Qualifying High Temperature Conductors for Use on Overhead Transmission Lines*; CIGRE: Paris, France, 2010.

5. *ASTM D696-08e1: Standard Test Method for Coefficient of Linear Thermal Expansion of Plastics Between $-30\,^{\circ}C$ and $30\,^{\circ}C$ with a Vitreous Silica Dilatometer*; ASTM International: West Conshohocken, PA, USA, 2008.

6. *ASTM E228-11: Standard Test Method for Linear Thermal Expansion of Solid Materials with a Push-Rod Dilatometer*; American National Standards Institute: New York, NY, USA, 2007.

7. *ASTM E831-12: Standard Test Method for Linear Thermal Expansion of Solid Materials by Thermomechanical Analysis*; ASTM International: West Conshohocken, PA, USA, 2012.

8. *IEC 61597: Overhead Electrical Conductors—Calculation Methods for Stranded Bare Conductors*; IEC: Geneva, Switzerland, 1995.

9. Powerline Conductor Accelerated Testing Facility (PCAT) at Oak Ridge National Laboratory. Available online: https://web.ornl.gov/sci/renewables/docs/factsheets/PCAT-factsheet.pdf (accessed on 13 December 2019).

10. Fernandez, E.; Albizu, I.; Bedialauneta, M.T.; de Arriba, S.; Mazon, A.J. System for ampacity monitoring and low sag overhead conductor evaluation. In Proceedings of the IEEE MELECON Conference, Yasmine Hammamet, Tunisia, 25–28 March 2012; IEEE: Piscataway, NJ, USA, 2012.

11. Clairmont, B. *High-Temperature Low-Sag Transmission Conductors*; EPRI: Palo Alto, CA, USA, 2008.

12. Custer, L. DOE-3M demonstration project. In Proceedings of the IEEE TP & C Line Design Meeting, Albuquerque, NM, USA, 15–19 October 2006.

13. Goel, A. Hydro one networks. In Proceedings of the IEEE TP & C Line Design Meeting, Las Vegas, NV, USA, 1 February 2005.

14. Kupke, S. Pilot project—High temperature low sag conductors. In Proceedings of the CIGRE WG B2.42, Stockholm, Sweden, 21 May 2010.

15. *CIGRE Brochure (Ref. No. 324): Sag-Tension Calculation Methods for Overhead Lines*; CIGRE: Paris, France, 2007.

16. Olivieri, C.; de Paulis, F.; Orlandi, A.; Giannuzzi, G.; Salvati, R.; Zaottini, R.; Morandini, C.; Mocarelli, L. Remote monitoring of joints status on in-service high-voltage overhead lines. *Energies* **2019**, *12*, 1004. [CrossRef]

17. *EN 50182: Conductors for Overhead Lines—Round Wire Concentric Lay Stranded Conductors*; CENELEC: Brussels, Belgium, 2001.

18. *IEC 62420: Concentric Lay Stranded Overhead Electrical Conductors Containing One or More Gap(s)*; IEC: Geneva, Switzerland, 2008.

19. *EN 50540: Conductors for Overhead Lines—Aluminium Conductors Steel Supported (ACSS)*; CENELEC: Brussels, Belgium, 2010.

20. *A Method of Stress-Strain Testing of Aluminum Conductor and a Test for Determining the Long Time Tensile Creep of Aluminum Conductors in Overhead Lines*; Electrical Technical Committee of the Aluminum Association: Arlington, TX, USA, 1999.

Article

Multiresolution GPC-Structured Control of a Single-Loop Cold-Flow Chemical Looping Testbed

Shu Zhang [1,†], Joseph Bentsman [1,*], Xinsheng Lou [2,‡], Carl Neuschaefer [2], Yongseok Lee [1] and Hamza El-Kebir [3]

[1]   Department of Mechanical Science and Engineering, University of Illinois at Urbana-Champaign, 1206 W Green St., Urbana, IL 61801, USA

[2]   Alstom Thermal Power, Windsor, CT 06095, USA

[3]   Department of Aerospace Engineering, University of Illinois at Urbana-Champaign, 104 S Wright St., Urbana, IL 61801, USA

*   Correspondence: jbentsma@illinois.edu; Tel.: +01-217-244-1076

†   Current address: Citadel LLC, New York, NY 10022, USA.

‡   Current address: GE Steam Power, Windsor, CT 06095, USA.

Received: 10 February 2020; Accepted: 25 March 2020; Published: 7 April 2020

Abstract: Chemical looping is a near-zero emission process for generating power from coal. It is based on a multi-phase gas-solid flow and has extremely challenging nonlinear, multi-scale dynamics with jumps, producing large dynamic model uncertainty, which renders traditional robust control techniques, such as linear parameter varying H_∞ design, largely inapplicable. This process complexity is addressed in the present work through the temporal and the spatiotemporal multiresolution modeling along with the corresponding model-based control laws. Namely, the nonlinear autoregressive with exogenous input model structure, nonlinear in the wavelet basis, but linear in parameters, is used to identify the dominant temporal chemical looping process dynamics. The control inputs and the wavelet model parameters are calculated by optimizing a quadratic cost function using a gradient descent method. The respective identification and tracking error convergence of the proposed self-tuning identification and control schemes, the latter using the unconstrained generalized predictive control structure, is separately ascertained through the Lyapunov stability theorem. The rate constraint on the control signal in the temporal control law is then imposed and the control topology is augmented by an additional control loop with self-tuning deadbeat controller which uses the spatiotemporal wavelet riser dynamics representation. The novelty of this work is three-fold: (1) developing the self-tuning controller design methodology that consists in embedding the real-time tunable temporal highly nonlinear, but linearly parametrizable, multiresolution system representations into the classical rate-constrained generalized predictive quadratic optimal control structure, (2) augmenting the temporal multiresolution loop by a more complex spatiotemporal multiresolution self-tuning deadbeat control loop, and (3) demonstrating the effectiveness of the proposed methodology in producing fast recursive real-time algorithms for controlling highly uncertain nonlinear multiscale processes. The latter is shown through the data from the implemented temporal and augmented spatiotemporal solutions of a difficult chemical looping cold flow tracking control problem.

Keywords: chemical looping; wavelets; NARMA model; generalized predictive control (GPC)

---

1. Introduction

The current transition to clean power generation involves both the use of renewables, such as hydrokinetics [1], and cleaner coal-based techniques. The latter are projected to still supply power for the foreseeable future due to the abundance of coal in many industrialized and developing countries; however, they will be required to meet the hard caps on carbon emissions. Chemical looping (CL) is a near-zero emission coal-based technology in which multiple interacting loops of flowing, reactive, gas/solid mixtures produce energy via solid-oxygen carrier based combustion [2–4]. Chemical looping has remained an area of active research focused on improving its economic viability and reducing environmental footprint [5,6]. To reduce waste stream volumes and required energy, advanced optimizing control systems for the chemical looping process are required. However, the process exhibits extremely challenging nonlinear multi-scale dynamics that are hard to characterize and depend on a particular design. These features render traditional robust control techniques marginally successful in experimental trials.

The goal of the present paper is to present the development of the real-time computational control-oriented models and the corresponding model-based control design strategies found to provide the desired chemical looping tracking performance. In particular, we demonstrate a novel model-based process control methodology to control the pressure drop in the riser of a single loop chemical looping process, where the air flow rate is the controlled variable. This control approach was implemented and successfully tested on an industrial single loop cold gas/solid flow chemical looping testbed, where the previously available techniques had exhibited difficulties.

Prior to being able to control the process, it is imperative to characterize the system's response to control inputs. Classically, this would be done by devising a physical model of the system from first principles, but this often yields limited practical utility for increasingly complex nonlinear models when viewed from the perspective of process control design. To meet this challenge, an alternative technique, identification of a model constructed on the basis of the wavelet multiresolution analysis (MRA), is used in the present work. MRA has become one of the major tools in neural networks [7–10] and nonlinear system modeling [11–18]. Wavelet-based multiresolution decomposition has been proven to constitute a universal approximator for a wide range of function spaces in terms of linear combination of scaling and wavelet functions. Wavelet approximation has no smoothness requirement on the target function, making it an appropriate candidate for identification of complex nonlinear systems with multiscale dynamics, such as those encountered in chemical looping processes. Several controller designs incorporating wavelet system representations have been proposed in the literature. Reference [18] proposed adaptive adjustment of the model resolution and the corresponding structure of the nonlinear adaptive controller. However, no optimality in controller synthesis was introduced and no testing was done on the real multiresolution system. In [19] an optimal model predictive multiresolution control law with constraints was derived. However, the controller was given as a sequence of computational steps with no clear analytical formula for controller implementation and therefore no guarantee of the acceptable real-time performance; also no adaptation was included. In [20] utilization of wavelets in generalized predictive control (GPC)l has been proposed for reduction of the computational demands on the constrained GPC, but the application was not addressing multiresolution nonlinear system modeling and was restricted to linear systems only. Thus, there has been a clearly identifiable gap in producing an optimal adaptive control law with rate constraints and guaranteed real-time performance for systems with nonlinear multiscale dynamics.

The present work fills this gap through the development of the self-tuning wavelet MRA-based topology that combines temporal and spatiotemporal loops into a single closed loop control system. The GPC structure is employed for embedding into it the identified temporal nonlinear multiscale model due to its real-timable recursive receding horizon calculation, local optimality by the virtue of being a variant of LQG [21], relatively easy incorporation of rate constraints on the control signal, tunable robustness properties (not pursued in this paper), and natural embedding of the integrator to address setpoint step changes in the chemical looping based power generation.

The paper completes the brief presentation of the results of chemical looping project at Alstom Thermal Power given in a conference publication [22] through the addition of the temporal controller derivation, and presents the previously unreleased experimental data along with the corresponding discussion, as well as the derivation and implementation of an additional spatiotemporal controller for the fast process dynamics related to the riser. The material presented is based in part on two unpublished documents, an internal technical report [23] and the PhD thesis [24] of the first author; however, the detailed controller derivation was not given in either and is presented here for the first time.

Embedding highly nonlinear, but linear in parameters, adaptable multiresolution model into GPC is a novel idea requiring nontrivial analytical effort presented in this paper. Several researchers have successfully applied GPC proposed by Clark et al. [25,26] to many control areas [27–31]. GPC, however, has limitations, some of which have been discussed by Grimble in [32]. Since GPC uses a linear dynamic model to make predictions of process outputs over the prediction horizon, its performance will significantly degrade when the real process has severe nonlinearities and runs in a wide range of operating conditions, as is the case for a chemical looping process. Therefore, it is imperative to incorporate a high fidelity nonlinear dynamic model into the GPC scheme. Accordingly, we embed a wavelet MRA model of the nonlinear single loop cold flow of the chemical looping process into the GPC scheme. Specifically, first, a single-input-single-output (SISO) nonlinear autoregressive exogenous model (NARX) based on wavelet MRA is identified on-line using the chemical looping process test rig. Then, a GPC-type performance index is formulated, which incorporates the MRA model, and a gradient descent (GD) algorithm is developed for tuning both the weighting parameters of the wavelet MRA model and the control sequence in the GPC scheme. Further, the Lyapunov function-based theorems are proven to separately guarantee the convergence of the wavelet MRA identified model and the stability of the proposed GPC scheme without constraint and provide a guidance on both controller and identifier performance tuning. A rate constraint is then imposed on the control signal to smooth out the CL process transients. The resulting controller is shown to yield a satisfactory closed loop performance over a broad operating range, effectively meeting the challenge of handling the chemical looping process complexity.

The resulting cold flow testbed control system was then further improved by augmenting the temporal closed loop structure described above with the additional spatiotemporal control of fast dynamics of the riser loop, which were not captured in the original low-frequency wavelet MRA system model. The response time of the 2-partial differential equations (2-PDE) riser model used for this purpose is much shorter than that of the identified NARX model, for which the sampling time is 1 s. Therefore, the control-oriented riser representation was obtained through the use of the 2-PDE riser model as follows: The model was simulated to obtain the riser impulse response, the latter was employed to approximate the faster dynamics of the system, and the result was used in a convolution to obtain a model of the transients. To simplify the calculations, the impulse response was approximated using Gaussian spatial and temporal wavelets. Simulation and experimental results verified the validity of the spatiotemporal wavelet-based control system topology augmentation.

The paper is organized as follow: Section 2.1 provides the nomenclature for the main variables and symbols used in the paper. Section 2.2 introduces the chemical looping process model. A NARX model representation and a wavelet MRA representation are given in Section 2.3. Section 2.4 provides derivation of a wavelet MRA-based GPC strategy for solving the tracking problem for a single loop cold flow system. The convergence of the prediction error of the wavelet MRA model identification algorithm and the tracking error of the proposed GPC control strategy are separately proven in Section 2.5. An input-constrained GPC scheme is presented in Section 2.6. Experimental results are discussed in Section 3. The closed loop topology augmentation with the spatiotemporal model-based control to account for the pressure drop DP47 over the riser related to the fluidizing air flows is presented in Section 4. The discussion of the results is presented in Section 5. A conclusion is provided in Section 6.

2. Materials and Methods

2.1. Nomenclature

$\phi_{j,k}(x)$: orthonormal basis for V_j; $\psi_{j,k}(x)$: orthonormal basis for W_j; $y(t)$: system output; $u(t)$: system input; $w(t)$: system noise; $e(t)$: model output estimation error; n_y: maximum lag in the output; n_u: maximum lag in the input; θ: weighting parameter vector trained on-line; g_i: multivariable scaling or wavelet basis function of past inputs and outputs; γ_θ: adaptation gain for the control input vector; U: control input vector; Δu_{target}: unconstraint control signal calculated by the predictive control law; μ: design parameter; ε : voidage; u_s: solid velocity; U_g: superficial gas velocity; $S_{1,w}(t)$: control command calculated by wavelet adaptive GPC control; $y_r(t)$: reference signal; $\nabla_\theta J_1(n)$: gradient of loss function J_1 with respect to θ; $\nabla_U J_2(n)$: gradient of the loss function J_2 with respect to U; $\nabla_\theta e(n)$: sensitivity derivative at time n.

2.2. Chemical Looping Process

The modeling and control methodologies proposed in this paper focus on the hybrid combustion-gasification chemical looping (CL) process initially developed by Alstom Power. Chemical looping is a two-step process which first separates oxygen (O_2) from nitrogen (N_2) in an air stream in an air reactor. The O_2 is transferred to a solid oxygen carrier. Next, the oxygen is carried by the solid oxide and is then used to gasify or combust solid fuel in a separate fuel reactor. As shown in Figure 1, a metal or calcium material (oxygen carrier) is burned in air forming a hot oxide (MeO_x or CaO_x) in the air reactor (oxidizer). The oxygen in the hot metal oxide is used to gasify coal in the fuel reactor (reducer), thereby reducing the oxide for continuous reuse in the chemical looping cycle. CL coal power technology is an entirely new, ultra clean, low cost, high efficiency coal power plant technology for the future power market. The concept promises to be the technological link from today's steam cycle power plants to tomorrow's clean coal power plants, capable of high efficiency and CO_2 capture.

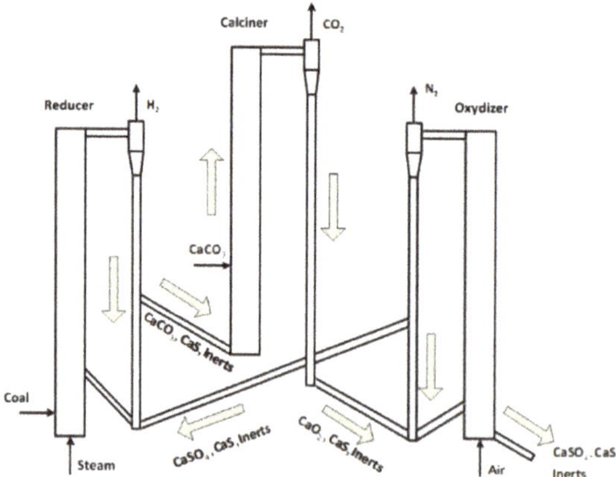

Figure 1. Alstom's combustion-gasification process.

The CL process with its multi-phase flows and complicated chemical reactions is characterized by process nonlinearities and time delays due to mass transport and chemical reactions. The specific operational characteristics are new and are still being studied. Hence, there is a need for further investigation and the potential for advanced control solutions. In this paper, we have focused on developing a control-oriented model for a single loop cold gas/solid flow test rig which omits all chemical reactions and interactions with other loops.

The block diagram of a single loop cold flow CL process is shown in Figure 2. It consists of a lower level pipeline, a riser pipeline, an upper level horizontal pipeline, a cyclone, a dip leg, seal pot control valves (SPCV), and a solid return leg. The lower level pipeline accepts air flow and solids returned from both seal pot control valves and solids which are added manually. In the riser the air-solid mixture (two-phase) flows upwards, turns into the horizontal pipeline, and then enters the cyclone. The cyclone separates the solid particles from the air. The separated solids then drop into the dip leg and enter the SPCV. The SPCV splits the solids between the return leg in its own loop and the return leg in another loop. The SPCV also maintains a pressure control boundary.

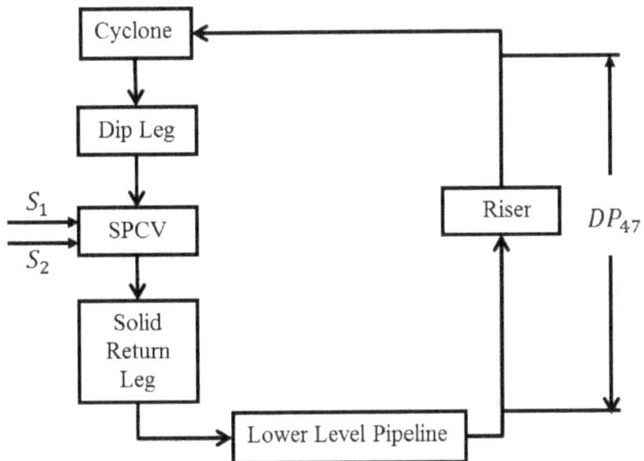

Figure 2. Block diagram for a single-loop cold flow CL test rig.

In our test rig, the manipulated variables (MV) include $S1, S2$—two fluidizing air flow rates into the SPCV, which change pressures in the SPCV and the flow conditions upstream and downstream of the SPCV. The controlled variable (CV) of interest is DP47, which stands for the pressure drop measured across the riser—a substantive indicator of solid/gas flow transport stability along the whole loop. The performance of the test rig implementing the controller to track a reference command was evaluated both under step-changes and cycling operation.

A wavelet MRA modelling and its embedding into a GPC-based predictive controller are described in the next two subsections.

2.3. Wavelet MRA Model Structure

Wavelet multiresolution analysis [14] is a function approximation tool representing function details at different scales of resolution in both the time and the frequency domains in terms of shifted and dilated scaling and wavelet functions. In general, MRA consists of a sequence of successive approximation closed subspaces $V_j \in L_2(R)$, $j \in Z$ satisfying:

$$\cdots V_{-1} \subset V_0 \subset V_1 \cdots , \tag{1}$$

with the following properties:

$$\cup_{j \in Z} V_j \text{ is dense in } L_2(R); \; \cap_{j \in Z} V_j = \{0\}, \tag{2}$$

$$f(x) \in V_j \Leftrightarrow f(2x) \in V_{j+1} \tag{3}$$

$$f(x) \in V_j \Leftrightarrow f(x - 2^{-j}k) \in V_j, \; k \in Z, \tag{4}$$

$$V_j = \text{span}\{\phi_{j,k}, k \in Z\}, \tag{5}$$

where Z is the set of all integers, $\phi_{j,k}(x) = 2^{j/2}\phi(2^j x - k)$ is an orthonormal basis for V_j and $L_2(R)$ is the space of square integrable functions of scalar real variable.

If we define W_j to be the orthogonal complement of V_j in V_{j+1}, then:

$$V_{j+1} = V_j \oplus W_j, \ V_j \perp W_j, \tag{6}$$

$$W_j = \text{span}\{\psi_{j,k}, k \in Z\}, \tag{7}$$

where $\psi_{j,k}(x) = 2^{j/2}\psi(2^j x - k)$ is an orthonormal basis for W_j. It follows from Equations (1) and (6) that, any V_j can be written for any $l < j$ as:

$$V_{j+1} = V_l \oplus W_l \oplus W_{l+1} \oplus W_{l+2} \oplus \cdots \oplus W_j, \tag{8}$$

where all the subspaces are orthogonal. Then this implies that:

$$L_2(R) = \oplus_{j \in Z} W_j \tag{9}$$

The functions $\phi_{j,k}$ and $\psi_{j,k}$ will be referred to as scaling and wavelet functions respectively. According to Equations (8) and (9), any $f(x) \in L_2(R)$ can be represented as:

$$f(x) = \sum_n (f, \phi_{J,n})\phi_{J,n} + \sum_{j \geq J, n} (f, \psi_{j,n})\psi_{j,n} \tag{10}$$

The approximation starts from some lower resolution level J and can be truncated at certain higher resolution level N when:

$$\| f(x) - [\sum_n (f, \phi_{J,n})\phi_{J,n}(x) + \sum_{j=J}^{N}\sum_n (f, \psi_{j,n})\psi_{j,n}(x)] \| < \varepsilon, \tag{11}$$

for any predefined small error $\varepsilon > 0$.

Multivariable wavelet bases can be constructed from the tensor product of a radial basis function of a one-dimensional wavelet as described for images in [33]. Because wavelet MRA can approximate any finite energy nonlinear function to any desired accuracy level, in this paper, the wavelet MRA will be used to build the nonlinear empirical model for a single loop cold flow CL process, as shown in the next subsection.

The NARX Model Structure

Many systems in a variety of applications contain nonlinearities which render linear model incapable of capturing the complex dynamic system behavior. Therefore, it is of interest to develop for these applications sufficiently accurate nonlinear dynamical models. An NARX model [34] is a well-established input/output representation for nonlinear system identification. Under some mild assumptions, a discrete-time stochastic nonlinear SISO system can be expressed as:

$$y(t) = f(y(t-1), \cdots, y(t-n_y), u(t-1), \cdots, u(t-n_u)) + w(t), \tag{12}$$

where $y(t)$, $u(t)$, $w(t)$ are the system output, input, and noise, and t is discrete time, respectively, n_y and n_u are the maximum lags in the output and input, $w(t)$ is assumed to be a zero mean, independent, and bounded noise variable, and $f(\cdot)$ is some nonlinear function. Unless some prior knowledge of the system dynamics is available, most methods use nonparametric regression to estimate the nonlinear function f from the data. In our case, f is implemented as a linear expansion in terms of the scaling and wavelet functions of regressors g_i such that

$$f = \sum_{i=1}^{m} \theta_i g_i \tag{13}$$

minimizes a pre-specified approximation adequacy criterion, where $\theta = \{\theta_i\}$ is a parameter vector trained on-line, $g_i \in \{\phi_{j,k}, \psi_{j,k}\}$ is a multivariable scaling or wavelet basis function of past inputs and outputs, and m is the number of basis functions to meet some given modeling accuracy requirement.

In the next subsection, the NARX model structure introduced above is embedded into the parameter adaptation law and the GPC performance criterion, and the self-tuning MRA-based control law is derived.

2.4. Wavelet MRA-Based GPC Scheme

The basic methodology of GPC is to calculate the current control actions on-line at each sampling instant in order to solve a finite horizon, open-loop, optimal control problem where the first control in the optimal control sequence is applied to the plant. In this section, we present both the online wavelet MRA system identification algorithm and the GPC based predictive control strategy for the stable tracking problem of a single loop CL system. To clearly illustrate the idea of the proposed control scheme, we derive the algorithm for a SISO nonlinear dynamic system. The extension to a multi-input-multi-output setting is straightforward.

Referring to Figure 2 and its description, let DP47 be the actual system output y and S1 be the control input u, while S2 is set to a constant value. Let $\hat{y}$ denote the approximated system output. Then, the identified wavelet MRA based model is defined as follows:

$$\hat{y}(t) = f(y(t-1), \cdots, y(t-n_y), u(t-1), \cdots, u(t-n_u)), \tag{14}$$

where f is defined in Equation (13). Then, the error between the real plant output y and the estimated output $\hat{y}$ is defined as:

$$e(n) = y(n) - \hat{y}(n). \tag{15}$$

The weighting parameters θ in Equation (13) are trained online to minimize the loss function defined as:

$$J_1(n) = \frac{1}{2}e^2(n), \tag{16}$$

where n indicates discrete time. To make J_1 small, we employ a parameter adaptation law in the form of a gradient descent (GD) algorithm, which adjusts the weighting gains θ to keep the gradient of J_1 negative, that is:

$$\theta(n+1) = \theta(n) - \gamma_\theta \nabla_\theta J_1(n) = \theta(n) - \gamma_\theta e(n) \nabla_\theta e(n), \tag{17}$$

where γ_θ is the adaptation gain, $\nabla_\theta J_1(n)$ is the gradient of J_1 with respect to θ at discrete time n, and $\nabla_\theta e(n)$ is the so-called sensitivity derivative at time n indicating how the error is influenced by the weighting parameters θ. From Equations (13)–(15), the sensitivity derivative $\nabla_\theta e$ can be derived as follows:

$$\nabla_\theta e = -\nabla_\theta \hat{y} = -\nabla_\theta f = -g \Rightarrow \theta(n+1) = \theta(n) + \gamma_\theta e(n)g(n). \tag{18}$$

Suppose the future set-point signals $y_m(n+k), k = 1, 2, \cdots$ are available. In the context of GPC, define another loss function as follows:

$$J_2 = \frac{1}{2}\left\{ \sum_{k=N_1}^{N_2} (y_m(n+k) - \hat{y}(n+k))^2 + \sum_{k=1}^{N_u} \rho_k \Delta u(n+k-1)^2 \right\}, \tag{19}$$

where N_1 and N_2 are the minimum and the maximum output prediction horizons, respectively, N_u is the control horizon, Δ is the difference operator, $\Delta u(n) = u(n) - u(n-1)$, and ρ_k is the k-th control weighting factor. Assuming $N_1 = 1$, $N_2 = L = N_p$, and identical control weighing factor $\rho_k = \rho$, Equation (19) can be rewritten in the vector form as:

$$J_2 = \frac{1}{2}\left\{ \| Y_m(n+1) - \hat{Y}(n+1) \|^2 + \rho\| \Delta U(n) \|^2 \right\}, \tag{20}$$

where:

$$Y_m(n+1) = [y_m(n+1), y_m(n+2), \cdots, y_m(n+L)]^T,$$

$$\hat{Y}(n+1) = [\hat{y}(n+1), \hat{y}(n+2), \cdots, \hat{y}(n+L)]^T,$$

$$U(n) = [u(n), u(n+1), \cdots, u(n+N_u-1)]^T,$$

$$\Delta U(n) = [\Delta u(n), \Delta u(n+1), \cdots, \Delta u(n+N_u-1)]^T,$$

and $\| \cdot \|$ is the L_2 norm of the n-dimensional real vectors.

The loss function J_2 is now minimized to drive the system output y to the reference signal y_m given that the wavelet MRA identifier accurately approximates the real process dynamics on-line. At each sampling instant, an optimal control sequence is calculated using future predicted output values of the identified model, but only the first one is applied to the system. To minimize J_2, the GD method is implemented again to recursively calculate the N_u-dimensional control increment sequence ΔU as follows:

$$\Delta U(n) = -\gamma_u \nabla_U J_2(n), \tag{21}$$

where γ_u is the adaptation gain for the control input vector U. Noting that for any vector $y(x)$, $\nabla_x \| y \|^2 = 2(\nabla_x y)y$, from Equations (20) and (21), the gradient of the loss function J_2 with respect to U can be obtained as:

$$\begin{aligned}
\nabla_U J_2(n) &= \tfrac{1}{2}\nabla_U\Big\{\| Y_m(n+1) - \hat{Y}(n+1) \|^2 + \rho\| \Delta U(n) \|^2\Big\} \\
&= \nabla_U\Big\{Y_m(n+1) - \hat{Y}(n+1)\Big\}\Big\{Y_m(n+1) - \hat{Y}(n+1)\Big\} + \rho\nabla_U\{\Delta U(n)\}\{\Delta U(n)\}.
\end{aligned}$$

The first part of the expression above is evaluated as follows. First, we note that since $y_m(n)$ are the reference signals, which are preset constants, we have $\partial y_m(n+1)/\partial u(n) = 0$. Then, since $\hat{y}(n+k) = \theta^T u$, $\hat{y}$ depends only on the past u, we have:

$$\frac{\partial y(n+k)}{\partial u(n+l)} = \begin{cases} \dfrac{\partial \hat{y}(n+k)}{\partial u(n+l)}, & \text{when } k > l, \\[2mm] 0, & \text{when } k \le l. \end{cases}$$

This yields:

$$\nabla_U\Big\{Y_m(n+1) - \hat{Y}(n+1)\Big\} =$$

$$\begin{bmatrix}
\frac{\partial}{\partial u(n)}\{y_m(n+1)-\hat{y}(n+1)\} & \frac{\partial}{\partial u(n)}\{y_m(n+2)-\hat{y}(n+2)\} & \cdots & \frac{\partial}{\partial u(n)}\{y_m(n+L)-\hat{y}(n+L)\} \\
\frac{\partial}{\partial u(n+1)}\{y_m(n+1)-\hat{y}(n+1)\} & \frac{\partial}{\partial u(n+1)}\{y_m(n+2)-\hat{y}(n+2)\} & \cdots & \frac{\partial}{\partial u(n+1)}\{y_m(n+L)-\hat{y}(n+L)\} \\
\vdots & \vdots & \ddots & \vdots \\
\frac{\partial}{\partial u(n+N_u-1)}\{y_m(n+1)-\hat{y}(n+1)\} & \frac{\partial}{\partial u(n+N_u-1)}\{y_m(n+2)-\hat{y}(n+2)\} & \cdots & \frac{\partial}{\partial u(n+N_u-1)}\{y_m(n+L)-\hat{y}(n+L)\}
\end{bmatrix}_{N_u \times N_p}$$

$$= \begin{bmatrix}
-\frac{\partial \hat{y}(n+1)}{\partial u(n)} & -\frac{\partial \hat{y}(n+2)}{\partial u(n)} & \cdots & -\frac{\partial \hat{y}(n+N_u)}{\partial u(n)} & \cdots & -\frac{\partial \hat{y}(n+L)}{\partial u(n)} \\
0 & -\frac{\partial \hat{y}(n+2)}{\partial u(n+1)} & \cdots & -\frac{\partial \hat{y}(n+N_u)}{\partial u(n+1)} & \cdots & -\frac{\partial \hat{y}(n+L)}{\partial u(n+1)} \\
& & & \vdots & & \\
0 & \cdots & 0 & -\frac{\partial \hat{y}(n+N_u)}{\partial u(n+N_u-1)} & \cdots & -\frac{\partial \hat{y}(n+L)}{\partial u(n+N_u-1)}
\end{bmatrix}_{N_u \times N_p}$$

The second part of $\nabla_U J_2(n)$ is evaluated by taking into account the relation $\Delta u(n) = u(n) - u(n-1)$, so that $\partial \Delta u(n)/\partial u(n) = 1$ and $\partial \Delta u(n)/\partial u(n-1) = -1$. The latter yields:

$$\nabla_U\{\Delta U(n)\} = \begin{bmatrix} \frac{\partial \Delta u(n)}{\partial u(n)} & \frac{\partial \Delta u(n+1)}{\partial u(n)} & \frac{\partial \Delta u(n+2)}{\partial u(n)} & \cdots & \frac{\partial \Delta u(n+N_u-1)}{\partial u(n)} \\ \frac{\partial \Delta u(n)}{\partial u(n+1)} & \frac{\partial \Delta u(n+1)}{\partial u(n+1)} & \frac{\partial \Delta u(n+2)}{\partial u(n+1)} & \cdots & \frac{\partial \Delta u(n+N_u-1)}{\partial u(n+1)} \\ \vdots & \vdots & \vdots & \ddots & \vdots \\ \frac{\partial \Delta u(n)}{\partial u(n+N_u-1)} & \cdots & \frac{\partial \Delta u(n+N_u-3)}{\partial u(n+N_u-1)} & \frac{\partial \Delta u(n+N_u-2)}{\partial u(n+N_u-1)} & \frac{\partial \Delta u(n+N_u-1)}{\partial u(n+N_u-1)} \end{bmatrix}_{N_u \times N_u}$$

$$= \begin{bmatrix} 1 & -1 & 0 & \cdots & 0 \\ 0 & 1 & -1 & \cdots & 0 \\ \vdots & \vdots & \vdots & \vdots & \vdots \\ 0 & \cdots & 0 & 0 & 1 \end{bmatrix}_{N_u \times N_u} . \tag{22}$$

Combining the above expressions into:

$$\nabla_U J_2(n) = -G\big(Y_m(n+1) - \hat{Y}(n+1)\big) + \rho H \Delta U(n),$$

and substituting into Equation (21) as:

$$\Delta U(n) = -\gamma_u \nabla_U J_2(n) = \gamma_u G\big(Y_m(n+1) - \hat{Y}(n+1)\big) - \gamma_u \rho H \Delta U(n), \tag{23}$$

where:

$$G = \begin{bmatrix} \frac{\partial \hat{y}(n+1)}{\partial u(n)} & \frac{\partial \hat{y}(n+2)}{\partial u(n)} & \cdots & \frac{\partial \hat{y}(n+N_u)}{\partial u(n)} & \cdots & \frac{\partial \hat{y}(n+L)}{\partial u(n)} \\ 0 & \frac{\partial \hat{y}(n+2)}{\partial u(n+1)} & \cdots & \frac{\partial \hat{y}(n+N_u)}{\partial u(n+1)} & \cdots & \frac{\partial \hat{y}(n+L)}{\partial u(n+1)} \\ & & & \vdots & & \\ 0 & \cdots & 0 & \frac{\partial \hat{y}(n+N_u)}{\partial u(n+N_u-1)} & \cdots & \frac{\partial \hat{y}(n+L)}{\partial u(n+N_u-1)} \end{bmatrix}_{N_u \times N_p} \tag{24}$$

and:

$$H = \begin{bmatrix} 1 & -1 & 0 & \cdots & 0 \\ 0 & 1 & -1 & \cdots & 0 \\ \vdots & \vdots & \vdots & \vdots & \vdots \\ 0 & \cdots & 0 & 0 & 1 \end{bmatrix}_{N_u \times N_u} \tag{25}$$

yields the control law of the form:

$$\Delta U(n) = (I + \gamma_u \rho H)^{-1} \gamma_u G\big(Y_m(n+1) - \hat{Y}(n+1)\big), \tag{26}$$

where I is the $N_u \times N_u$ identity matrix. G can be computed from the chosen wavelet MRA model structure. The proposed wavelet MRA model-based GPC control schematic is shown in Figure 3. As a result, the tracking problem for a single loop cold flow system can be solved by the wavelet MRA-based GPC control strategy using the convergence tuning guidelines developed in the next section.

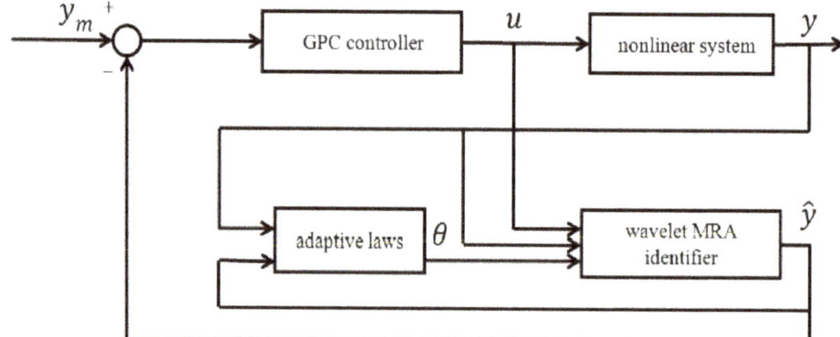

Figure 3. Schematic of the wavelet MRA-based self-tuning GPC control system.

2.5. Convergence and Stability

In this section, we show the output error convergence of the wavelet MRA model identification algorithm and the tracking error convergence of the proposed GPC-based control strategy. These proofs serve to show that the system identification scheme will converge to the true system model of the preselected resolution, while the predictive control scheme will provide tracking of the desired output by the system output. The adaptive identification and control laws have one parameter each in the form of the adaptation gains chosen by the user. It has been shown [35] that adaptation gains are crucial to the stability and performance of an adaptive control system. Therefore, we have provided analytic guidelines for selecting these gains. The validity of such separate convergence analysis is certainly limited under significant coupling between identification and control (e.g., for aggressively chosen gains), and the coupled analysis will be reported elsewhere. However, the results are well supported by the actual implementation and testing.

2.5.1. Convergence of Wavelet MRA Identifier

Define a discrete-type Lyapunov function as:

$$V_1(n) = \frac{1}{2}e^2(n), \tag{27}$$

where $e(n)$ defined in Equation (15) represents the output modeling error. Then, the increment of the Lyapunov function is given by:

$$\Delta V_1(n) = V_1(n+1) - V_1(n) = \frac{1}{2}\left(e^2(n+1) - e^2(n)\right). \tag{28}$$

The error difference can be represented using the Jacobian matrix by:

$$\Delta e(n) = e(n+1) - e(n) = \left[\frac{\partial e(n)}{\partial \theta(n)}\right]\Delta \theta(n) \tag{29}$$

where $\Delta\theta(n) = \{\Delta\theta_i(n)\}_{i=1}^{m}$ represents a change in the arbitrary component of the weighting gain vector θ. From Equation (18), $\Delta\theta_i(n)$ can be obtained by:

$$\Delta\theta(n) = \gamma_\theta e(n)g(n), \tag{30}$$

$$\left[\frac{\partial e(n)}{\partial \theta(n)}\right] = -\left[\frac{\partial \hat{y}(n)}{\partial \theta(n)}\right] = -g^T(n), \tag{31}$$

where $g(n) = \{g_i(n)\}_{i=1}^{m}$.

Theorem 1. *Let γ_θ be the adaptation gain for the weights of the wavelet MRA identified model and g_{max} be defined as $g_{max} := \max_n \| g(n) \|$, where g is the wavelet MRA basis function, n is the discrete time index, and $\| \cdot \|$ is the L_2 norm of a real vector. Then convergence is guaranteed if γ_θ is chosen as:*

$$0 < \gamma_\theta < \frac{2}{g_{max}^2} \tag{32}$$

Proof. From Equations (28)–(31), $\Delta V_1(n)$ can be represented as:

$$\Delta V_1(n) = \Delta e(n)\left[e(n) + \frac{1}{2}\Delta e(n)\right] = \left[\frac{\partial e(n)}{\partial \theta}\right]\gamma_\theta e(n) g(n) \times \left\{e(n) + \frac{1}{2}\left[\frac{\partial e(n)}{\partial \theta}\right]\gamma_\theta e(n) g(n)\right\}$$

$$= -\gamma_\theta e^2(n)\| g(n) \|^2 + \frac{1}{2}\gamma_\theta^2 e^2(n)\| g(n) \|^4 = -\lambda e^2(n), \tag{33}$$

where:

$$\lambda = \frac{1}{2}\gamma_\theta\| g(n) \|^2(2 - \gamma_\theta\| g(n) \|^2) \geq \frac{1}{2}\gamma_\theta\| g(n) \|^2(2 - \gamma_\theta g_{max}^2) > 0. \tag{34}$$

From Equation (32) we obtain $V_1(n) \geq 0$ and $\Delta V_1(n) < 0$, then the convergence of the weighting parameters of the identified wavelet MRA model is guaranteed. $\square$

2.5.2. Stability Analysis of Wavelet MRA-Based GPC

Define a second discrete Lyapunov function as:

$$V_2(n) = \frac{1}{2}\| E(n+1) \|^2, \tag{35}$$

where $E(n+1) = Y_m(n+1) - \hat{Y}(n+1)$. Then the change of the Lyapunov function is obtained as:

$$\Delta V_2(n) = V_2(n+1) - V_2(n) = \frac{1}{2}(\| E(n+2) \|^2 - \| E(n+1) \|^2). \tag{36}$$

Similarly to Equation (29), the error difference can be represented using the Jacobian matrix by:

$$\Delta E(n+1) = E(n+2) - E(n+1) = \left[\frac{\partial E(n+1)}{\partial U(n)}\right]\Delta U(n), \tag{37}$$

where $\Delta U(n)$ is defined in Equation (26) and $\frac{\partial E(n+1)}{\partial U(n)} = -G^T$. Then Equation (37) can be expressed as:

$$\Delta E(n+1) = -G^T(I + \gamma_u\rho H)^{-1}\gamma_u GE(n+1). \tag{38}$$

Theorem 2. *Let γ_u be the adaptation gain for the GPC control input sequence. Assume a control weighting factor $\rho > 0$. Then the stable tracking convergence of the wavelet MRA based GPC control system is guaranteed if:*

$$0 < \gamma_u < \frac{2}{\lambda_{max}(GG^T)}, \tag{39}$$

where $\lambda_{max}(\cdot)$ is the maximum eigenvalue of the matrix.

Proof. From Equations (36)–(38), $\Delta V_2(n)$ can be represented as:

$$
\begin{aligned}
\Delta V_2(n) &= \tfrac{1}{2}\Big[(E(n+1)+\Delta E(n+1))^T(E(n+1)+\Delta E(n+1)) - E(n+1)^T E(n+1)\Big] \\
&= \Delta E^T(n+1)\Big[E(n+1) + \tfrac{1}{2}\Delta E(n+1)\Big] \\
&= -(GE)^T \gamma_u ((I+\gamma_u\rho H)^{-1})^T \Big[I - \tfrac{1}{2}GG^T(I+\gamma_u\rho H)^{-1}\gamma_u\Big]GE \\
&= -(GE)^T R_1 R_2 GE,
\end{aligned}
\tag{40}
$$

where:

$$
R_1 = \gamma_u\big((I+\gamma_u\rho H)^{-1}\big)^T,
\tag{41}
$$

$$
R_2 = I - \frac{1}{2}\gamma_u GG^T(I+\gamma_u\rho H)^{-1}.
\tag{42}
$$

$\square$

If R_1 and R_2 are both positive definite matrices, it follows that $\Delta V_2(n) < 0$. Together with $V_2(n) > 0$, the stable tracking of the reference signals is guaranteed.

From Equation (25) it can be shown that the eigenvalues of H are $\lambda_H = \{1,\cdots,1\}_{N_u\times 1}$. Then the eigenvalues of R_1 can be derived as:

$$
\lambda_{R_1} = \Big\{\gamma_u(1+\gamma_u\rho)^{-1},\cdots,\gamma_u(1+\gamma_u\rho)^{-1}\Big\}_{N_u\times 1}
\tag{43}
$$

Hence, all eigenvalues of R_1 are positive if $\gamma_u > 0$. It follows that $R_1 > 0$.
If $0 < \gamma_u < \frac{2}{\lambda_{\max}(GG^T)}$, then:

$$
I - \frac{1}{2}\gamma_u GG^T > 0.
\tag{44}
$$

From Equation (25) we have:

$$
\gamma_u\rho H > 0.
\tag{45}
$$

Then from Equations (44) and (45)

$$
I - \frac{1}{2}\gamma_u GG^T + \gamma_u\rho H > 0.
\tag{46}
$$

Similarly to the way it was done for Equation (43), we can prove that $I + \gamma_u\rho H > 0$. Then Equation (46) can be rewritten as:

$$
(I+\gamma_u\rho H)\Big(I - \frac{1}{2}\gamma_u GG^T(I+\gamma_u\rho H)^{-1}\Big) > 0.
\tag{47}
$$

Since $I + \gamma_u\rho H > 0$, $\big(I - \tfrac{1}{2}\gamma_u GG^T(I+\gamma_u\rho H)^{-1}\big) > 0$ follows. Now we have $V_2(n) \geq 0$ and $\Delta V_2(n) < 0$. With this, the convergence of the prediction error of the wavelet MRA model identification algorithm and the tracking error of the proposed GPC control strategy have been separately proven.

2.6. Wavelet MRA GPC with Input Constraints

The stability analysis in Section 2.5 does not account for constraints. In practice, all process inputs are subject to certain constraints due to the actuation limits. In [36], two types of constraints are considered in the GPC design procedure, namely the rate and the magnitude limits on the input control signal, given, respectively, by:

$$
\Delta u_{\min} \leq u(n+k) - u(n+k-1) \leq \Delta u_{\max},
\tag{48}
$$

$$
u_{\min} \leq u(n+k) \leq u_{\max},
\tag{49}
$$

where $0 \leq k \leq N_u - 1$. When constraints are included, the stability properties obtained above must be reanalyzed. The stability analysis for constrained wavelet MRA–GPC architecture will be addressed elsewhere. Taking into account the CL process actuator constraints, the control input u is subject to an input magnitude constraint saturation:

$$u^*(n) = \text{sat}[u(n)] = \begin{cases} u_{\min} \text{ if } u(n) < u_{\min} \\ u_{\max} \text{ if } u(n) > u_{\max} \\ \quad u(n) \text{ otherwise} \end{cases} . \tag{50}$$

In the experiments, the latter constraints were seldomly attained, whereas control rate constraints of the form of Equation (48) had to be introduced to achieve good experimental results, as presented in the next section.

3. Results

3.1. MRA Temporal Modeling of the Chemical Looping Process Testbed

This section describes implementation of the proposed wavelet MRA model-based GPC scheme on the single loop gas/solid cold flow CL process testbed developed at Alstom Power Inc. to carry out experiments without consideration of the oxidation reaction. The experimental facility is shown in Figure 4.

Figure 4. Experimental facility for control testing.

The controllers were developed in MATLAB/SIMULINK, compiled in C and run on the proprietary ASTOM processing platform. The software used for wavelet identification was MATLAB Wavenet. The system output y was selected to be the riser pressure drop DP47 (inch H_2O). Fluidizing air flow $S1$ (standard cubic feet per hour, scfh), was used as the single control input u, while the other air flow $S2$ (scfh) was set to a constant value of about 20 scfh. The characterization of the complex dynamic behavior, to be obtained through the identification procedure, was chosen as a SISO NARX wavelet multiresolution network model of the form:

$$\hat{y}(t) = f(y(t-1), \cdots, y(t-n_y), u(t-1), \cdots, u(t-n_u)) = \sum_{i=1}^{m} \theta_i g_i, \tag{51}$$

where f is the unknown nonlinear mapping to be identified, $u(t)$ and $y(t)$ are the sampled input and output sequences, n_y and n_u are the maximum lags in the output and the input to be determined, respectively; $\theta = \{\theta_i\}$ is the parameter vector trained on-line, $g_i \in \{\phi_{j,k}, \psi_{j,k}\}$ is a multivariable scaling or wavelet basis function of past inputs and outputs, and m is the number of required basis functions to meet satisfactory modeling accuracy requirements.

First, several offline experimental tests were performed to understand the process better and to leverage the test results in tuning the identification structure and the model parameters. The input signal S_1 was generated in the form of a pseudo random binary signal (PRBS) changed about a nominal value, and the pressure drop across the riser DP47 was measured as the output. All the sequences used in the experiment were generated by MATLAB commands. The experimental results of the PRBS test are shown in Figures 5 and 6 below, where the sampling period is 1 s.

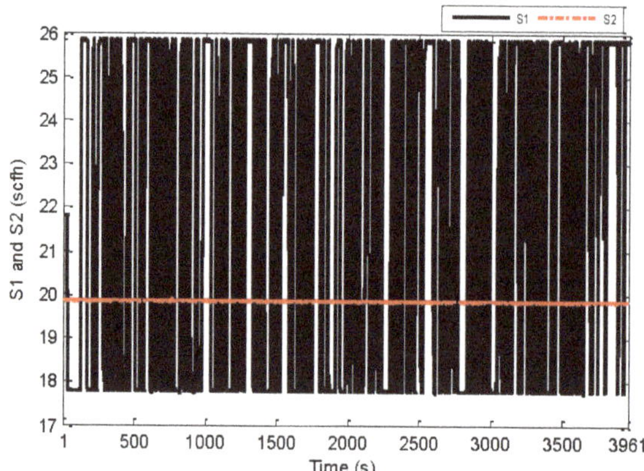

Figure 5. PRBS test-input S1 and S2 (scfh).

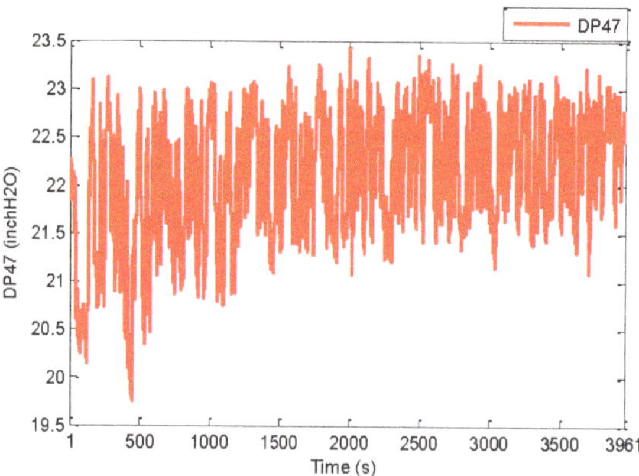

Figure 6. PRBS test-Output DP47 (inch H$_2$O).

The data set consists of 3961 input and output samples. The NARX multiresolution wavelet MRA model was used to approximate the nonlinear relationship between S1 and DP47 based on the experimental data. The regressor set was specified as:

$$y(t-1), y(t-2), u(t-1), u(t-2), \cdots, u(t-8). \tag{52}$$

Hence $n_y = 2$, $n_u = 8$. For MRA model we chose radial Marr scaling and wavelet functions [12]:

$$\phi(x) = \exp(-0.5\| x \|^2), \ \psi(x) = (\dim(x) - \| x \|^2) \exp(-0.5\| x \|^2). \tag{53}$$

The initial coarse layer index J was chosen to be 3, with the number of basis functions doubling when resolution was incremented by 1 starting with 10. The final resolution adopted was $K = 6$. Figure 7 below shows how the model predicted output compares with the experimental results.

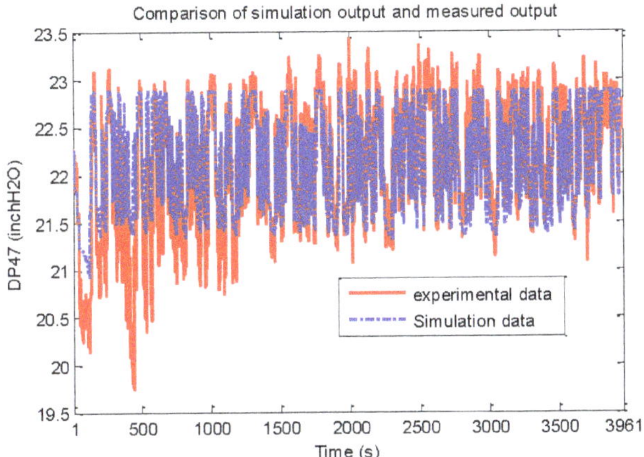

Figure 7. PRBS test-Simulation data vs. experimental data.

The one-step-ahead predicted output and the test data set are shown in Figure 8. From these figures it is seen that the NARX wavelet MRA model obtained predicted the system outputs well. The model was found to be sufficiently accurate and no finer resolution levels were needed to be added to the model structure.

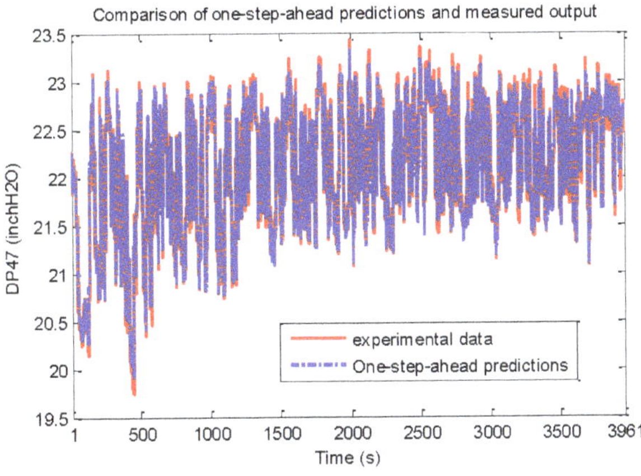

Figure 8. PRBS test-One-step-ahead predictions vs. experimental data.

3.2. MRA–GPC Scheme Implementation on a Chemical Looping Process Testbed

In order to make the wavelet MRA GPC applicable to the CL process, the control input u was subjected to rate constraints of the form:

$$\Delta u = \Delta u_{\text{target}} \times \exp(1 - \mu \parallel \Delta u_{\text{target}} \parallel), \tag{54}$$

where Δu_{target} is the unconstraint control signal calculated by the predictive control law and $\mu > 0$ is a design parameter to adjust the rate of the control signal. The effectiveness of such input constrained wavelet predictive controller on the CL process is demonstrated next through experimental results.

The control objective of the GPC design is to ensure that the output of the system y asymptotically tracks the reference signal y_m. The cost function to be minimized is defined in (19). The design parameters for GPC configuration were chosen as $N_1 = 1$, $N_2 = 10$, $N_u = 8$, $\rho = 1$. The adaptation gains derived from Theorems 1 and 2, were chosen as $\gamma_\theta = 0.01$, $\gamma_u = 0.1$. The system was initially stable around level of $y_0 = 13$ inch H_2O. Two setpoint step change experiments were performed consecutively. After 5 min, the setpoint was first increased to 16 inch H_2O and stayed at the latter value for about 7 min. Then it went back to the original level of 13 inch H_2O. The air flow $S2$ was set to a constant value of about 20 scfh. The tracking response of the system output and the corresponding control efforts are shown in Figures 9 and 10, respectively. It can be seen from these figures that the proposed wavelet MRA-based GPC method effectively tracks the setpoint changes for a single loop CL process.

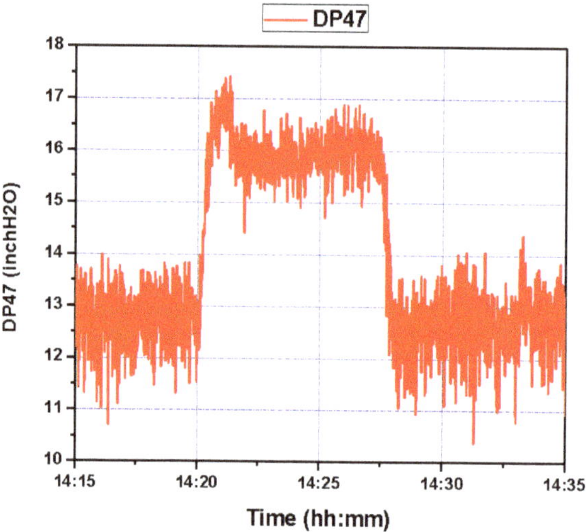

Figure 9. Pressure difference response of riser (DP47) during step setpoint changes.

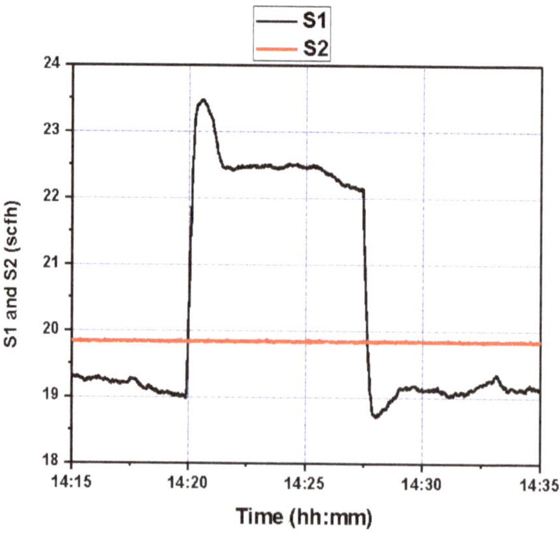

Figure 10. Fluidizing air flow control (S1 and S2) during step setpoint changes.

As can be seen in Figure 9, starting at a setpoint of 13 inH$_2$O for the duration of five minutes, responds to a setpoint change at 14:20 within roughly two minutes. While there is an initial overshoot, this overshoot has a magnitude of one inH$_2$O, and is quickly subdued; a similar phenomenon is seen at 14:27, when the setpoint is restored to 13 inH$_2$O. This points to an adequate response to step-changes in DP47 setpoint, which the process control methodology is capable of handling due to a sufficiently accurate wavelet MRA model, and effective GPC tuning; the latter can be seen in Figure 10, where controller rates are limited to within feasible bounds, and control efforts are limited, making for a subdued control input history.

In the second test, reference signal was set to be a sinusoidal of the form $y_m(t) = 13 + 2\sin(2\pi \times 0.01 \times t)$, while $S2$ (scfh) was still held at a constant value of about 20. The tracking response of system output and the corresponding control effort, shown in Figures 11 and 12, respectively, demonstrate that the controller satisfies the tracking performance requirement, with a time delay between the control signal and system output potentially addressed through prediction adjustment.

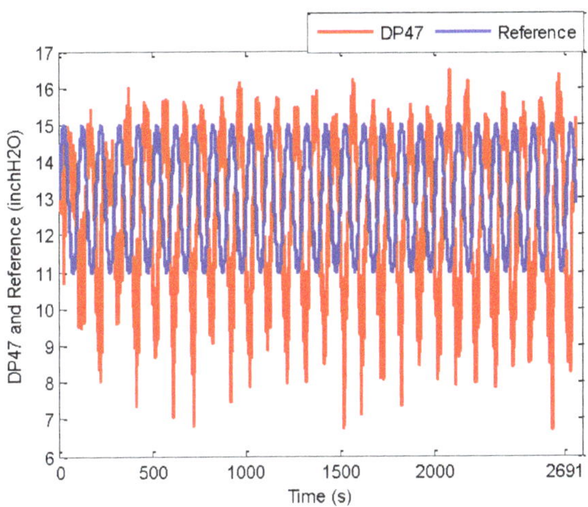

Figure 11. Pressure difference response of riser during sinusoidal setpoint changes.

In particular, a relatively timely response to reference signal changes is clearly seen in Figure 11; the controller effectuates the reference signal in about 50 s, with little overshoot, as was seen in the previous setpoint step-change experiment. However, besides the phase difference between the reference and true output, large values of undershoot are seen. This can be attributed to the presence of overshoot; as GPC reacts to samples 10 s ahead of time, any overshoot is met with an overaggressive response to lower it (see Figure 12), resulting in excessive undershoot. This issue may be addressed by increasing the control weighing factor to penalize excessive actuation.

The next section presents the derivation and implementation of the spatiotemporal control law for the fast riser dynamics to augment the temporal controller described above and tighten the closed loop tracking performance.

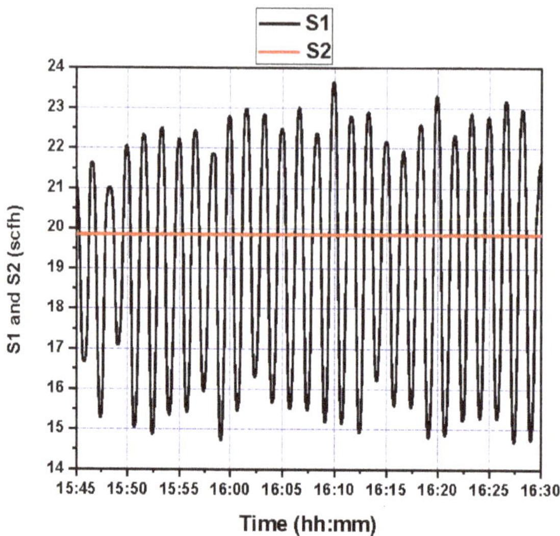

Figure 12. Fluidizing air flow control (S1 and S2) during setpoint changes.

4. Spatiotemporal Wavelet Decomposition

Since the empirically identified wavelet temporal model was obtained using data collected at a 1 s sampling rate, some of the fast dynamics of the plant that gave rise to jumps were not recorded. The fast dynamics comes primarily from the spatially distributed riser geometry. Hence, we simulated the impulse response of the 2-PDE riser model [37], approximated the faster riser dynamics with the transient spatiotemporal NARMA-L1 [38] model, and used the result in a convolution to obtain a spatiotemporal model of the transients. We then put the empirical temporal NARX model and the fast dynamics spatiotemporal model in parallel. Finally, we combined the temporal GPC control and the spatiotemporal deadbeat control, each based on its respective model, into the closed loop dual-model self-tuning configuration shown in Figure 13. In this configuration, the dynamic inter-loop coupling is rather minimal due to significantly differing time scales of each elf-tuning loop, ideally requiring two-sampling-rates hardware/software setting, not available for this experiment.

The nonlinear 2-PDE model governing the evolution of the variables (voidage and solid velocity) in the riser can be represented [37] as:

$$\begin{aligned}
\frac{\partial \varepsilon}{\partial t} &= (1 - \varepsilon)\frac{\partial u_s}{\partial x} - u_s \frac{\partial \varepsilon}{\partial x}, \\
\frac{\partial u_s}{\partial t} &= -u_s \frac{\partial u_s}{\partial x} + C_1 \varepsilon^{-6.7} - C_2 \varepsilon^{-5.7} u_s + C_3 \varepsilon^{-4.7} u_s^2 + C_4 (1 - \varepsilon)^{-0.54} - C_5,
\end{aligned} \tag{55}$$

where ε is the voidage and u_s is the solid velocity. The other parameters are defined in [37]. From simulations of this model, we could obtain a response $h(x, t)$ to an impulse actuation in solid velocity with area of 0.1.

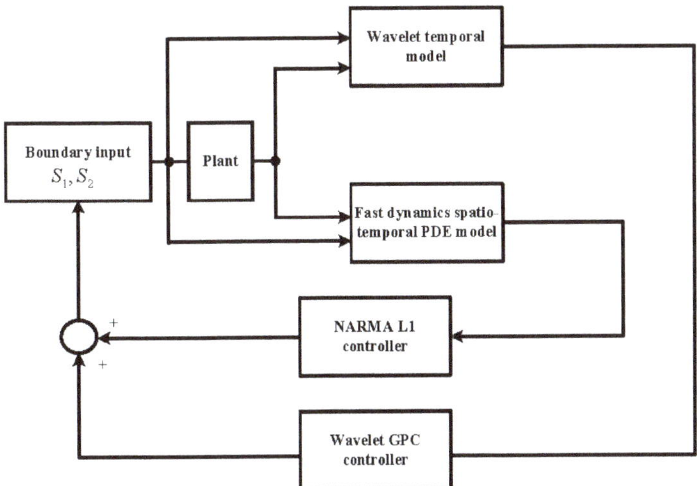

Figure 13. Block diagram of controller implementation with fast dynamics.

Then the response to an arbitrary inlet solid velocity $u(t)$ can be calculated as:

$$y(x,t) = \int_{-\infty}^{t} h(x,\tau) \cdot 10u(t-\tau)d\tau, \tag{56}$$

where the scaling factor is necessary since the simulated input was not 1. The simulated impulse responses are shown in Figure 14.

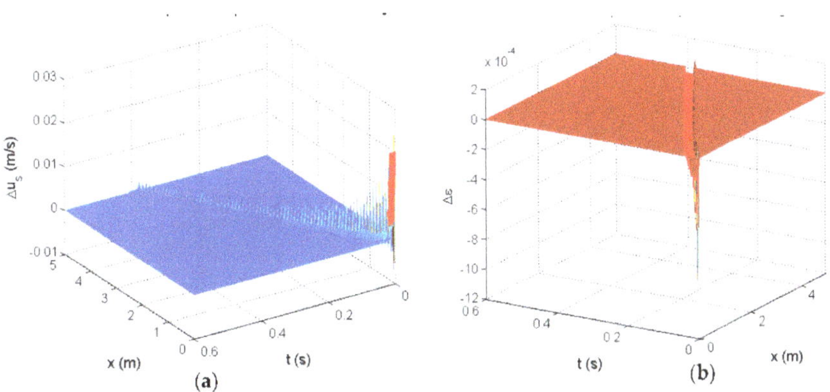

Figure 14. Simulated impulse response of the 2-PDE riser model: (**a**)—the solid velocity response, (**b**)—the voidage response.

Since the impulse response is uniformly zero after 0.6 s, Equation (56) can be limited to:

$$y(x,t) = \int_{t-0.6}^{t} h(x,\tau) \cdot 10u(t-\tau)d\tau. \tag{57}$$

To obtain a low-order high fidelity finite-dimensional representation of the impulse response, a wavelet decomposition [39,40] was used to approximate $h(x,t)$. That is, the impulse response $h(x,t)$ was approximated as:

$$h(x,t) = \sum_{m=1}^{m_{max}} \sum_{n=1}^{n_{max}} \beta_m(x)c_{m,n}\alpha_n(t), \tag{58}$$

where $\{\beta_m(x)\}$ and $\{\alpha_n(t)\}$ are wavelet basis functions.

Figure 15 is the resulting wavelet approximation of the impulse response. Here we chose Gaussian wavelet functions specifically. In this case, 23 spatial and 22 temporal wavelets were used. The coefficients $c_{m,n}$ were determined using a least-squares regression.

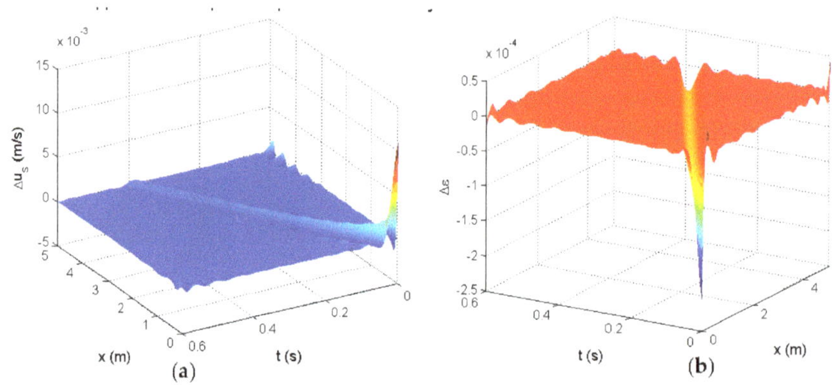

Figure 15. Wavelet-approximated impulse response $h(x,t)$: (**a**)—the solid velocity response approximation, (**b**)—the voidage response approximation.

The following notation is used to divide the impulse response into separate parts for voidage ε and velocity u_s:

$$
\begin{aligned}
h_{u_s}(x,t) &= \sum_{m=1}^{m_{max}} \sum_{n=1}^{n_{max}} \beta_m(x)c_{m,n}\alpha_n(t), \\
h_{\varepsilon}(x,t) &= \sum_{m=1}^{m_{max}} \sum_{n=1}^{n_{max}} \beta_m(x)d_{m,n}\alpha_n(t).
\end{aligned}
\tag{59}
$$

Then using the convolution:

$$\Delta u_s(x,t) = h_{u_s}(x,\tau) * 10(u-\tau) = 10 \int_0^{0.6} \sum_{m=1}^{m_{max}} \sum_{n=1}^{n_{max}} \beta_m(x)c_{m,n}\alpha_n(\tau)u(t-\tau)d\tau. \tag{60}$$

Since the online measurements were only available at 1 s intervals, it was assumed that:

$$u(t-\tau) = (1-\tau)u(t) + \tau u(t-1),\ 0 \leq \tau \leq 1, \tag{61}$$

to linearly interpolate between the measurements. Then:

$$
\begin{aligned}
\Delta u_s(x,t) &= 10\int_0^{0.6} \sum_{m=1}^{m_{max}} \sum_{n=1}^{n_{max}} \beta_m c_{m,n}\alpha_n(\tau)[(1-\tau)u(t)+\tau u(t-1)]d\tau \\
&= 10 \sum_{m=1}^{m_{max}} \sum_{n=1}^{n_{max}} \beta_m(x)c_{m,n}\int_0^{0.6} \alpha_n(\tau)[(1-\tau)u(t)+\tau u(t-1)]d\tau \\
&= 10\left[\sum_{m=1}^{m_{max}} \sum_{n=1}^{n_{max}} \beta_m(x)c_{m,n}\int_0^{0.6}(1-\tau)\alpha_n(\tau)d\tau\right]u(t) + 10\left[\sum_{m=1}^{m_{max}} \sum_{n=1}^{n_{max}} \beta_m(x)c_{m,n}\int_0^{0.6}\tau\alpha_n(\tau)d\tau\right]u(t-1).
\end{aligned}
\tag{62}
$$

Denote:

$$\alpha_{n,0} = \int_0^{0.6} (1-\tau)\alpha_n(\tau)d\tau,$$
$$\alpha_{n,1} = \int_0^{0.6} \tau\alpha_n(\tau)d\tau, \tag{63}$$

and:

$$\gamma_0(x) = 10 \sum_{m=1}^{m_{max}} \sum_{n=1}^{n_{max}} \beta_m(x)c_{m,n}\alpha_{n,0},$$
$$\gamma_1(x) = 10 \sum_{m=1}^{m_{max}} \sum_{n=1}^{n_{max}} \beta_m(x)c_{m,n}\alpha_{n,1}. \tag{64}$$

Then Equation (62) simplifies to:

$$\Delta u_s(x,t) = \gamma_0(x)u(t) + \gamma_1(x)u(t-1). \tag{65}$$

Similarly, we have:

$$\Delta\varepsilon(x,t) = \eta_0(x)u(t) + \eta_1(x)u(t-1), \tag{66}$$

where:

$$\eta_0(x) = 10 \sum_{m=1}^{m_{max}} \sum_{n=1}^{n_{max}} \beta_m(x)d_{m,n}\alpha_{n,0},$$
$$\eta_1(x) = 10 \sum_{m=1}^{m_{max}} \sum_{n=1}^{n_{max}} \beta_m(x)d_{m,n}\alpha_{n,1}. \tag{67}$$

Omitting for brevity several routine manipulations (available in [23]), the output DP47 representing the pressure drop across the riser can be given as:

$$P(5) - P(0) = \left[\rho_g\varepsilon(0)u_g^2(0) + \rho_s(1-\varepsilon(0))u_s^2(0)\right]$$
$$-\left[\rho_g\varepsilon(5)u_g^2(5) + \rho_s(1-\varepsilon(5))u_s^2(5)\right]$$
$$-\int_0^5 g\left[\rho_g\varepsilon(y) + \rho_s(1-\varepsilon(y))\right]dy, \tag{68}$$

where the riser length of 5 m is used, constant g is the gravity acceleration, u is the velocity, ρ is the density, the subscripts s and g refer to solid and gas, respectively, and $u_g = \frac{U_g}{\varepsilon(x)}$ where U_g is the superficial gas velocity. Expanding Equation (68) yields:

$$P(5) - P(0) = \left[\rho_g\frac{u_g^2}{\varepsilon(0)} + \rho_s(1-\varepsilon(0))u_s^2(0)\right]$$
$$-\left[\rho_g\frac{u_g^2}{\varepsilon(5)} + \rho_s(1-\varepsilon(5))u_s^2(5)\right]$$
$$-\int_0^5 g\left[\rho_g\varepsilon(y) + \rho_s(1-\varepsilon(y))\right]dy$$
$$= \rho_g\frac{u_g^2}{\varepsilon_{ss}(0)+\Delta\varepsilon(0)} + \rho_s(1-\varepsilon_{ss}(0)-\Delta\varepsilon(0))(u_s(0)+\Delta u_s(0))^2$$
$$-\rho_g\frac{u_g^2}{\varepsilon_{ss}(5)+\Delta\varepsilon(5)} - \rho_s(1-\varepsilon_{ss}(5)-\Delta\varepsilon(5))(u_s(5)+\Delta u_s(5))^2$$
$$-g\rho_g\int_0^5 (\varepsilon_{ss}(y)+\Delta\varepsilon(y))dy - g\rho_s\int_0^5 (1-\varepsilon_{ss}(y)-\Delta\varepsilon(y))dy, \tag{69}$$

where the subscript ss designates the steady state. Now, substituting the wavelet model gives:

$$P(5) - P(0) = \rho_g\frac{u_g^2}{\varepsilon_{ss}(0)+\eta_0(0)u(t)+\eta_1(0)u(t-1)}$$
$$+\rho_s[1-\varepsilon_{ss}(0)-\eta_0(0)u(t)-\eta_1(0)u(t-1)][u_s(0)+\gamma_0(0)u(t)+\gamma_1(0)u(t-1)]^2$$
$$-\rho_g\frac{u_g^2}{\varepsilon_{ss}(5)+\eta_0(5)u(t)+\eta_1(5)u(t-1)}$$
$$-\rho_s[1-\varepsilon_{ss}(5)-\eta_0(5)u(t)-\eta_1(5)u(t-1)][u_s(5)+\gamma_0(5)u(t)+\gamma_1(5)u(t-1)]^2$$
$$-g\rho_g\int_0^5 (\varepsilon_{ss}(y)+\eta_0(y)u(t)+\eta_1(y)u(t-1))dy$$
$$-g\rho_s\int_0^5 (1-\varepsilon_{ss}(y)-\eta_0(y)u(t)-\eta_1(y)u(t-1))dy. \tag{70}$$

Our goal was to use the model given by Equation (70) to account for the spatiotemporal behavior of the CL system to the extent allowed by the available sampling rate, and also to develop the control setting to be ready to employ much higher sampling rates for performance improvement, once they become available on the test rig through the processor upgrades to the GPUs and FPGAs. It is also of interest to calculate the steady-state pressure drop:

$$
\begin{aligned}
\Delta P_0 = \ & \rho_g \frac{u_g^2}{\varepsilon_{ss}(0)+\eta_0(0)u(t)+\eta_1(0)u(t)} \\
& + \rho_s \left[1 - \varepsilon_{ss}(0) - \eta_0(0)u(t) - \eta_1(0)u(t)\right]\left[u_s(0) + \gamma_0(0)u(t) + \gamma_1(0)u(t)\right]^2 \\
& - \rho_g \frac{u_g^2}{\varepsilon_{ss}(5)+\eta_0(5)u(t)+\eta_1(5)u(t)} \\
& - \rho_s \left[1 - \varepsilon_{ss}(5) - \eta_0(5)u(t) - \eta_1(5)u(t)\right]\left[u_s(5) + \gamma_0(5)u(t) + \gamma_1(5)u(t)\right]^2 \\
& - g\rho_g \int_0^5 \left(\varepsilon_{ss}(y) + \eta_0(y)u(t) + \eta_1(y)u(t)\right) dy \\
& - g\rho_s \int_0^5 \left(1 - \varepsilon_{ss}(y) - \eta_0(y)u(t) - \eta_1(y)u(t)\right) dy.
\end{aligned}
\tag{71}
$$

This is the pressure drop predicted by this model for constant input $u(t)$, as opposed to the linear interpolation described above. We can then use this model to approximate the transient difference, and the NARX wavelet model to approximate the steady state. The difference between the transient pressure drop $\Delta P(t)$ and the eventual steady pressure drop $\Delta P_0(t)$ is then equal to:

$$
\Delta P - \Delta P_0 = P(5) - P(0) - \Delta P_0.
\tag{72}
$$

Linearizing Equation (72) about $u(t) = u(t-1)$ gives:

$$
\Delta P - \Delta P_0 \approx f(u(t-1))(u(t) - u(t-1)),
\tag{73}
$$

where:

$$
\begin{aligned}
f(y) = \ & k_1 \frac{(k_2+k_3 y)}{(k_2+k_4 y)^4} + k_5(k_6 + k_7 y)^2 \\
& + k_8(1 - k_2 - k_4 y)(k_6 + k_9 y) + k_{10} \frac{(k_{11}+k_{12}y)}{(k_{11}+k_{13}y)^4} \\
& + k_{14}(k_{15} + k_{16}y)^2 + k_{17}(1 - k_{11} - k_{13}y)(k_{15} + k_{18}y) + k_{19},
\end{aligned}
\tag{74}
$$

and:

$$
\begin{aligned}
& k_1 = \rho_g U_g \eta_1(0), \ k_2 = \varepsilon_{ss}(0), \ k_3 = \eta_0(0) + 2\eta_1(0), \\
& k_4 = \eta_0(0) + \eta_1(0), \ k_5 = \rho_s \eta_1(0), \ k_6 = u_{s,ss}(0), \\
& k_7 = \gamma_0(0) + \gamma_1(0), \ k_8 = -\rho_s \gamma_1(0), \ k_9 = \gamma_0(0) + 2\gamma_1(0), \\
& k_{10} = -\rho_g U_g \eta_1(5), \ k_{11} = \varepsilon_{ss}(5), \ k_{12} = \eta_0(5) + 2\eta_1(5), \\
& k_{13} = \eta_0(5) + \eta_1(5), \ k_{14} = -\rho_s \eta_1(5), \ k_{15} = u_{s,ss}(5), \\
& k_{16} = \gamma_0(5) + \gamma_1(5), \ k_{17} = \rho_s \gamma_1(5), \ k_{18} = \gamma_0(5) + 2\gamma_1(5), \\
& k_{19} = g(\rho_g - \rho_s) \int_0^5 \eta_1(y) \, dy.
\end{aligned}
\tag{75}
$$

The input to the computational model was in terms of the velocity boundary condition, so that $u(t) = \Delta u_s(0,t)$. This can be connected to the inputs S_1 and S_2 via a quadratic model fitted to the test data where:

$$
\begin{aligned}
u(t) \approx \ & \frac{1}{\varepsilon_0}\left(2a_1 S_1(t-1) + a_3 S_2(t) + a_4\right)(S_1(t) - S_1(t-1)) \\
& + \frac{1}{\varepsilon_0}\left(a_1 S_1^2(t-1) + a_2 S_2^2(t) + a_3 S_1(t-1)S_2(t) + a_4 S_1(t-1) + a_5 S_2(t) + a_6\right) - u_{s,ss}(0) \\
= \ & \frac{1}{\varepsilon_0}\left(a_1 S_1(t-1) + a_3 S_2(t) + a_4\right)S_1(t) \\
& + \frac{1}{\varepsilon_0}\left(a_2 S_2^2(t) + a_5 S_2(t) + a_6\right) - u_{s,ss}(0).
\end{aligned}
\tag{76}
$$

Then:

$$\begin{aligned}\Delta P - \Delta P_0 \;&\approx\; \tfrac{1}{\varepsilon_0} f(u(t-1))(2a_1 S_1(t-1) + a_3 S_2(t) + a_4) S_1(t) \\ &\quad + f(u(t-1))\Big[\tfrac{1}{\varepsilon_0}\big(a_2 S_2^2(t) + a_5 S_2(t) + a_6\big) - u_{s,ss}(0) - u(t-1)\Big] \\ &= g_{\Delta P}(S_1(t-1), S_2(t-1), S_2(t))S_1(t) + f_{\Delta P}(S_1(t-1), S_2(t-1), S_2(t)).\end{aligned} \tag{77}$$

The NARX wavelet MRA model takes the form as in Equation (51):

$$y_w(t) = f\big(y(t-1), \cdots, y(t-n_y), S_{1\_w}(t-1), \cdots, S_{1\_w}(t-n_u)\big), \tag{78}$$

where $S_{1,w}(t)$ is the control command calculated by wavelet adaptive GPC control. Then, the fast transient behavior model of Equation (77) can be combined with Equation (78) to obtain a spatiotemporal multiscale dynamic network representation of the entire CL process:

$$y(t) \approx y_w(t) - f_{\Delta P} - g_{\Delta P} S_1(t). \tag{79}$$

The sign change is necessary because the pressure drop across the riser is negative in the model above, i.e., $P(5) - P(0) < 0$. Then, the deadbeat predictive controller taking account of fast dynamics is:

$$S_{1,\text{fast}}(t) = \frac{y_r(t) - y_w(t) + f_{\Delta P}}{g_{\Delta P}}, \tag{80}$$

where $y_r(t)$ is the reference signal. Hence, the final spatiotemporal wavelet controller $S_1(t)$ implemented on the real CL process is given by:

$$S_1(t) = S_{1,w}(t) + S_{1,\text{fast}}(t). \tag{81}$$

This controller was also implemented in the single loop cold flow CL test rig. S_1 was taken as the single input and DP47 as the output, while S_2 was mostly steady, but with jumps. The reference signal was set to 16 initially and then reduced to 13 around time 17:01. The tracking response of system output and the corresponding control efforts are shown in Figures 16 and 17, respectively. The controller is seen to stabilize the system quite well under difficult operating conditions. The pressure drop DP47 over the riser related to the fluidizing air flows was discussed based on the closed loop topology augmentation with the spatiotemporal model-based control.

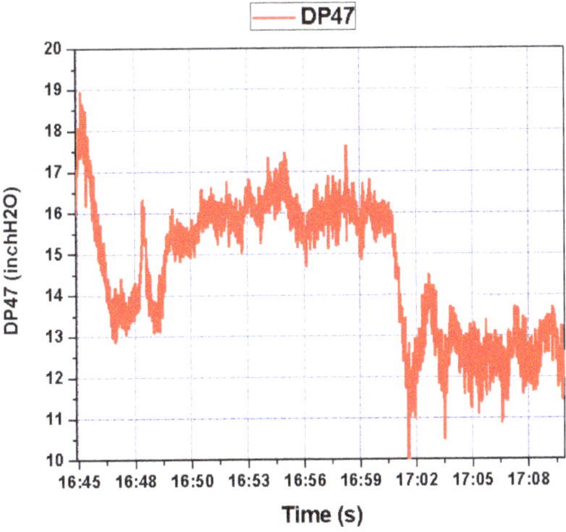

Figure 16. Pressure difference response of riser (DP47) during step setpoint changes.

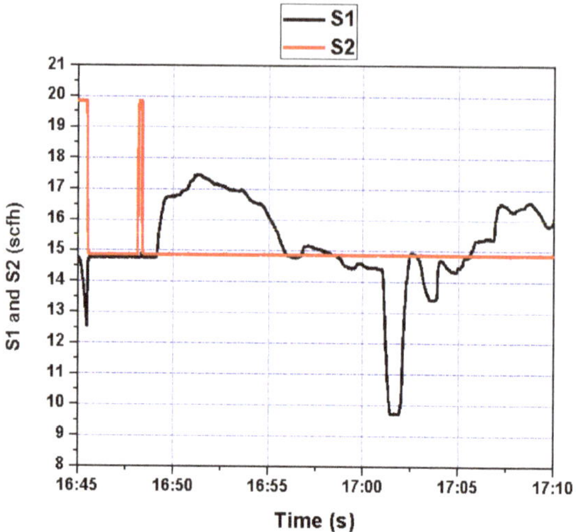

Figure 17. Fluidizing air flow control (S1 and S2) during setpoint changes.

5. Discussion

The control objective of developing the tracking controller for the CL test rig at ALSTOM Power was addressed at the start of the project through both first-principles model development and empirical system identification from the input/output experimental data record. The first approach resulted in the analytical development and the numerical simulation of the 1D, 2D, and 3D partial differential equation (PDE) networks—systems of coupled PDEs, each describing the testbed subsystem. On the basis of these models, approximately tuned to the process through the experimental data, first, the linear finite dimensional models were developed, after which robust controllers based on the H_∞ approach were designed. The latter were implemented on the test rig; however, no satisfactory performance was obtained. The empirical approach initially involved a polynomial NARX model test fit with the use of model predictive control (MPC) with constraints attained through the QP (quadratic programming) based control signal calculation. At the same time the hypothesis was posed that the process is highly multiscale and that a multiscale controller design should be attempted. The experimental data was then subjected to multi-resolution decomposition and was indeed found to be highly multiscale [23]. After some debate, it was suggested to refocus the effort from model-based robust controller improvement and traditional constrained MPC to empirical multiresolution controller development. The initial step was to fit a multiresolution nonlinear, but linear in parameters, temporal model to the input/output data, with the subsequent steps involving the self-tuning GPC-type controller development, as shown in Figure 3. After a number of prolonged experiments, the resulting controller was outfitted with rate constraints and tuned to work on the test rig, showing performance noticeably superior to that of the other techniques employed. Further on, since the underlying process dynamics was exhibited by the intrinsically spatially distributed processes, a spatiotemporal multiresolution control component was developed for the riser dynamics and combined with the temporal topology. The latter part was implemented and tested, producing reasonable overall performance, but it could not be fully utilized due to the computational real-time controller limitations. Subsequently, the results presented in this paper were acknowledged by ALSTOM as making a real breakthrough in the project. The findings and their implications in the broad context imply that if a robust linear controller does not work well on a system, the system is likely rather complex, and the nonlinear multiresolution modeling in combination with linear-type control structures under constraints could offer effective configurations for high performance system control.

The future theoretical effort in the temporal implementation in Section 3 should involve developing the proof of the simultaneous error convergence of the coupled identifier/controller system under control rate constraints—a challenging analytical task. Future research should also address spatiotemporal controller development and implementation in more generality than that presented in Section 4. In the present work, the latter controller had limited efficacy due to the low sampling/computation rate of the test rig control processor, but the approach presented stands ready for further theoretical advancement and applications involving more advanced hardware.

Due to the combination of the coarser temporal and the finer spatiotemporal control in the overall control configuration, the results presented have a broad appeal in other applications involving complex multiscale spatiotemporal dynamics, such as, for example, robotic electrosurgery [41]. In this application, the motion of the cutting probe is strongly coupled to the spatiotemporal electrosurgical impact of the latter on the target tissues, and the near-field probe-tissue interaction process is best described by a complex time-fractional PDE [42]. These features make the technique proposed a good match for this application.

6. Conclusions

In this paper, the following results have been presented:

- closing the gap between the actual system output data and that predicted by the first-principles model of the complex chemical looping process through the empirically identified wavelet multiresolution analysis model initially trained on-line to estimate the nonlinear dynamic characteristics;
- embedding the multiresolution model into the generalized predictive control structure to obtain self-tuning control strategy for stable tracking of a chemical looping process with complex process dynamics under actuator rate constraints;
- showing boundedness of the adaptation gains for identification and control laws using the Lyapunov function theorems, and providing a guidance for attainment of asymptotic stability of the closed-loop system through the choice of these adaptation gains;
- using a spatiotemporal wavelet decomposition of the impulse response of the chemical looping process riser for designing a deadbeat predictive controller for further enhancement of the closed loop performance to account for fast system dynamics.
- experimentally confirming the effectiveness of the proposed controller design methods though their implementation on the novel single loop chemical looping cold flow testbed with complex dynamics,

Limitations of the present study lie in the insufficient spatiotemporal modeling and controller design for the riser and the inability to fully utilize the designed spatiotemporal controller for it because of the insufficient real-time performance of the control processor and the data acquisition system. Future efforts should focus on the advancement of the spatiotemporal part of the design, the overall controller robustness evaluation and enhancement, the development of the rigorous convergence proofs for the coupled identification/control laws under control rate constraints for the temporal and the entire spatiotemporal control topologies, and the implementation of the controller using advanced processors. The proposed techniques is planned to be applied by the authors to other areas, such as robotic electrosurgery.

Author Contributions: Conceptualization, S.Z., J.B., X.L. and C.N.; methodology, S.Z. and J.B.; software, S.Z. and X.L.; validation, S.Z., J.B., X.L., and C.N.; formal analysis, S.Z., J.B., Y.L., and H.E.-K.; investigation, S.Z., J.B., X.L., and C.N.; resources, J.B., X.L., and C.N.; data curation, S.Z. and X.L.; writing—original draft preparation, S.Z.; writing—review and editing, S.Z., J.B., X.L., C.N., Y.L., H.E.-K.; visualization, S.Z., J.B., X.L., and H.E.-K.; supervision, J.B., X.L., and C.N.; project administration, J.B., X.L., and C.N.; funding acquisition, C.N. All authors have read and agreed to the published version of the manuscript.

Funding: This work was funded by DOE/ALSTOM, contract number DE-FC26-07NT43095 and, in part, by by the National Institute of Biomedical Imaging and Bioengineering of the National Institutes of Health under award number R01EB029766. The content is solely the responsibility of the authors and does not necessarily represent the official views of the National Institutes of Health.

Acknowledgments: The authors gratefully acknowledge the support of ALSTOM Power Plant Labs and DOE NETL. In particular, an invaluable organizational support of Ray P. Chamberland of ALSTOM and programmatic support of Robert R. Romanosky of DOE NETL is greatly appreciated. The authors are also greatly indebted to the DOE NETL project manager Susan M. Maley for her significant help with assessment of the state-of-the-art in mathematical modeling and numerical simulation of controlled CFBRs (continuous fluidized bed reactors). Cyrus Taft is gratefully acknowledged for facilitating participation of the University of Illinois team in the project. The University of Illinois team members Vivek Natarajan, currently at IIT Bombay, Bryan Petrus, currently at Nucor Steel, and Dong Ye, currently at Microsoft, are gratefully acknowledged for making innumerable contributions to all aspects of the project.

Conflicts of Interest: The authors declare no conflict of interest. The funders had no role in the design of the study; in the collection, analyses, or interpretation of data; in the writing of the manuscript, or in the decision to publish the results.

References

1.	Karthikeyan, A.; Previsic, M.; Scruggs, J.; Chertok, A. Non-linear Model Predictive Control of Wave Energy Converters with Realistic Power Take-off Configurations and Loss Model. In Proceedings of the 3rd IEEE Conference on Control Technology and Applications (CCTA 2019), Hong Kong, China, 19–21 August 2019; pp. 270–277.

2.	Lou, X.; Neuschaefer, C.; Lei, H. Simulation and Advanced Controls for Alstom's Chemical Looping Process. In Proceedings of the 33rd International Technical Conference on Coal Utilization and Fuel Systems, Clearwater, FL, USA, 1–5 June 2008.

3.	Lou, X.; Neuschaefer, C.; Lei, H. Simulation and Advanced Controls for Hybrid Combustion-Gasification Chemical Looping Process. In Proceedings of the 51st ISA Power Industry Division Symposium and 18th Annual Joint ISA POWID/EPRI Controls and Instrumentation Conference, Scottsdale, Arizona, 8–13 June 2008.

4.	Fan, L.-S. *Chemical Looping Systems for Fossil Energy Conversions*; Wiley: Hoboken, NJ, USA, 2010.

5.	Iloeje, C.O.; Zhao, Z.; Ghoniem, A.F. Design and techno-economic optimization of a rotary chemical looping combustion power plant with CO_2 capture. *Appl. Energy* **2018**, *231*, 1179–1190. [CrossRef]

6.	Chen, P.; Sun, X.; Gao, M.; Ma, J.; Guo, Q. Transformation and migration of cadmium during chemical-looping combustion/gasification of municipal solid waste. *Chem. Eng. J.* **2019**, *365*, 389–399. [CrossRef]

7.	Vidal, R.; Bruna, J.; Giryes, R.; Soatto, S. Mathematics of Deep Learning. *arXiv* **2017**, arXiv:1712.04741v1.

8.	Bruna, J.; Mallat, S. Invariant Scattering Convolution Networks. *IEEE Trans. Pattern Anal. Mach. Intell.* **2013**, *35*, 1872–1886. [CrossRef] [PubMed]

9.	Alexandridis, A.K.; Zapranis, A.D. Wavelet neural networks: A practical guide. *Neural Netw.* **2013**, *42*, 1–27. [CrossRef] [PubMed]

10.	Doucoure, B.; Agbossou, K.; Cardenas, A. Time series prediction using artificial wavelet neural network and multi-resolution analysis: Application to wind speed data. *Renew. Energy* **2016**, *92*, 202–211. [CrossRef]

11.	Zhang, Q.; Benveniste, A. Wavelet networks. *IEEE Trans. Neural Netw.* **1992**, *3*, 889–898. [CrossRef]

12.	Zhang, Q. Using wavelet network in nonparametric estimation. *IEEE Trans. Neural Netw.* **1997**, *8*, 227–236. [CrossRef]

13.	Billings, S.A.; Wei, H.L. A new class of wavelet networks for nonlinear system identification. *IEEE Trans. Neural Netw.* **2005**, *16*, 862–874. [CrossRef]

14.	Coca, D.; Billings, A. Non-linear system identification using wavelet multiresolution models. *Int. J. Control* **2001**, *74*, 1718–1736. [CrossRef]

15.	Khelifa, S.; Gourine, B.; Rami, A.; Taibi, H. Assessment of nonlinear trends and seasonal variations in global sea level using singular spectrum analysis and wavelet multiresolution analysis. *Arab. J. Geosci.* **2016**, *9*, 560. [CrossRef]

16.	Sudre, J.; Yahia, H.; Pont, O.; Garçon, V. Ocean Turbulent Dynamics at Superresolution from Optimal Multiresolution Analysis and Multiplicative. *IEEE Trans. Geosci. Remote Sens.* **2015**, *53*, 6274–6285. [CrossRef]

17.	Hsu, C.; Lin, C.; Lee, T. Wavelet adaptive backstepping control for a class of nonlinear systems. *IEEE Trans. Neural Netw.* **2006**, *17*, 1175–1183. [PubMed]

18. Xu, J.; Tan, Y. Nonlinear adaptive wavelet control using constructive wavelet networks. *IEEE Trans. Neural Netw.* **2007**, *18*, 115–127. [CrossRef] [PubMed]

19. Summers, S.; Jones, C.N.; Lygeros, J.; Morari, M. A Multiresolution Approximation Method for Fast Explicit Model Predictive Control. *IEEE Trans. Autom. Control* **2011**, *56*, 2530–2541. [CrossRef]

20. Li, S.; Xi, Y. Application of wavelet to constrained generalized predictive control. In Proceedings of the 15th IEEE international Symposium on Intelligent Control (ISIC 2000), Rio, Patras, Greece, 17–19 July 2000; pp. 247–252.

21. Bitmead, R.; Gevers, M.; Wertz, V. *Adaptive Optimal Control: The Thinking Man's GPC*; Prentice Hall of Australia: Brunswick, Victoria, Australia, 1990.

22. Zhang, S.; Bentsman, J.; Lou, X.; Neuschaefer, C. Wavelet Multiresolution Model based Generalized Predictive Control for Hybrid Combustion-Gasification Chemical Looping Process. In Proceedings of the 51st IEEE Conference on Decision and Control, Maui, Hawaii, USA, 10–13 December 2012; pp. 2409–2414.

23. Bentsman, J.; Zhang, S.; Natarajan, V.; Petrus, B.; Ye, D. *Development of Computational Approaches for Simulation and Advanced Controls for Hybrid Combustion-Gasification Chemical Looping*; Technical Report; Department of Energy National Energy Technology Laboratory: Pittsburgh, PA, USA, 2011.

24. Zhang, S. Wavelet Adaptive and Predictive Control with Applications to Chemical Looping System. Ph.D. Thesis, University of Illinois at Urbana-Champaign, Urbana, IL, USA, 2014.

25. Clark, D.; Mohtadi, C.; Tuffs, P. Generalized predictive control-Part I. the basic algorithm. *Automatica* **1987**, *23*, 137–148. [CrossRef]

26. Clark, D.; Mohtadi, C.; Tuffs, P. Generalized predictive control-Part II. Extensions and interpretations. *Automatica* **1987**, *23*, 149–160. [CrossRef]

27. Matko, D.; Biasizzo, K.K. Generalized predictive control of a thermal plant using fuzzy model. In Proceedings of the 2000 American Control Conference, Chicago, IL, USA, 28–30 June 2000; Volume 3, pp. 2053–2057.

28. Gu, D.; Hu, H. Wavelet neural network-based predictive control for mobile robots. In Proceedings of the 2000 IEEE International Conference on Systems, Man and Cybernetics. 'Cybernetics Evolving to Systems, Humans, Organizations, and Their Complex Interactions', Nashville, TN, USA, 8–11 October 2000; Volume 5, pp. 3544–3549.

29. Yoo, S.J.; Choi, Y.H.; Park, J.B. Generalized predictive control based on self-recurrent wavelet neural network for stable path tracking of mobile robots: Adaptive learning rates approach. *IEEE Trans. Circuits Syst.* **2006**, *53*, 1381–1394.

30. Tao, J.; Dehmer, M.; Xie, G.; Zhou, Q. A Generalized Predictive Control-Based Path Following Method for Parafoil Systems in Wind Environments. *IEEE Access* **2019**, *7*, 42586–42595. [CrossRef]

31. Li, S.; Jiang, P.; Han, K. Generalized Predictive Control of an Intensified Continuous Reactor. In Proceedings of the 37th Chinese Control Coference, Wuhan, China, 25–27 July 2018.

32. Grimble, M.J. Generalized predictive optimal control: An introduction to the advantages and limitations. *Intern. J. Syst. Sci.* **1992**, *23*, 85–98. [CrossRef]

33. Mallat, J. A Theory for Multiresolution Signal Decomposition: The Wavelet Representation. *IEEE Trans. Pattern Anal. Mach. Intell.* **1989**, *11*, 674–693. [CrossRef]

34. Chen, S.; Billings, S.A.; Luo, W. Orthogonal least squares methods and their application to nonlinear system identification. *Int. J. Control* **1989**, *50*, 1873–1896. [CrossRef]

35. Astrom, K.J.; Wittenmark, B. *Adaptive Control*, 2nd ed.; Addison-Wesley: Reading, MA, USA, 1995.

36. Tsang, T.; Clarke, D. Generalized predictive control with input constraints. *IEE Proc.* **1988**, *135*, 451–460. [CrossRef]

37. Huang, Y.; Turtona, R.; Park, J.; Famourib, P.; Boylec, E.J. Dynamic model of the riser in circulating fluidized bed. *Powder Technol.* **2006**, *163*, 23–31. [CrossRef]

38. Narendra, K.S.; Mukhopadhyay, S. Adaptive control using neural networks and approximate models. *IEEE Trans. Neural Netw.* **1997**, *8*, 475–485. [CrossRef] [PubMed]

39. Zhao, H.; Bentsman, J. Biorthogonal wavelet based identification of fast linear time varying systems-part 1: System representations. *J. Dyn. Syst. Meas. Control* **2001**, *123*, 585–592. [CrossRef]

40. Zhao, H.; Bentsman, J. Biorthogonal wavelet based identification of fast linear time varying systems-part 2: Algorithms and performance analysis. *J. Dyn. Syst. Meas. Control* **2001**, *123*, 593–600. [CrossRef]

41. Opfermann, J.D.; Leonard, S.; Decker, R.S.; Uebele, N.A.; Bayne, C.E.; Joshi, A.S.; Krieger, A. Semi-Autonomous Electrosurgery for Tumor Resection Using a Multi-Degree of Freedom Electrosurgical Tool and Visual Servoin. In Proceedings of the 2017 IEEE/RSJ International Conference on Intelligent Robots and Systems (IROS), Vancouver, BC, Canada, 24–28 September 2017; pp. 3653–3660.
42. Madhukar, A.; Park, Y.; Kim, W.; Sunaryanto, H.J.; Berlin, R.; Chamorro, L.P.; Bentsman, J.; Ostoja-Starzewski, M. Heat conduction in porcine muscle and blood: Experiments and time-fractional telegraph equation model. *J. R. Soc. Interface* **2019**, *16*, 1–8. [CrossRef]

Article

Multi-Objective Optimal Operation for Steam Power Scheduling Based on Economic and Exergetic Analysis

Yu Huang [1], Weizhen Hou [1], Yiran Huang [2,*], Jiayu Li [1], Qixian Li [1], Dongfeng Wang [1] and Yan Zhang [1]

[1] Department of Automation, North China Electric Power University, Baoding 071003, China; huangyufish@ncepu.edu.cn (Y.H.); 2182216069@ncepu.edu.cn (W.H.); 2182216097@ncepu.edu.cn (J.L.); 2192216022@ncepu.edu.cn (Q.L.); wangdongfeng@ncepu.edu.cn (D.W.); zhangyan@ncepu.edu.cn (Y.Z.)

[2] Department of Engineering Electronics and Communication Engineering, North China Electric Power University, Baoding 071003, China

* Correspondence: hyr_2000@sina.com; Tel.: +86-1378-5216-978

Received: 18 January 2020; Accepted: 31 March 2020; Published: 13 April 2020

Abstract: Steam supply scheduling (SSS) plays an important role in providing uninterrupted reliable energy to meet the heat and electricity demand in both the industrial and residential sectors. However, the system complexity makes it challenging to operate efficiently. Besides, the operational objectives in terms of economic cost and thermodynamic efficiency are usually contradictory, making the online scheduling even more intractable. To this end, the thermodynamic efficiency is evaluated based on exergetic analysis in this paper, and an economic-exergetic optimal scheduling model is formulated into a mixed-integer linear programming (MILP) problem. Moreover, the ε-constraint method is used to obtain the Pareto front of the multi-objective optimization model, and fuzzy satisfying approach is introduced to decide the unique operation strategy of the SSS. In the single-period case results, compared with the optimal scheduling which only takes the economic index as the objective function, the operation cost of the multi-objective optimization is increased by 4.59%, and the exergy efficiency is increased by 9.3%. Compared with the optimal scheduling which only takes the exergetic index as the objective function, the operation cost of the multi-objective optimization is decreased by 19.83%, and the exergy efficiency is decreased by 2.39%. Furthermore, results of single-period and multi-period multi-objective optimal scheduling verify the effectiveness of the model and the solution proposed in this study.

Keywords: steam supply scheduling; exergetic analysis; multi-objective; ε-constraint method

1. Introduction

As the material basis of human survival and development, energy plays an increasingly important role in promoting social and economic development as well as in improving people's living standards [1]. The energy situation and environmental problems have recently attracted worldwide attention. Steam supply scheduling (SSS) consumes primary energy to provide energy for an enterprise and simultaneously produces a substantial number of pollutants. This paper focuses on optimizing operation of the SSS to reduce the operation cost and improve the thermodynamic efficiency, thus the economic–exergetic operation of the system can be realized [2,3].

Scholars have conducted in-depth studies on the operation optimization of SSS and have made some achievements. Grossman proposed a mixed-integer linear programming (MILP) model framework for utility systems [4], and a mixed-integer nonlinear programming (MINLP) problem based on a successive MILP approach was solved [5]. Based on utility systems modeling, numerous scholars have focused on optimal operation strategy so as to achieve cost minimization and energy distribution. In order to reduce operation cost, a model was established which integrated the start and stop of utility operating units under different process requirements [6]. In [7], the multi-period with different electricity or steam demand was introduced into the operation strategy of the utility system, and the optimal choice of units for each period was determined by using the two-stage approach. Based on the linear single-device models in public utility systems, the influence of the change of external electricity price on the system operation scheduling was studied by taking into account the steam equilibrium, fuel supply and devices operation constraints [8]. A multi-period MILP model for byproduct gases, steam and power distribution optimization was proposed in steam power plants, and the experimental results showed that the proposed model could effectively reduce system cost [9]. Given the uncertainty of device efficiency and process demand, a data-driven method was proposed to achieve the tradeoff between optimality and robustness of operational decisions in utility system optimization, and the experiments demonstrated the effectiveness of the method [10,11]. Besides, a method for the simultaneous synthesis of heat exchanger networks and utility systems was presented, and the two-stage algorithm was used to identify the best tradeoff between utility systems and heat exchanger networks costs [12,13]. Due to the escalating environmental crisis, several scholars have conducted extensive researches on the environmental issue. Central utility systems with adjoining waste-to-energy networks were integrated to form an ecological friendly energy management system, and the feasibility of the combination of the two networks was demonstrated from environmental and economic perspectives through experiments [14]. By taking into account the impact of pollutants on environment, the utility system consisting of boilers, gas turbines with heat recovery steam generators, ST and CT has been developed [15]. Considering the environmental performance of the entire site utility system, the structural design was optimized to minimize the total annual cost [16]. In addition, the multimodal genetic algorithm was used in the exergoenvironmental analysis of a combined heat and power plant [17]. In general, most studies have focused on the economic and environmental operation that consider energy quantity saving in the SSS, without considering the quality distinction between different energy resources.

In 1953, Rant put forward the concept of exergy, which is a physical quantity that synthesizes the first and the second law of thermodynamics to measure working ability. This concept can focus on the quality and quantity of energy [18–20], which provides a highly efficient method to evaluate the energy efficiency of the system [21–23]. Certain scholars have recently studied the energy system by means of the exergetic analysis. A kind of solid oxide fuel cell integrated with gas and steam trigeneration systems was optimized, and the energy, exergy and economy of the system were analysed in [24]. In addition, the influence of parameter changes on system performance was further studied. An existing CHP system was analysed in terms of energy, exergy and environmental (3E) aspects [25]. In order to analyse the performance and optimize parameter of the geothermal power plant, a system optimization model was formulated to maximize the exergy efficiency, which was solved through the gravitational search algorithm [26]. A study was conducted to examine the energy and exergy performance as well as multi-objective optimization of an exhaust air heat recovery system, which could provide reference for system planning [27]. In [28], considering the total cost, carbon dioxide emission and exergetic destruction, a multi-objective optimization of district heating system was carried out and the Pareto front was obtained with the weighting method. By means of exergy, exergoeconomic and exergoenvironmental analysis, the optimal integration of steam and power system with a steam power plant as the source and a utility system as the sink was investigated, and the experimental results reflected that the integration of steam power plant and utility system is a favorable option [29]. However, most studies have applied exergetic analysis to the performance evaluation and parameter optimization of the energy system, while there are relatively few quality studies on the energy such as heat and electricity in the SSS.

In this paper, exergy is introduced into the operation optimization of SSS. Firstly, the multi-objective mixed-integer linear programming (MOMILP) model of SSS is established by using the exergetic analysis method to reduce the operation cost and exergy input for SSS. At the same time, the Pareto front of the multi-objective optimization model is obtained with the ε-constraint approach, and the compromise solution on the Pareto front was acquired with the fuzzy satisfying approach. Finally, the effectiveness of the proposed model and solution method was verified by the results of single-period and multi-period multi-objective optimal scheduling.

This paper is organized as follows: the MOMILP model of SSS is developed in Section 2. In Section 3, the multi-objective operation strategy of SSS is presented to obtain the Pareto front, and a tradeoff is conducted between these different objectives. Case studies are analysed in Section 4, and the conclusion is summarised in Section 5.

2. Problem Formulation of Multi-objective Optimal Operation of SSS

The SSS converts primary energy (fuel coal, fuel gas, fuel oil) into secondary energy (electricity, steam and hot water) to provide enterprises with the required process steam, thermal energy and electricity, and its typical structure is illustrated in Figure 1. To realize the economic and efficient operation of the SSS, the mathematical model of each equipment is built and the concept of exergy is adopted to evaluate all types of energy. Subsequently, the MOMILP optimal model of the SSS is formulated.

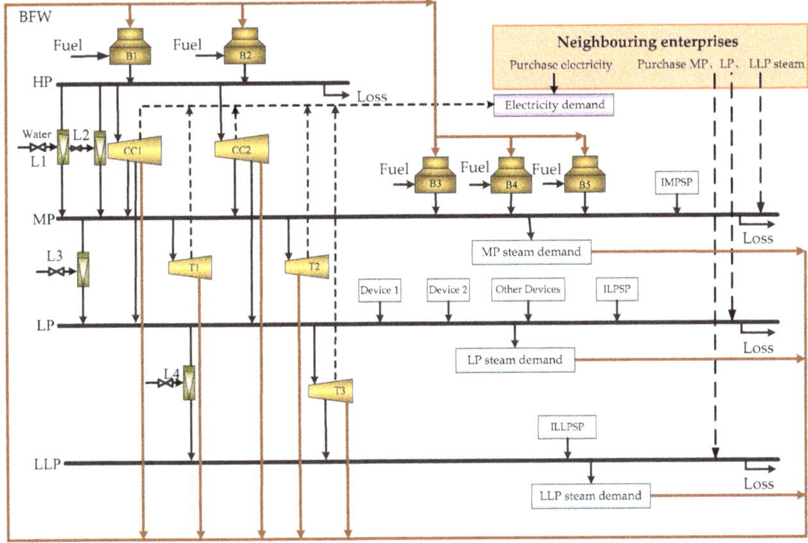

Figure 1. Steam supply scheduling (SSS) of a petrochemical enterprise.

2.1. Objectives

2.1.1. Economic Objective

Generally, the economic objective of SSS operation is to minimize the total cost of the entire period, including fuel consumption cost, electricity or steam purchase cost, equipment operation and maintenance cost, depreciation cost and equipment start/stop cost. The specific expression is as follows:

$$
\begin{aligned}
f_1 = &\sum_{Bn} \sum_i \sum_t F_{Bn,t,i} p_i + \sum_r \sum_t PS_{r,t} p_r + \sum_t PE_t p_e \\
&+ \sum_n \sum_t CM_n Y_{n,t} + \sum_n \sum_t CST_n Z_{n,t} + \sum_n \sum_t CSP_n ZS_{n,t}
\end{aligned}
\tag{1}
$$

2.1.2. Exergetic Objective

Exergy represents the maximum amount of useful work that can be obtained from a given form of energy, that is, the quality of energy. For different kinds of energy, the quality of the same quantity of energy is not necessarily the same. Hence, the quality of different energy of the SSS is evaluated with the exergetic analysis method, and then the maximum exergy efficiency can be achieved. Among them, exergy efficiency can be defined as the ratio of the total output exergy to the total input exergy [21], and the formula is as follows:

$$\psi = \frac{\sum\limits_{t} \dot{Ex}_{out,t}}{\sum\limits_{t} \dot{Ex}_{in,t}} \tag{2}$$

Figure 1 shows that the energy types on the load side include electric energy and heat energy, and the input energy includes fossil energy (fuel coal, fuel oil, fuel gas), purchased steam and electricity. Therefore, the load demand exergy and the input exergy of the SSS are expressed as follows:

$$\begin{cases} \dot{Ex}_{out,t} = \dot{Ex}^{e}_{out,t} + \dot{Ex}^{h}_{out,t} \\ \dot{Ex}_{in,t} = \dot{Ex}^{coal}_{in,t} + \dot{Ex}^{oil}_{in,t} + \dot{Ex}^{gas}_{in,t} + \dot{Ex}^{e}_{in,t} + \dot{Ex}^{h}_{in,t} \end{cases} \tag{3}$$

Fuel coal, fuel oil and fuel gas are all chemical fuels, and their specific exergy is generally expressed by a lower heating value (LHV) and an exergy factor [30,31]. The exergy factors of different types of fuels are slightly different, but basically the same [32]. The exergy in chemical fuels can be expressed by the following formula:

$$\begin{cases} \alpha_i = \gamma_{i,LHV}\zeta_i \\ \dot{Ex}^{i}_{in,t} = \alpha_i F_{Bn,t,i}/3.6 \end{cases} \tag{4}$$

The exergy of electricity is equal to electricity because it can be completely converted into work. However, the work done by heat energy is limited by the Carnot factor, and its heat exergy is equal to the work done by the Carnot cycle. Therefore, the heat and electricity exergy of SSS are described as follows:

$$\dot{Ex}^{h}_{out,t} = \sum_{r}\left(1 - \frac{T_a}{T^h_r}\right)DS_{r,t}h_r/3.6 \tag{5}$$

$$\dot{Ex}^{h}_{in,t} = \sum_{r}\left(1 - \frac{T_a}{T^h_r}\right)PS_{r,t}h_r/3.6 \tag{6}$$

$$\dot{Ex}^{e}_{out,t} = DE_t \tag{7}$$

$$\dot{Ex}^{e}_{in,t} = PE_t \tag{8}$$

The energy demand of SSS is predictable, that is, the total output exergy can be calculated. Therefore, the exergetic objective could be converted from maximum exergy efficiency to minimum exergy input of the SSS. The formula is as follows:

$$f_2 = \sum_{t} \dot{Ex}_{in,t} \tag{9}$$

2.2. Constraints

2.2.1. Device Constraints

Industrial boilers generally convert the chemical energy in fuel into heat energy, and then transfer this heat energy to water through different heating surfaces, and finally produce the high-pressure (HP) or medium-pressure (MP) steam required by the system. The model is represented by Equation (10). Equation (11) indicates that the boiler load should be placed within a certain safety range:

$$F_{Bn,t,i}\gamma_{i,LHV} = a_{Bn}Y_{Bn,t} + b_{Bn}G_{Bn,t} \tag{10}$$

$$G_{Bn}^{L}Y_{Bn,t} \leq G_{Bn,t} \leq G_{Bn}^{U}Y_{Bn,t} \tag{11}$$

The operation characteristics of a simple ST, including backpressure and CT, can be expressed by the linear relationship between steam intake and power output, and the steam intake and output power of ST should be placed within a certain safety range. The general constraints are detailed as follows:

$$P_{Tn,t} = d_{Tn}Y_{Tn,t} + e_{Tn}G_{Tn,in,t} \tag{12}$$

$$P_{Tn}^{L}Y_{Tn,t} \leq P_{Tn,t} \leq P_{Tn}^{U}Y_{Tn,t} \tag{13}$$

$$G_{Tn,in}^{L}Y_{Tn,t} \leq G_{Tn,in,t} \leq G_{Tn,in}^{U}Y_{Tn,t} \tag{14}$$

The double extraction CT generates steam with different pressure, and at the same time it can generate the power required to meet the external electricity load. Its output power is related to steam intake, industrial sectors or heating steam extractions. The model generally uses a linear function to represent the relationship among steam intake, adjustable extraction and power of the double extraction CT. It can be described as follows:

$$P_{cn,t} = g_{cn}G_{cn,in,t} - \sum_{oj} h_{cn,oj}G_{cn,oj,t}^{out} - f_{cn}Y_{cn,t} \tag{15}$$

$$P_{cn}^{L}Y_{cn,t} \leq P_{cn,t} \leq P_{cn}^{U}Y_{cn,t} \tag{16}$$

$$G_{cn,in}^{L}Y_{cn,t} \leq G_{cn,in,t} \leq G_{cn,in}^{U}Y_{cn,t} \tag{17}$$

$$G_{cn,oj}^{out,L}Y_{cn,t} \leq G_{cn,oj,t}^{out} \leq G_{cn,oj}^{out,U}Y_{cn,t} \tag{18}$$

The main function of the pressure reducer and attemperator is to adjust the steam from high temperature and high pressure to the relatively low temperature and low pressure required by the system. The general form of the model is shown in Equation (19) [10,33]. For a given steam system, once the steam pressure and temperature of each level steam are specified, the enthalpy $h_{Ln,out}$, $h_{Ln,in}$, and $h_{Ln,w}$ are constant. By defining a parameter $\eta_{Ln.}$, the expression is shown in Equation (20):

$$G_{Ln,in,t} = \frac{h_{Ln,out,t} - h_{Ln,w,t}}{h_{Ln,in,t} - h_{Ln,w,t}}G_{Ln,out,t} \tag{19}$$

$$G_{Ln,in,t} = \eta_{Ln}G_{Ln,out,t} \tag{20}$$

2.2.2. Balance Constraints

Electricity balance and steam balance should be considered to meet the energy demand and energy distribution among various energy devices in the SSS should also be considered. Therefore, the steam and electricity balance model are expressed as follows:

$$DS_{r,t} = PS_{r,t} + \sum_{n} G_{Bn,t}Y_{Bn,t} + \sum_{n} G_{cn,oj,t}^{out}Y_{cn,t} + \sum_{n} G_{Ln,out,t}Y_{Ln,t} - \sum_{n} G_{Tn,in,t}Y_{Tn,t} - \sum_{n} G_{Ln,in,t}Y_{Ln,t} - Loss_{r} \tag{21}$$

$$DE_{t} = PE_{t} + \sum_{n} P_{Tn,t}Y_{Tn,t} + \sum_{n} P_{cn,t}Y_{cn,t} \tag{22}$$

2.2.3. Logical Constraints on Device Start and Stop

Due to the different demand for steam and power in different periods, the equipment is easy to start and stop in the adjacent two periods, thus resulting in the start and stop cost of the equipment. In this study, Equations (23) and (24) are used to represent the start and stop logic of the equipment:

$$Z_{n,t} = Y_{n,t} - Y_{n,t-1}, Y_{n,0} = 0 \tag{23}$$

$$ZS_{n,t} = Y_{n,t} - Y_{n,t+1}, Y_{n,t+1} = 0 \tag{24}$$

In summary, the MOMILP optimal model of SSS can be represented as follows:

$$\begin{aligned} &\min \text{obj}_1 = f_1(x_t, y_t) \\ &\min \text{obj}_2 = f_2(x_t, y_t) \\ &\text{subject to} \begin{cases} g(x_t, y_t) \leq 0 \\ h(x_t, y_t) = 0 \\ x_t \in (x^L, x^U) \\ y_t \in (0, 1) \end{cases} \end{aligned} \tag{25}$$

3. Solution and Decision of the Optimal Condition

In order to achieve optimal operation of the economy and exergetics of the SSS, this section uses the ε-constraint method to solve the proposed multi-objective problem, so as to obtain the Pareto front. The Pareto curve is used to determine the optimal solution with the fuzzy satisfying approach.

3.1. ε-Constraint Based Solution

In the optimization problem of SSS, the two objectives of reducing the operation cost and exergy input are considered to affect each other, that is, it is difficult for both sides to reach the optimal simultaneously. Therefore, the ε-constraint method is used to solve this multi-objective optimal problem. The ε-constraint method preserves one of the objectives in the objective function and transforms the rest of the objective functions into constraints. Thereby the multi-objective optimization problem is transformed into a series of single-objective optimization problems, which can be solved by modifying the value range of the constraints condition step by step. The details are presented as follows:

$$\begin{aligned} &\min f_1 \\ &\begin{cases} \text{subject to } f_2 \leq \varepsilon \\ \text{Equations } (1) - (24) \end{cases} \end{aligned} \tag{26}$$

where the value of ε is considered to be expressed by the following equation:

$$\varepsilon = f_{2,\max} - (f_{2,\max} - f_{2,\min})(a - 1)/(a_{\max} - 1) \tag{27}$$

where $a = 1, 2, \ldots, a_{max}$, a_{max} is the maximum number of cycles; $f_{2,\min}$ and $f_{2,\max}$ are the maximum and minimum values of f_2 obtained when f_2 and f_1 are considered as a single-objective function, respectively. Furthermore, the economic objective is taken as f_1 as shown in Equation (1) and the exergetic objective as f_2 as shown in Equation (9) in this paper.

3.2. Decision based on Fuzzy Satisfying Approach

To achieve the coordination and unification of the multi-objective of SSS, the fuzzy satisfying approach [34] is introduced to help the operator establish a trade-off between the economic objective and the exergetic objective. The target values of each operation strategy are normalized according to

(28). Subsequently, the membership function value of each operation strategy is calculated according to (29), and the best operation strategy is selected according to Equation (30):

$$\mu_k^a = \begin{cases} 1 & f_k^a \leq f_{k,\min} \\ \dfrac{f_{k,\max} - f_k^a}{f_{k,\max} - f_{k,\min}} & f_{k,\min} \leq f_k^a \leq f_{k,\max} \\ 0 & f_k^a \geq f_{k,\max} \end{cases} \tag{28}$$

$$\mu^a = \min(\mu_1^a, \mu_2^a) \tag{29}$$

$$\mu^{\max} = \max(\mu^1, \mu^2, \ldots \mu^{a_{\max}}) \tag{30}$$

To make the fuzzy satisfying approach clearer, a small example is given in Table 1, and the multi-objective optimization strategy selected is scheme 7, which is the bold part of Table 1. In addition, Figure 2 summarizes the specific process of the multi-objective operation optimization.

Table 1. An example of the fuzzy satisfying approach.

a	f_1	f_2	μ_1	μ_2	$\mu^{\max}$
1	57883.09281	575814.554	1	0	0
2	58242.90797	560822.484	0.956601	0.105264	0.105264
3	58697.95742	545830.368	0.901716	0.210528	0.210528
4	59153.00687	530838.253	0.846831	0.315792	0.315792
5	59608.05632	515846.137	0.791946	0.421056	0.421056
6	60063.10577	500854.021	0.73706	0.52632	0.52632
7	**60566.37947**	**485861.905**	**0.676359**	**0.631584**	**0.631584**
8	61116.28338	470869.789	0.610033	0.736848	0.610033
9	61666.18728	455877.674	0.543707	0.842112	0.543707
10	66174.01862	433390.6	0	1	0

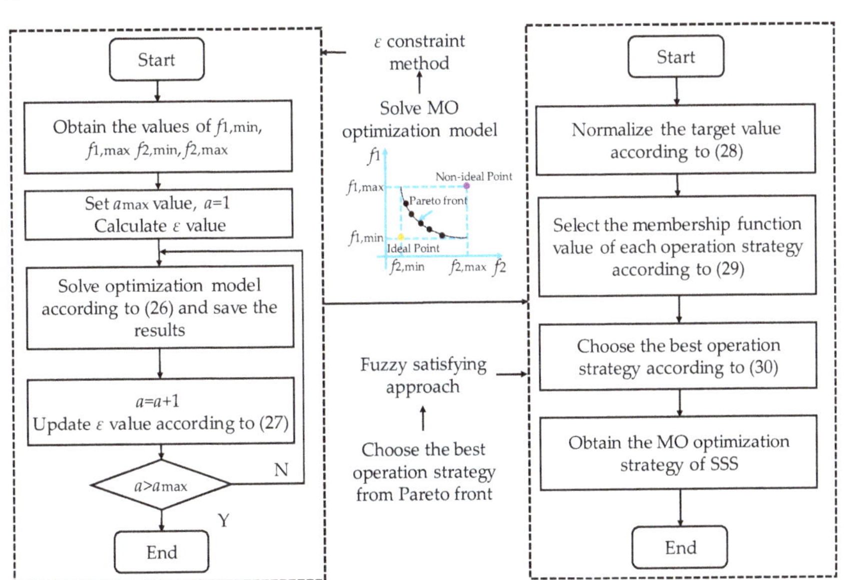

Figure 2. Flowchart of the multi-objective optimization solution.

4. Case Study

To verify the effectiveness of the MILP model of SSS with economic and exergetic objectives, the optimal model of single-period and multi-period of SSS are solved and results are analyzed in this section. Moreover, it is necessary to declare the case studied in this paper does not consider optimization situation of neighbouring enterprises, which is a partial optimization.

4.1. Case Description

This study takes the SSS of petrochemical enterprises as an example (Figure 1), which includes four different levels of steam, namely, high-pressure steam (HP, 9.5 MPa and 535 °C), medium pressure steam (MP, 3.5 MPa and 425 °C), low-pressure steam (LP, 1 MPa and 300 °C) and low and low-pressure steam (LLP, 0.3 MPa and 200 °C). B1–B2 are coal-fired boilers that produce HP steam with a blowdown rate of 8%; B3–B5 are dual fuel boilers, which burn oil and gas to produce MP steam. The amount of gas is determined by the processing unit, and the maximum available gas import capacity is 12 t/h. Double extraction CT (CC1, CC2) produces MP and LP steam as well as electric energy. T1 and T2 steam turbines generate power, and T3 produces LLP steam and electric energy. As can be seen from Figure 1, the condensate is recycled and converted into boiler feed water. Furthermore, the minimum value of condensing steam amount of steam turbine is 63 t/h, the maximum value of condensing steam amount is 142 t/h, and the condensing pressure is 5.9 kPa. Pressure reducer and attemperator (L1, L2, L3 and L4) can convert high-temperature and high-pressure steam into relatively low-level steam. The study allows the maximum electricity import capacity from neighbouring enterprises of 50,000 kW. The maximum MP steam, LP steam and LLP steam import capacity from neighbouring enterprises of 100, 50 and 50 t/h. The effects of device 1, device 2 and other devices on the system are not considered in this study, and the loss of the system is neither considered, that is, $loss_r = 0$. Table 2 indicates the model parameters of boiler and steam turbine. Tables 3 and 4 list the equipment parameters, start/stop costs and equipment operation costs of boilers and ST. Table 5 indicates the unit price of resources. Table 6 shows the parameters of the resource, and Table 7 lists the start and stop time of the equipment.

Table 2. Model parameters of boiler and steam turbine (ST).

Parameter	Value	Parameter	Value
a_{B1}	95.8	f_{c1}	−179
a_{B2}	95.8	f_{c2}	−179
a_{B3}	19.47	g_{T3}	73.37
a_{B4}	5.818	g_{c1}	252
a_{B5}	5.159	g_{c2}	252
b_{B1}	0.8488	$h_{c1,o1}$	−235
b_{B2}	0.8488	$h_{c1,o2}$	−102
b_{B3}	0.931	$h_{c2,o1}$	−235
b_{B4}	1.021	h_{c2o2}	−102
b_{B5}	1.032	h_{T3}	−23.3
d_{T1}	−1459	η_{L1}	0.933
d_{T2}	−1650	η_{L2}	0.933
e_{T1}	86.2	η_{L3}	0.913
e_{T2}	88.6	η_{L4}	0.923
f_{T3}	−116.7		

Table 3. Boiler equipment information.

Boiler	B1	B2	B3	B4	B5
Rated evaporation (t/h)	320	320	140	75	65
Minimum evaporation (t/h)	160	160	80	35	50
Start and stop cost (¥)	10,000	10,000	13,000	13,000	13,000
Equipment operation cost (¥/h)	100	100	200	185	170
Steam temperature (°C)	535	535	425	425	425
Steam pressure (MPa)	9.5	9.5	3.5	3.5	3.5
Feedwater temperature (°C)	211.23	211.23	163	163	163
Feedwater pressure (MPa)	15.08	15.08	3.5	3.5	3.5

Table 4. Steam turbine equipment information.

Equipment	Start and Stop Cost (¥)	Equipment Operation Cost (¥/h)	Power Rating (kW)		Maximum Steam Intake (t/h)	Maximum First Extraction Steam (t/h)	Maximum Second Extraction Steam (t/h)
			Maximum	Minimum			
CC1	12,000	200	50,000	20,000	380	190	180
CC2	12,000	200	50,000	20,000	380	190	180
T1	6000	160	6500	1500	85	—	—
T2	6000	140	6500	1500	85	—	—
T3	6000	110	6000	3500	100	—	—

Table 5. The unit price of resources.

Fuel Coal (¥/t)	Fuel Gas (¥/t)	Fuel Oil (¥/t)	Electricity (¥/kWh)	MP Steam (¥/t)	LP Steam (¥/t)	LP Steam (¥/t)
750	1200	3500	0.45	180	135	80

Table 6. Parameters of the resource.

Fuel	Lower Heating Value (kJ/kg)	Exergy Factor	/	Enthalpy (kJ/kg)	Industrial Steam Production	Value (t/h)
Coal	24,440	1.08	MP	3280.7	IMPSP	57
Gas	32,503	1	LP	3051.7	ILPSP	76.8
Oil	40,245	1.06	LLP	2865.9	ILLPSP	34.9
			Water of Ln	632.2		

Table 7. Start and stop time of equipment.

Equipment	B1	B2	B3	B4	B5	CC1	CC2	T1	T2	T3
Start time (h)	10	10	7.5	7.5	7.5	3.3	3.3	3.3	3.3	3.3
Stop time (h)	10	10	7.5	7.5	7.5	3.3	3.3	3.3	3.3	3.3

4.2. Single-Period Case

Table 8 reports the demand for steam and electricity over a single period time, without considering the start and stop costs of the equipment in the economic objective. During the solution process, the maximum number of cycles n in the ε-constraint method is set to 20.

Table 8. Single-period steam and electricity demands.

Steam (t/h)			Electricity (kW)
MP	LP	LLP	
180	125	150	64,400

Figure 3 shows the single-period Pareto front for the SSS, the points on it are all optimal values, which can provide different operation strategies for operators. Furthermore, the multi-objective optimal operation strategy can be obtained with the fuzzy satisfying approach, which is the point marked on the Pareto curve in Figure 3. Tables 9 and 10 show the optimal scheduling results of boilers load and purchased resources for SSS.

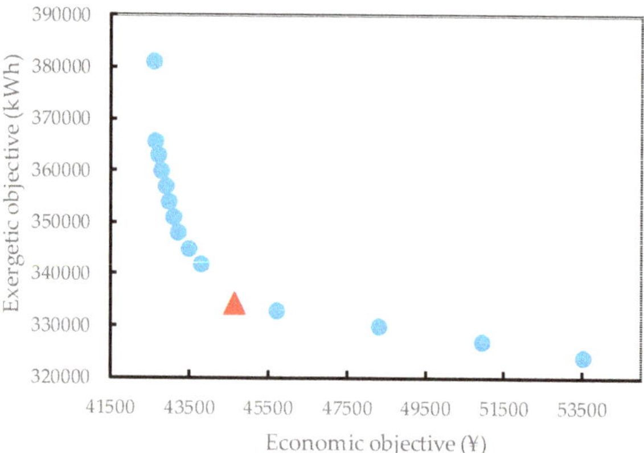

Figure 3. Pareto front of single- period for SSS.

Table 9. Optimal scheduling results of boilers load for SSS.

Boilers Load (t/h)/Situation	Multi-Objective Operation (Standard)	Economic Operation	Exergetic Operation
B1 steam production	320	0	0
B1 coal consumption	41.33	0	0
B2 steam production	0	320	247.03
B2 coal consumption	0	41.33	34.36
B3 steam production	0	103.00	0
B3 oil consumption	0	0	0
B3 gas consumption	0	8.55	0
B4 steam production	0	0	0
B4 oil consumption	0	0	0
B4 gas consumption	0	0	0
B5 steam production	0	0	0
B5 oil consumption	0	0	0
B5 gas consumption	0	0	0

Table 10. Optimal scheduling results of purchased resources for SSS.

Purchased Resources/Situation	Multi-Objective Operation (Standard)	Economic Operation	Exergetic Operation
PMP steam (t/h)	21.74	0	100
PLP steam (t/h)	0	0	0
PLLP steam (t/h)	0	0	0
PE (kW)	19,838.62	743.04	19,838.62

Evidently, compared with the multi-objective operation, the energy conversion equipment such as boilers and ST meets the demand for steam and most electricity in the economic operation. Consequently, less steam and electricity are purchased. By contrast, exergetic operation purchases more steam from the neighbouring enterprises. The multi-objective operation establishes a tradeoff between the economic objective and exergetic objective to satisfy the multi-objective optimal operation by coordinating the consumption of different types of energy (fossil energy, heat energy and electric energy).

Table 11 indicates the operation cost, input exergy and exergy efficiency in multi-objective optimization and single-objective optimization. Based on the results of multi-objective optimization, the growth rate of operation cost, input exergy and exergy efficiency in economic operation and exergetic operation are calculated. Results reveal that compared with the multi-objective optimal operation, the operation cost of the economic operation is decreased by 4.59%, while the input exergy is increased by 13.97%. On the contrary, for the exergetic operation, its input exergy is decreased by 3.06%, while its operation cost is increased by 19.83%.

Table 11. Comparison of multi-objective optimization results with single-objective optimization results.

Situation	Indicator	Value	Growth Rates
Multi-objective operation (standard)	Operation cost/¥	44,667.83	0
	Input exergy/kW	334,216	0
	Exergy efficiency/%	75.83%	0
Economic operation	Operation cost/¥	42,615.71	−4.59%
	Input exergy/kW	380,921.40	13.97%
	Exergy efficiency/%	66.53%	−9.3%
Exergetic operation	Operation cost/¥	53,526.94	19.83%
	Input exergy/kW	323,989.8	−3.06%
	Exergy efficiency/%	78.22%	2.39%

Compared with SSS optimal scheduling which only takes economic or exergetic as the objective function, from the above calculated data, it can see that the multi-objective operation can comprehensively consider energy efficiency from the point of view of economic and exergetic, make a tradeoff between the economic index and exergetic index, and pay attention to the quality and quantity of energy simultaneously, so as to achieve the purpose of reducing cost and increasing efficiency. Furthermore, this paper is in line with the sustainable energy development strategy of the world today.

4.3. Multi-Period Case

A multi-period case is established in this section to further verify the effectiveness of the proposed multi-objective model and solution method. The multi-period model includes six periods, each with a duration of 720 h, which is consistent with the solution method and the ε-constraint parameter setting in Section 4.2. Table 12 indicates the steam and electricity demands of the six periods. The optimal scheduling results are detailed as follows.

Table 12. Multi-period steam and electricity demands.

Period	Steam (t/h)			Electricity (kW)
	MP	LP	LLP	
1	180	125	150	64,400
2	350	165	75	73,700
3	320	185	60	85,700
4	250	200	95	58,700
5	200	175	105	67,600
6	335	300	120	78,300

Figure 4 depicts the steam distribution among the equipment. It can be seen that under the premise of fully considering the steam purchase, the boiler and ST jointly produce steam, and the system supplements the regulation of pressure reducer and attemperator, thus the integrated operation of steam production and supply at all levels can be realized.

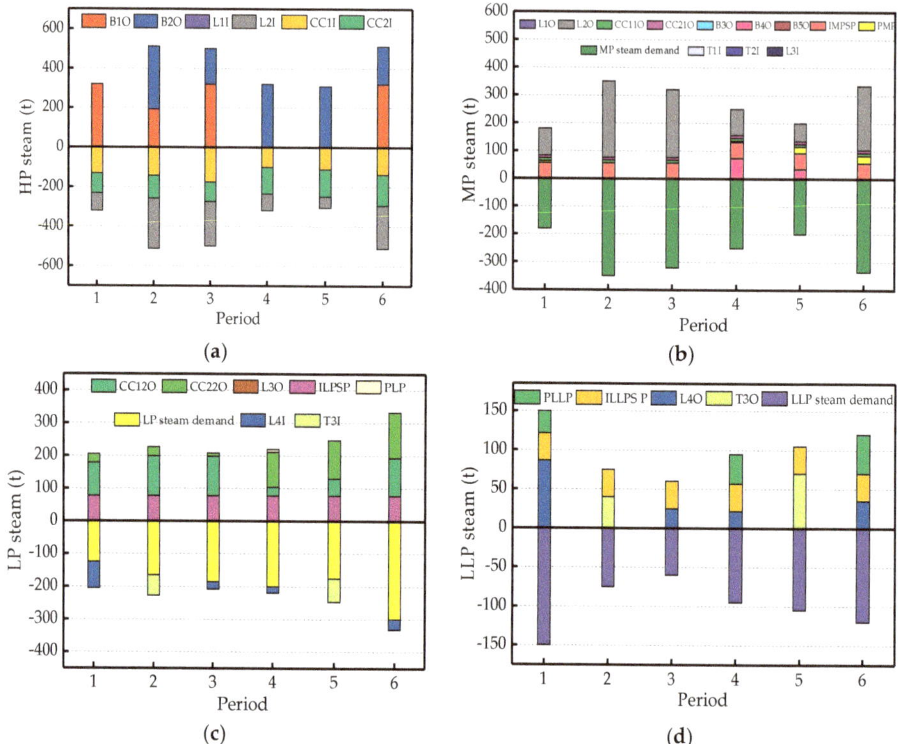

Figure 4. Optimal scheduling results for steam: (**a**) high-pressure steam; (**b**) medium pressure steam; (**c**) low-pressure steam; (**d**) low and low-pressure steam. (I, O indicate input and output of equipment, respectively.).

Table 13 shows the optimal scheduling results of the start and stop of multi-period operation equipment. Figure 5 describes the optimal scheduling results of fuel consumed in the system. Evidently, changes in steam and electricity demand lead to the inevitably start and stop of equipment, thus changing fuel consumption. Due to the relative high steam and electricity demand compared with other periods, the B1 and B2 are in operation in periods 2, 3 and 6. Furthermore, since the MP steam demand of period 1 is lower than that of periods 4 and 5, boilers producing MP steam are closed in period 1. Moreover, considering that the energy of the system is converted from the HP steam generated by B1 and B2, a large amount of coal is consumed.

Table 13. Optimal scheduling results of start/stop of multi-period operation equipment.

Equipment/Period	1	2	3	4	5	6
B1	1	1	1	0	0	1
B2	0	1	1	1	1	1
B3	0	0	0	0	0	0
B4	0	0	0	1	1	0
B5	0	0	0	0	0	0
CC1	1	1	1	1	1	1
CC2	1	1	1	1	1	1
T1	0	0	0	0	0	0
T2	0	0	0	0	0	0
T3	0	1	0	0	1	0

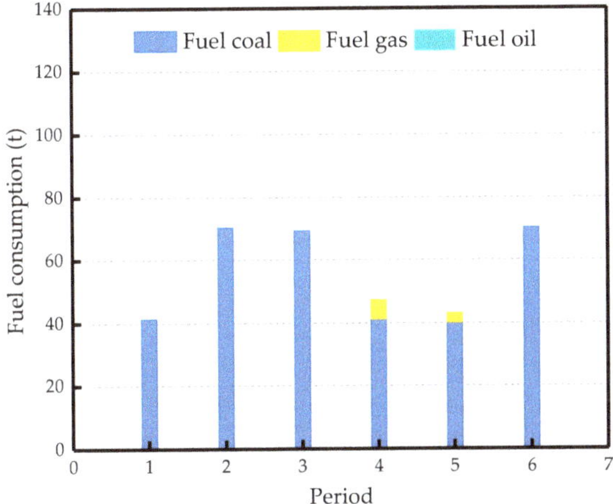

Figure 5. Optimal scheduling results for fuel.

Figure 6 shows the optimal scheduling results of electricity. Under the premise of purchasing electricity (Figure 6 PE), it can be observed that the coordinated operation of T1, T2, T3, CC1 and CC2 can meet the electricity demand. Besides, T1 and T2 are stopped in each period and double extraction CT is used more frequently during the operation process. On the one hand, this is because CC1 and CC2 can satisfy most of the electricity demand. Moreover, CC1 and CC2 can generate electricity and produce both MP and LP steam to meet the steam demand by consuming HP steam. On the other, T1 and T2 only generate electricity. In order to save operation costs, it is not necessary to maintain the operation of all units. In addition, T3 is used more frequently than T1 and T2, partly because T3 can generate electricity and LLP steam simultaneously. On the other hand, it can be seen from Table 4 that the operation cost of T3 is lower than that of T1 and T2. Accordingly, this operation strategy can save economic costs. Furthermore, in order to balance both economic and exergetic objectives, the system neither over purchases energy, nor blindly consumes chemical fuel to meet the electricity demand, thus realizing the primary energy saving and improving the thermodynamic efficiency of the system in multi-period operation.

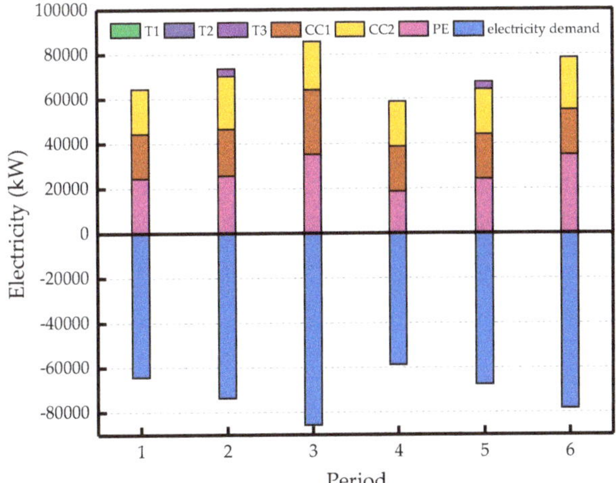

Figure 6. Optimal scheduling results for electricity.

Therefore, the multi-period case study in this section can provide guiding significance for the actual operating system, and the corresponding unit output plan can be made from the two aspects of system economy and thermodynamic efficiency. Furthermore, this study can help reduce greenhouse gas emissions and improve the thermodynamic efficiency under the premise of meeting the power and thermal demand of enterprises.

5. Conclusions

In order to achieve a good balance between enhancing energy efficiency and reducing system cost, this paper adopts the exergetic analysis method in thermodynamics to evaluate the effective energy contained in different kinds of energy. At the same time, the exergetic objective function is built. Considering the cost of electricity and steam, combined with the mathematical model of each equipment, an SSS optimal model based on economic index and exergetic index is further built. Utilizing the ε-constraint method to obtain the Pareto front of multi-objective optimization problems, the fuzzy satisfying approach is introduced to determine the optimal operation strategy. Taking the single-period operation as an example, it can be seen that the multi-objective optimization operation strategy can consider the economic and exergetic of the system by comparing with the single- objective optimization results. Meanwhile, the single-objective optimization only takes the economic or exergetic index as the objective function. Moreover, it can be verified by the results of multi-period scheduling that the multi-objective model and solution is effective. In addition, to deal with the multi-objective problem, the fuzzy satisfying approach is introduced to obtain the optimal results. However, the optimal results may rely on the fuzzy satisfying approach. Therefore, to get better multi-objective optimal results, our future work will focus on the effectiveness of various multi-objective optimal methods.

Author Contributions: Y.H. (Yiran Huang) supervised and revised the manuscript from the initial submission to the final version. Y.H. (Yu Huang) formulated the research problem; Y.H. (Yu Huang), W.H. and Q.L. proposed the mathematical model and solution scheme; W.H. and J.L. wrote the code and solved the optimization model; D.W. and Y.Z. revised the manuscript. All of the authors were involved in writing this paper. All authors have read and agreed to the published version of the manuscript.

Funding: This research was supported by 'the Fundamental Research Funds for the Central Universities 2018MS093 and 2019MS099'.

Conflicts of Interest: The authors declare no conflict of interest.

Nomenclature

a_{Bn}, b_{Bn}	model comodel coefficients of Bn
CM_n	operation and maintenance of equipment (¥)
CST_n	start cost of equipment (¥)
CSP_n	stop cost of equipment (¥)
DE_t	demand electricity of the SSS (kW)
$DS_{r,t}$	demand steam of the SSS (t/h)
d_{Tn}, e_{Tn}	model coefficients of Tn
$\dot{Ex}_{out,t}$	exergy output of SSS (kW)
$\dot{Ex}_{in,t}$	exergy input of SSS (kW)
$\dot{Ex}^e_{out,t}$	electricity exergy output (kW)
$\dot{Ex}^h_{out,t}$	heat exergy output (kW)
$\dot{Ex}^{coal}_{in,t}$	fuel coal exergy input (kW)
$\dot{Ex}^{oil}_{in,t}$	fuel oil exergy input (kW)
$\dot{Ex}^{gas}_{in,t}$	fuel gas exergy input (kW)
$\dot{Ex}^e_{in,t}$	electricity exergy input (kW)
$\dot{Ex}^h_{in,t}$	heat exergy input (kW)
$F_{Bn,t,i}$	fuel consumed by the boiler (t/h)

$f_{cn},\ g_{cn},\ h_{cn,oj}$	model coefficients of cn
f_1	economic objective
f_2	exergetic objective
$G_{Bn,t}$	generated steam by Bn (t/h)
$G_{Tn,in,t}$	flow rates of the inlet for Tn (t/h)
$G_{cn,in,t}$	flow rates of the inlet for cn (t/h)
$G_{Ln,in,t}, G_{Ln,out,t}$	flow rates of the inlet and outlet steam for Ln (t/h)
$G^{out}_{cn,oj,t}$	adjustable steam extraction of cn (t/h)
$G^{L}_{Bn},\ G^{U}_{Bn}$	upper and lower limits of the steam generated by Bn (t/h)
$G^{U}_{Tn,in},\ G^{L}_{Tn,in}$	upper and lower limits of flow rates of the inlet for Tn (t/h)
$G^{out,U}_{cn,oj},\ G^{out,L}_{cn,oj}$	upper and lower limits of adjustable steam extraction of cn (t/h)
$G^{U}_{cn,in},\ G^{L}_{cn,in}$	upper and lower limits of flow rates of the inlet for $Tn\ cn$ (t/h)
$g(x_t,y_t)$	inequality constraint equation
h_r	steam enthalpy (kJ/kg)
$h(x_t,y_t)$	equality constraint equation
$h_{Ln,out,t}$	enthalpy of outlet steam for Ln (kJ/kg)
$h_{Ln,in,t}$	enthalpy of inlet steam for Ln (kJ/kg)
$h_{Ln,w,t}$	enthalpy of water for Ln (kJ/kg)
$Loss_r$	loss stream at all levels
p_i	price of fuel (¥/t)
p_r	price of purchased steam (¥/t)
p_e	price of purchased electricity (¥/kW)
$PS_{r,t}$	purchased steam of the SSS (t/h)
PE_t	purchased electricity of the SSS (kW)
$P_{cn,t}$	generated electricity by cn (kW)
$P_{Tn,t}$	generated electricity by Tn (kW)
$P_{Tn}{}^{U},\ P_{Tn}{}^{L}$	upper and lower limits of electricity generated by Tn (kW)
$P^{U}_{cn},\ P^{L}_{cn}$	upper and lower limits of electricity generated by cn (kW)
T_a	ambient temperature (K)
$T_r{}^{h}$	steam temperature (K)
x_t	continuous variable
$x^{U},\ x^{L}$	upper and lower limits of x
y_t	binary variables of the running state of the equipment
$Y_{n,t}$	binary variables of the running state of the equipment
$Z_{n,t}$	binary variables
$ZS_{n,t}$	binary variables

Subscripts

a_{max}	maximum number of cycles
Bn	boiler equipment
cn	double extraction condensing turbines equipment
i	fuel type
k	the number of objective functions
Ln	pressure reducer and attemperator equipment
n	equipment
r	level for steam
Tn	simple steam turbine equipment

Symbols

α_i	specific exergy of fuel (kJ/kg)
ψ	exergy efficiency (%)
$\gamma_{i,LHV}$	lower heating value of fuel (kJ/kg)
ζ_i	exergy factor of fuel
η_{Ln}	conversion efficiency (%)

Abbreviations

B1~ B5	boiler 1~5
BFW	boiler feed water
CC1, CC2	double extraction condensing turbines 1~2
CC11	flow rates of first extraction steam for CC1
CC12	flow rates of second extraction steam for CC1
CC21	flow rates of first extraction steam for CC2
CC22	flow rates of second extraction steam for CC2
CT	condensing turbines
HP	high pressure
IMPSP	medium pressure produced in industrial
ILPSP	low pressure produced in industrial
ILLPSP	low and low-pressure steam produced in industrial
L1~ L4	pressure reducer and attemperator 1~4
LP	low pressure
LLP	low and low-pressure steam
LHV	lower heating value
MP	medium pressure
MILP	mixed-integer linear programming
MINLP	mixed-integer nonlinear programming
PMP	medium pressure purchased from neighbouring enterprises
PLP	low pressure purchased from neighbouring enterprises
PLLP	low and low-pressure steam purchased from neighbouring enterprises
ST	steam turbines
T1~T3	simple steam turbine 1~3

References

1. Luo, X.; Zhang, B.; Chen, Y.; Mo, S. Operational planning optimization of steam power plants considering equipment failure in petrochemical complex. *Appl. Energy* **2013**, *112*, 1247–1264. [CrossRef]
2. Luo, X.; Zhang, B.; Chen, Y.; Mo, S. Modeling and optimization of a utility system containing multiple extractions steam turbines. *Energy* **2011**, *36*, 3501–3512. [CrossRef]
3. Douglas, T.; Big-Alabo, A. A generic algorithm of sustainability (GAS) function for industrial complex steam turbine and utility system optimisation. *Energy* **2018**, *164*, 881–897. [CrossRef]
4. Iyer, R.R.; Grossmann, I.E. Synthesis and operational planning of utility systems for multiperiod operation. *Comput. Chem. Eng.* **1998**, *22*, 979–993. [CrossRef]
5. Varbanov, P.S.; Doyle, S.; Smith, R. Modelling and optimization of utility systems. *Chem. Eng. Res. Des.* **2004**, *82*, 561–578. [CrossRef]
6. Velasco-Garcia, P.; Varbanov, P.S.; Arellano-Garcia, H.; Wozny, G. Utility systems operation: Optimisation-based decision making. *Appl. Therm. Eng.* **2011**, *31*, 3196–3205. [CrossRef]
7. Iyer, R.R.; Grossmann, I.E. Optimal multiperiod operational planning for utility systems. *Comput. Chem. Eng.* **1997**, *21*, 787–800. [CrossRef]
8. Tao, T.; Ye, J.; Beixing, W. Modeling and optimization for steam power system in petrochemical plant under steam mutual supply condition. *Energy Conserv. Emiss. Reduct. Pet. Petrochem. Ind.* **2011**, *2*, 10–15.
9. Zeng, Y.; Xiao, X.; Li, J.; Sun, L.; Floudas, C.A.; Li, H. A novel multi-period mixed-integer linear optimization model for optimal distribution of byproduct gases, steam and power in an iron and steel plant. *Energy* **2018**, *143*, 881–899. [CrossRef]
10. Zhao, L.; You, F. A data-driven approach for industrial utility systems optimization under uncertainty. *Energy* **2019**, *182*, 559–569. [CrossRef]
11. Sun, L.; Shen, J.; Hua, Q.; Lee, K.Y. Data-driven oxygen excess ratio control for proton exchange membrane fuel cell. *Appl. Energy* **2018**, *231*, 866–875. [CrossRef]

12. Martelli, E.; Elsido, C.; Mian, A.; Marechal, F. Synthesis of Heat Exchanger Networks and Utility Systems: Sequential Initialization Procedure and Simultaneous MINLP Algorithm. In *Computer Aided Chemical Engineering*; Elsevier: London, UK, 2016; Volume 38, pp. 1449–1454.

13. Martelli, E.; Elsido, C.; Mian, A.; Marechal, F. MINLP model and two-stage algorithm for the simultaneous synthesis of heat exchanger networks, utility systems and heat recovery cycles. *Comput. Chem. Eng.* **2017**, *106*, 663–689. [CrossRef]

14. Hwangbo, S.; Sin, G.; Rhee, G.; Yoo, C. Development of an integrated network for waste-to-energy and central utility systems considering air pollutant emissions pinch analysis. *J. Clean. Prod.* **2020**, *252*, 119746. [CrossRef]

15. Sun, L.; Gai, L.; Smith, R. Site utility system optimization with operation adjustment under uncertainty. *Appl. Energy* **2017**, *186*, 450–456. [CrossRef]

16. Walmsley, T.G.; Jia, X.; Varbanov, P.S.; Klemeš, J.J.; Wang, Y. Total Site Utility Systems Structural Design Considering Environmental Impacts. In *Computer Aided Chemical Engineering*; Elsevier: London; UK, 2018; Volume 43, pp. 1305–1310.

17. Ahmadi, P.; Dincer, I. Exergoenvironmental analysis and optimization of a cogeneration plant system using Multimodal Genetic Algorithm (MGA). *Energy* **2010**, *35*, 5161–5172. [CrossRef]

18. Sevinchan, E.; Dincer, I.; Lang, H. Energy and exergy analyses of a biogas driven multigenerational system. *Energy* **2019**, *166*, 715–723. [CrossRef]

19. Morosuk, T.; Tsatsaronis, G. Advanced exergy-based methods used to understand and improve energy-conversion systems. *Energy* **2019**, *169*, 238–246. [CrossRef]

20. Elhelw, M.; Al Dahma, K.S.; el Hamid Attia, A. Utilizing exergy analysis in studying the performance of steam power plant at two different operation mode. *Appl. Therm. Eng.* **2019**, *150*, 285–293. [CrossRef]

21. Sansaniwal, S.K.; Sharma, V.; Mathur, J. Energy and exergy analyses of various typical solar energy applications: A comprehensive review. *Renew. Sustain. Energy Rev.* **2018**, *82*, 1576–1601. [CrossRef]

22. Noroozian, A.; Mohammadi, A.; Bidi, M.; Ahmadi, M.H. Energy, exergy and economic analyses of a novel system to recover waste heat and water in steam power plants. *Energy Convers. Manag.* **2017**, *144*, 351–360. [CrossRef]

23. Wang, J.; Chen, Y.; Lior, N.; Li, W. Energy, exergy and environmental analysis of a hybrid combined cooling heating and power system integrated with compound parabolic concentrated-photovoltaic thermal solar collectors. *Energy* **2019**, *185*, 463–476. [CrossRef]

24. Ehyaei, M.A.; Rosen, M.A. Optimization of a triple cycle based on a solid oxide fuel cell and gas and steam cycles with a multiobjective genetic algorithm and energy, exergy and economic analyses. *Energy Convers. Manag.* **2019**, *180*, 689–708. [CrossRef]

25. Ahmadi, G.; Toghraie, D.; Akbari, O. Energy, exergy and environmental (3E) analysis of the existing CHP system in a petrochemical plant. *Renew. Sustain. Energy Rev.* **2019**, *99*, 234–242. [CrossRef]

26. Özkaraca, O.; Keçebaş, A. Performance analysis and optimization for maximum exergy efficiency of a geothermal power plant using gravitational search algorithm. *Energy Convers. Manag.* **2019**, *185*, 155–168. [CrossRef]

27. Shahsavar, A.; Khanmohammadi, S. Feasibility of a hybrid BIPV/T and thermal wheel system for exhaust air heat recovery: Energy and exergy assessment and multi-objective optimization. *Appl. Therm. Eng.* **2019**, *146*, 104–122. [CrossRef]

28. Dorotić, H.; Pukšec, T.; Duić, N. Economical, environmental and exergetic multi-objective optimization of district heating systems on hourly level for a whole year. *Appl. Energy* **2019**, *251*, 113394. [CrossRef]

29. Manesh, M.K.; Navid, P.; Baghestani, M.; Abadi, S.K.; Rosen, M.A.; Blanco, A.M.; Amidpour, M. Exergoeconomic and exergoenvironmental evaluation of the coupling of a gas fired steam power plant with a total site utility system. *Energy Convers. Manag.* **2014**, *77*, 469–483. [CrossRef]

30. Huang, Y.; Li, S.; Ding, P.; Zhang, Y.; Yang, K.; Zhang, W. Optimal operation for economic and exergetic objectives of a multiple energy carrier system considering demand response program. *Energies* **2019**, *12*, 3995. [CrossRef]

31. Kotas, T.J. *The Exergy Method of Thermal Plant Analysis*; Elsevier: London, UK, 2013.

32. Ramirez–Elizondo, L.M.; Paap, G.C.; Ammerlaan, R.; Negenborn, R.R.; Toonssen, R. On the energy, exergy and cost optimisation of multi–energy–carrier power systems. *Int. J. Exergy* **2013**, *13*, 364–386. [CrossRef]

33. Li, Z.; Du, W.; Zhao, L.; Qian, F. Modeling and optimization of a steam system in a chemical plant containing multiple direct drive steam turbines. *Ind. Eng. Chem. Res.* **2014**, *53*, 11021–11032. [CrossRef]
34. Nojavan, S.; Majidi, M.; Najafi-Ghalelou, A.; Ghahramani, M.; Zare, K. A cost-emission model for fuel cell/PV/battery hybrid energy system in the presence of demand response program: ε-constraint method and fuzzy satisfying approach. *Energy Convers. Manag.* **2017**, *138*, 383–392. [CrossRef]

MDPI
St. Alban-Anlage 66
4052 Basel
Switzerland
Tel. +41 61 683 77 34
Fax +41 61 302 89 18
www.mdpi.com

Energies Editorial Office
E-mail: energies@mdpi.com
www.mdpi.com/journal/energies